College Geometry

An Introduction to the Modern Geometry of the Triangle and the Circle

Nathan Altshiller-Court

Second Edition
Revised and Enlarged

Dover Publications, Inc.
Mineola, New York

Copyright

Bibliographical Note

This Dover edition, first published in 2007, is an unabridged republication of the second edition of the work, originally published by Barnes & Noble, Inc., New York, in 1952.

Library of Congress Cataloging-in-Publication Data

Altshiller-Court, Nathan, b. 1881.
College geometry : an introduction to the modern geometry of the triangle and the circle / Nathan Altshiller-Court. — Dover ed.
p. cm.
Originally published: 2nd ed., rev. and enl. New York : Barnes & Noble, 1952.
Includes bibliographical references and index.
ISBN 0-486-45805-9
1. Geometry, Modern—Plane. I. Title.

QA474.C6 2007
516.22—dc22

2006102940

Manufactured in the United States of America
Dover Publications, Inc., 31 East 2nd Street, Mineola, N.Y. 11501

To My Wife

PREFACE

Before the first edition of this book appeared, a generation or more ago, modern geometry was practically nonexistent as a subject in the curriculum of American colleges and universities. Moreover, the educational experts, both in the academic world and in the editorial offices of publishing houses, were almost unanimous in their opinion that the colleges felt no need for this subject and would take no notice of it if an avenue of access to it were opened to them.

The academic climate confronting this second edition is radically different. College geometry has a firm footing in the vast majority of schools of collegiate level in this country, both large and small, including a considerable number of predominantly technical schools. Competent and often even enthusiastic personnel are available to teach the subject.

These changes naturally had to be considered in preparing a new edition. The plan of the book, which gained for it so many sincere friends and consistent users, has been retained in its entirety, but it was deemed necessary to rewrite practically all of the text and to broaden its horizon by adding a large amount of new material.

Construction problems continue to be stressed in the first part of the book, though some of the less important topics have been omitted in favor of other types of material. All other topics in the original edition have been amplified and new topics have been added. These changes are particularly evident in the chapter dealing with the recent geometry of the triangle. A new chapter on the quadrilateral has been included.

Many proofs have been simplified. For a considerable number of others, new proofs, shorter and more appealing, have been substituted. The illustrative examples have in most cases been replaced by new ones.

The harmonic ratio is now introduced much earlier in the course. This change offered an opportunity to simplify the presentation of some topics and enhance their interest.

The book has been enriched by the addition of numerous exercises of varying degrees of difficulty. A goodly portion of them are noteworthy propositions in their own right, which could, and perhaps should, have their place in the text, space permitting. Those who use the book for reference may be able to draw upon these exercises as a convenient source of instructional material.

N. A.-C.

Norman, Oklahoma

ACKNOWLEDGMENTS

It is with distinct pleasure that I acknowledge my indebtedness to my friends Dr. J. H. Butchart, Professor of Mathematics, Arizona State College, and Dr. L. Wayne Johnson, Professor and Head of the Department of Mathematics, Oklahoma A. and M. College. They read the manuscript with great care and contributed many important suggestions and excellent additions. I am deeply grateful for their valuable help.

I wish also to thank Dr. Butchart and my colleague Dr. Arthur Bernhart for their assistance in the taxing work of reading the proofs.

Finally, I wish to express my appreciation to the Editorial Department of Barnes and Noble, Inc., for the manner, both painstaking and generous, in which the manuscript was treated and for the inexhaustible patience exhibited while the book was going through the press.

N. A.-C.

CONTENTS

TO THE INSTRUCTOR

This book contains much more material than it is possible to cover conveniently with an average college class that meets, say, three times a week for one semester. Some instructors circumvent the difficulty by following the book from its beginning to whatever part can be reached during the term. A good deal can be said in favor of such a procedure.

However, a judicious selection of material in different parts of the book will give the student a better idea of the scope of modern geometry and will materially contribute to the broadening of his geometrical outlook.

The instructor preferring this alternative can make selections and omissions to suit his own needs and preferences. As a rough and tentative guide the following omissions are suggested. Chap. II, arts. 27, 28, 43, 44, 51–53. Chap. III, arts. 70, 71, 72, 78, 83, 84, 91–93, 102–104, 106–110, 114, 115, 130, 168, 181, 195–200, 203, 205, 211–217, 229–239. Chap. IV, Chap. V, arts. 284–287, 292, 298–305. Chap. VI, arts. 308, 323, 324, 337–344. Chap. VIII, arts. 362, 372, 398, 399, 418–420, 431, 432, 438, 439, 463–470, 479–490, 495, 499–517. Chap. IX, Chap. X, arts. 583–587, 591, 592, 595, 596, 614–624, 639–648, 658–665, 672–683.

The text of the book does not depend upon the exercises, so that the book can be read without reference to them. The fact that it is essential for the learner to work the exercises needs no argument.

The average student may be expected to solve a considerable part of the groups of problems which follow immediately the various subdivisions of the book. The supplementary exercises are intended as a challenge to the more industrious, more ambitious student. The lists of questions given under the headings of "Review Exercises" and "Miscellaneous Exercises" may appeal primarily to those who have an enduring interest, either professional or avocational, in the subject of modern geometry.

TO THE STUDENT

The text. Novices to the art of mathematical demonstrations may, and sometimes do, form the opinion that memory plays no role in mathematics. They assume that mathematical results are obtained by reasoning, and that they always may be restored by an appropriate argument. Obviously, such an opinion is superficial. A mathematical proof of a proposition is an attempt to show that this new proposition is a consequence of definitions and theorems already accepted as valid. If the reasoner does not have the appropriate propositions available in his mind, the task before him is well-nigh hopeless, if not outright impossible.

The student who embarks upon the study of college geometry should have accessible a book on high-school geometry, preferably his own text of those happy high-school days. Whenever a statement in *College Geometry* refers, explicitly or implicitly, to a proposition in the elementary text, the student will do well to locate that proposition and enter the precise reference in a notebook kept for the purpose, or in the margin of his college book. It would be of value to mark references to *College Geometry* on the margin of the corresponding propositions of the high-school book.

The cross references in this book are to the preceding parts of the text. Thus art. 189 harks back to art. 73. When reading art. 189, it may be worth while to make a record of this fact in connection with art. 73. Such a system of "forward" references may be a valuable help in reviewing the course and may facilitate the assimilation of the contents of the book.

Figures. The student will do well to cultivate the habit of drawing his own figures while reading the book, and to draw a separate figure for each proposition. A rough free-hand sketch is sufficient in most cases. Where a more complex figure is required, the corresponding figure in the book may be consulted as a guide to the disposition of the various parts and elements. Such practices help to fix the propositions in the reader's mind.

The Exercises. The purpose of exercises in the study of mathematics is usually two-fold. They provide the reader with a check on his mastery of the contents of the course, and also with an opportunity to test his ability to use the material by applying the methods presented in the book. These two phases are, of course, not unrelated.

It goes without saying that the student cannot possibly solve a problem if he does not know what the problem is. To argue the contrary would be nothing short of ridiculous. In the light of experience, however, it may be useful to insist on this point. We begin our problem-solving career with such simple statements that there is no doubt as to our understanding their contents. When, in the course of time, conditions change radically, we continue, by force of habit, to assume an instantaneous knowledge of the statement of the problem.

Like the problems in most books on geometry, nearly all the problems in *College Geometry* are verbal problems. Nevertheless, surprising as it may sound, it is often difficult to know what a given problem is. More or less effort may be required to determine its meaning. Clearly, this effort of understanding the problem must be made first, however, before any steps toward a solution are undertaken. In fact, the mastery of the meaning of the problem may be the principal part, and often is the most difficult part, of its solution.

To make sure that he understands the statement of the question, the student should repeat its text verbally, without using the book, or, still better, write the text in full, from memory. Moreover, he must have in mind so clearly the meaning of the spoken or written sentences that he will be able to explain a problem, in his own words, to anyone, equipped with the necessary information, who has never before heard of the problem.

Finally, it is essential to draw the figure the question deals with. A simple free-hand illustration will usually suffice. In some cases a carefully executed drawing may provide valuable suggestions.

Obviously, no infallible rule can be given which will lead to the solution of all problems. When the student has made sure of the meaning of the problem, has listed accurately the given elements of the problem and the elements wanted, and has before him an adequate figure, he will be well armed for his task, and with such help even a recalcitrant question may eventually become more manageable.

The student must not expect that a solution will invariably occur to him as soon as he has finished reading the text of a question. If it does, as it often may, well and good. But, in most cases, a question requires, above all, patience. A number of unsuccessful starts is not unusual, and need not cause discouragement. The successful solver of problems is the one whose determination — whose will to overcome obstacles — grows and increases with the resistance encountered. Then, after the light breaks through, and the goal has been reached, his is the reward of a gratifying sense of triumph, of achievement.

I

GEOMETRIC CONSTRUCTIONS

A. PRELIMINARIES

1. Notation. We shall frequently denote by:

$A, B, C, \ldots$ the vertices or the corresponding angles of a polygon;

$a, b, c, \ldots$ the sides of the polygon (in the case of a triangle, the small letter will denote the side opposite the vertex indicated by the same capital letter);

$2p$ the perimeter of a triangle;

h_a, h_b, h_c the altitudes and m_a, m_b, m_c the medians of a triangle ABC corresponding to the sides a, b, c;

t_a, t_b, t_c the internal, and t_a', t_b', t_c' the external, bisectors of the angles A, B, C;

R, r the radii of the circumscribed and inscribed circles (for the sake of brevity, we shall use the terms *circumcircle*, *circumradius*, *circumcenter*, and *incircle*, *inradius*, *incenter*);

(A, r) the circle having the point A for center and the segment r for radius;

$M = (PQ, RS)$ the point of intersection M of the two lines PQ and RS.

2. Basic Constructions. Frequent use will be made of the following constructions:

To divide a given segment into a given number of equal parts.

To divide a given segment in a given ratio (i) internally; (ii) externally (§ 54).

To construct the fourth proportional to three given segments.

To construct the mean proportional to two given segments.

To construct a square equivalent to a given (i) rectangle; (ii) triangle.

To construct a square equivalent to the sum of two, three, or more given squares.

To construct two segments given their sum and their difference.

To construct the tangents from a given point to a given circle.

To construct the internal and the external common tangents of two given circles.

EXERCISES

Construct a triangle, given:

1. a, b, c.
2. a, b, C.
3. a, B, C.
4. a, h_a, B.
5. a, b, m_a.
6. a, B, t_b.
7. A, h_a, t_a.

Construct a right triangle, with its right angle at A, given:

8. a, B.
9. b, C.
10. a, b.
11. b, c.

Construct a parallelogram $ABCD$, given:

12. AB, BC, AC.
13. AB, AC, B.
14. $AB, BD, \angle ABD$.

Construct a quadrilateral $ABCD$, given:

15. A, B, C, AB, AD.
16. AB, BC, CD, B, C.
17. A, B, C, AD, CD.

18. With a given radius to draw a circle tangent at a given point to a given (i) line; (ii) circle.

19. Through two given points to draw a circle (i) having a given radius; (ii) having its center on a given line.

20. To a given circle to draw a tangent having a given direction.

21. To divide a given segment internally and externally in the ratio of the squares of two given segments p, q. (*Hint.* If AD is the perpendicular to the hypotenuse BC of the right triangle ABC, $AB^2:AC^2 = BD:DC$.)

22. Construct a right triangle, given the hypotenuse and the ratio of the squares of the legs.

23. Given the segments a, p, q, construct the segment x so that $x^2:a^2 = p:q$.

24. Construct an equilateral triangle equivalent to a given triangle.

3. Suggestion. Most of the preceding problems are stated in conventional symbols. It is instructive to state them in words. For instance, Exercise 4 may be stated as follows: Construct a triangle given the base, the corresponding altitude, and one of the base angles.

B. GENERAL METHOD OF SOLUTION OF CONSTRUCTION PROBLEMS

4. Analytic Method. Some construction problems are direct applications of known propositions and their solutions are almost immediately apparent. *Example:* Construct an equilateral triangle.

If the solution of a problem is more involved, but the solution is known, it may be presented by starting with an operation which we know how to perform, followed by a series of operations of this kind, until the goal is reached.

This procedure is called the synthetic method of solution of problems. It is used to present the solutions of problems in textbooks.

However, this method cannot be followed when one is confronted with a problem the solution of which is not apparent, for it offers no clue as to what the first step shall be, and the possible first steps are far too numerous to be tried at random.

On the other hand, we do know definitely what the problem is — we know what figure we want to obtain in the end. It is therefore helpful to start with this very figure, provisionally taken for granted. By a careful and attentive study of this figure a way may be discovered leading to the desired solution. The procedure, which is called the analytic method of solving problems, consists, in outline, of the following steps:

ANALYSIS. Assuming the problem solved, draw a figure approximately satisfying the conditions of the problem and investigate how the given parts and the unknown parts of the figure are related to one another, until you discover a relation that may be used for the construction of the required figure.

CONSTRUCTION. Utilizing the information obtained in the analysis, carry out the actual construction.

PROOF. Show that the figure thus constructed satisfies all the requirements of the problem.

DISCUSSION. Discuss the problem as to the conditions of its possibility, the number of solutions, etc.

The following examples illustrate the method.

5. Problem. *Two points A, B are marked on two given parallel lines x, y. Through a given point C, not on either of these lines, to draw a secant $CA'B'$ meeting x, y in A', B' so that the segments AA', BB' shall be proportional to two given segments p, q.*

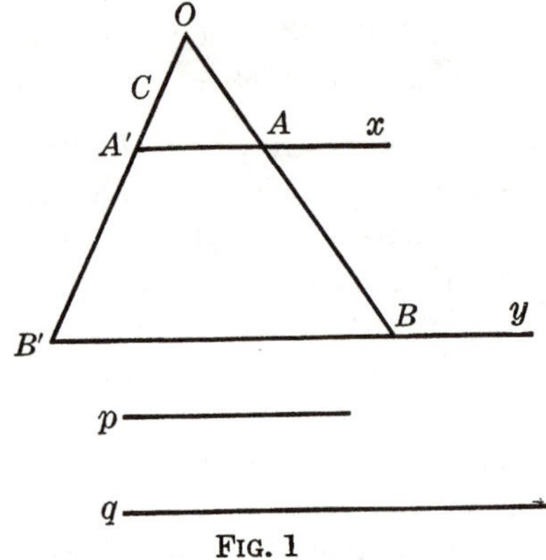

FIG. 1

ANALYSIS. Let $CA'B'$ be the required line, so that (Fig. 1):

$$AA':BB' = p:q,$$

and let $O = (AB, A'B')$. The two triangles OAA', OBB' are similar; hence:

$$AO:BO = AA':BB'.$$

But the latter ratio is known; hence the point O divides the given segment AB in the given ratio $p:q$. Thus we may construct O, and OC is the required line.

CONSTRUCTION. Construct the point O such that:

$$AO:BO = p:q.$$

The points O and C determine the required line.

PROOF. Left to the student.

DISCUSSION. There are two points, O and O', which divide the given segment AB in the given ratio $p:q$, one externally and the other internally, and we can always construct these two points; hence the problem has two solutions if neither of the lines CO, CO' is parallel to the lines x, y.

Consider the case when $p = q$.

6. Problem. *Through a given point, outside a given circle, to draw a secant so that the chord intercepted on it by the circle shall subtend at the center an angle equal to the acute angle between the required secant and the diameter, produced, passing through the given point.*

ANALYSIS. Let the required secant MBA (Fig. 2) through the given point M cut the given circle, center O, in A and B. The two triangles AOB, AOM have the angle A in common and, by assumption, angle AOB = angle M; hence the two triangles are equiangular. But the triangle AOB is isosceles; hence the triangle AOM is also isosceles, and $MA = MO$. Now the length MO is known; hence the distance MA of the point A from M is known, so that the point A may be constructed, and the secant MA may be drawn.

CONSTRUCTION. Draw the circle (M, MO). If A is a point common to the two circles, the line MA satisfies the conditions of the problem.

PROOF. Let the line MA meet the given circle again in B. The triangles AOB, AOM are isosceles, for $OA = OB$, $MA = MO$, as radii of the same circle, and the angle A is a common base angle in the two triangles; hence the angles AOB and M opposite the respective bases AB and AO in the two triangles are equal. Thus MA is the required line.

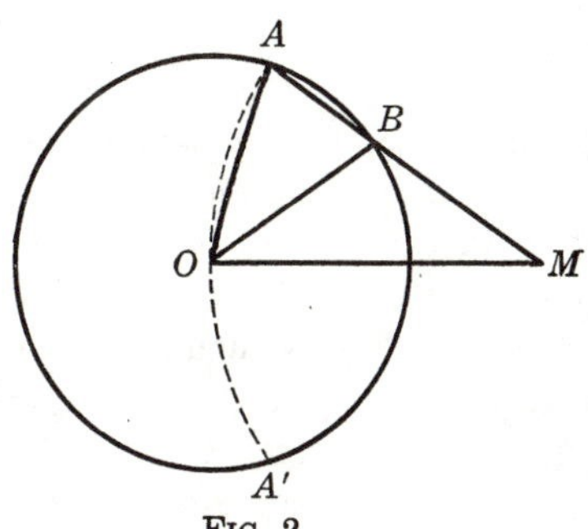

FIG. 2

DISCUSSION. We can always draw the circle (M, MO) which will cut the given circle in two points A and A'; hence the problem always has two solutions, symmetrical with respect to the line MO.

Could the line MBA be drawn so that the angle AOB would be equal to the obtuse angle between MBA and MO? If that were possible, we would have:

$$\angle AOB + \angle OMA = 180°;$$

hence:

$$\angle OMA = \angle OAB + \angle OBA.$$

But in the triangle OBM we have:

$$\angle M < \angle OBA.$$

We are thus led to a contradiction; hence a line satisfying the imposed condition cannot be drawn.

Consider the problem when the point M is given inside the given circle, or on the given circle.

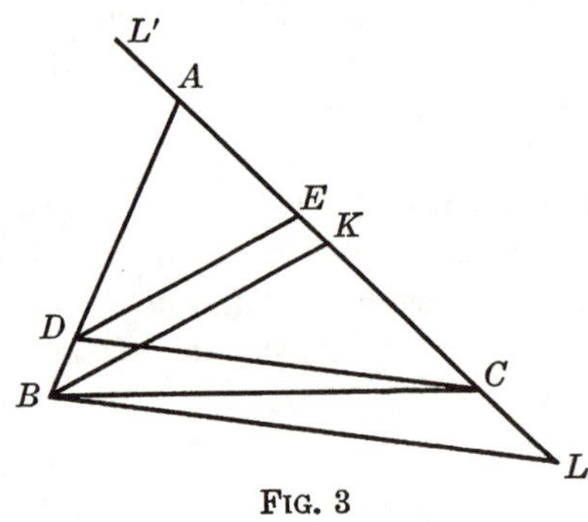

FIG. 3

7. Problem. *On the sides AB, AC, produced if necessary, of the triangle ABC to find two points D, E such that the segments AD, DE, and EC shall be equal* (Fig. 3).

ANALYSIS. Suppose that the points D, E satisfy the conditions of the problem, and let the parallels through B to the lines DE, DC meet AC in K, L. The two triangles ADE, ABK are similar, and since $AD = DE$, by assumption, we have $AB = BK$, and the point K is readily constructed.

Again, the triangles DEC, BKL are similar, and since $DE = EC$, by assumption, we have $BK = KL$; hence L is known.

CONSTRUCTION. Draw the circle (B, BA) cutting AC again in K. Draw the circle (K, BA) cutting AC in L. The parallel through C to BL meets the line AB in the first required point D, and the parallel through D to BK meets AC in the second required point E.

PROOF. The steps in the proof are the same as those in the analysis, but taken in reverse order.

DISCUSSION. The point K always has one and only one position. When K is constructed we find two positions for L, and we have two solutions, DE and $D'E'$, for the problem.

If A is a right angle the problem becomes trivial.

8. Problem. *On two given circles find two points a given distance apart and such that the line joining them shall have a given direction.*

ANALYSIS. Let P, Q be the required points on the two given circles (A), (B). Through the center A of (A) (Fig. 4) draw a parallel to PQ and lay off $AR = PQ$. In the parallelogram $APQR$ we have $RQ = AP$. But AP is a radius of a given circle; hence the length RQ is known. On the other hand the point R is known, for both the direction and the length of AR are given; hence the point Q may be constructed. The point P is then readily found.

CONSTRUCTION. Through the center A of one of the two given circles, (A), draw a line having the given direction and lay off AR equal to the given length, m. With R as center and radius equal to the radius of (A) draw a circle (A'), meeting the second given circle in a point Q. Through Q draw a parallel to AR and lay off $QP = AR$, so as to form a parallelogram in which AR is a side (and not a diagonal). The points P, Q satisfy the conditions of the problem.

PROOF. The segment PQ has the given length and the given direction, by construction. The point Q was taken on the circle (B). In order to show that P lies on the circle (A) it is enough to point out that in the parallelogram $ARQP$ we have $AP = RQ$, and RQ is equal to the radius of (A), by construction.

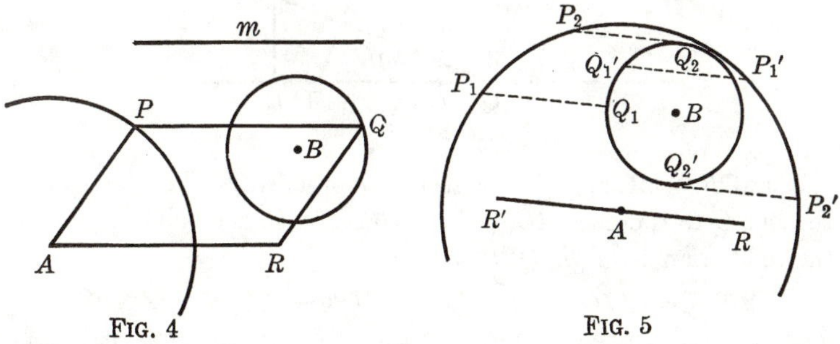

FIG. 4

FIG. 5

DISCUSSION. It is always possible to draw through A a line having the given direction. The point R may then be marked on either side of A, so that we always have for R two positions, R and R'. The circle with either of these points as center and radius equal to the radius of (A) may cut (B) in two points, or may be tangent to (B), or may not cut (B) at all. Consequently the problem may have four, three, two, one, or no solutions.

Figure 5 illustrates the case when there are four solutions.

9. Problem. *Construct a square so that each side, or the side produced, shall pass through a given point.*

ANALYSIS. Let $PQRS$ be a square whose sides PQ, QR, RS, SP pass, respectively, through the given points A, B, C, D (Fig. 6).

If the perpendicular from D upon AC meets the line QR in F, then $DF = AC$. Indeed, if we leave the line DF fixed while we revolve the square, and the line AC with it, around its center by an angle of 90° so that the sides PQ, QR, RS, SP shall occupy the present positions of QR, RS, SP, PQ, respectively, the line AC will become parallel to DF; hence $AC = DF$.

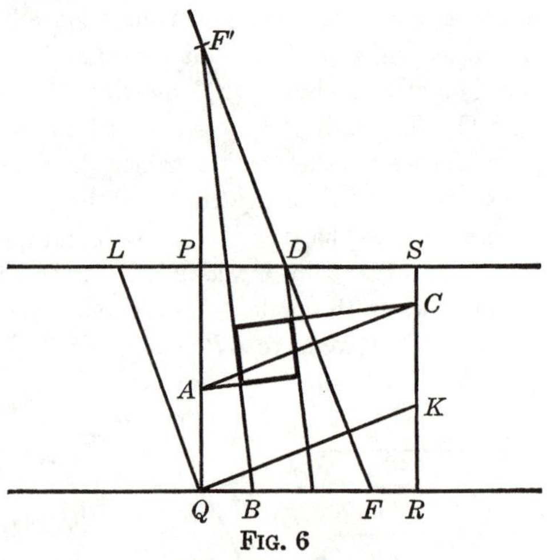

FIG. 6

That $DF = AC$ may also be seen in another way. Let the parallels through Q to the lines AC, DF meet RS, SP in the points K, L. In the right triangles PQL, QRK we have:

$$\angle PQL = \angle LQR - \angle PQR = \angle LQR - \angle LQK = \angle KQR.$$

Thus the two triangles are equiangular, and since $PQ = QR$, by assumption, they are congruent; hence $QL = QK$. But $QL = DF$, $QK = AC$; hence $AC = DF$.

The equality of these two segments suggests the following

CONSTRUCTION. From the given point D drop a perpendicular DF upon the given line AC and lay off $DF = AC$. Join F to the fourth given point B, and through D draw a parallel to BF. These two parallel lines and the perpendiculars to them through A and C form the required square $PQRS$.

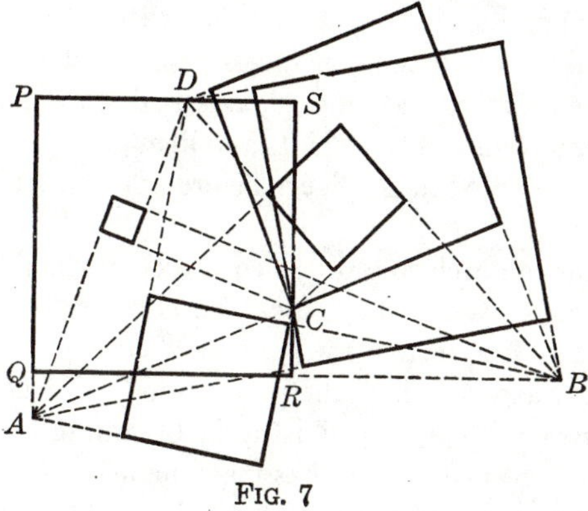

FIG. 7

PROOF. From the construction it follows immediately that $PQRS$ is a rectangle whose sides pass respectively through the given points A, B, C, D.

To show that this rectangle is a square we consider again the triangles PQL, QRK, and, as in the analysis, we show that they are equiangular. Now $DF = AC$, by construction; hence $QL = QK$; therefore the triangles are congruent, and $QP = QR$.

DISCUSSION. We obtain a second solution if we make the symmetric F' of F with respect to D play the role of the point F.

Moreover, the perpendicular from D may also be dropped upon either of the other two sides AB, BC of the triangle ABC, and we will obtain two more pairs of solutions. The problem thus has, in general, six different solutions. It is, of course, immaterial to which of the four given points we assign the role of the point D.

Should the point F happen to coincide with B, the direction of the line BF becomes indeterminate, that is, any line through B may be taken as a side of the required square, and the figure completed accordingly. Thus if the segment joining two of the four given points is equal to and perpendicular to the segment joining the remaining two points, the problem has an infinite number of solutions.

Figure 7 presents the case when the problem has six solutions.

10. Suggestions. The figure for the analysis may be drawn freehand and the given elements arranged as conveniently as possible to give a correct idea of the problem and to exhibit the existing interrelations between the elements involved.

The construction should be carried out with ruler and compasses. The given magnitudes, like segments and angles, should be set down before any construction is attempted, and used in the actual construction. If the problem involves points and lines given in position, these data should be marked in the figure before any constructional operations are begun.

In the discussion each step of the construction should be examined for the number of ways it may be carried out and for the number of lines or points of intersection the considered step will yield.

The different aspects which the problem may assume, as indicated by the discussion, should be illustrated by suitable figures. As a general rule a problem or a figure has many more possibilities than is at first apparent. A careful study of a problem will often reveal vistas which in a more casual treatment may readily escape notice.

EXERCISES

1. Through a given point to draw a line making equal angles with the sides of a given angle.
2. Through a given point to draw a line so that two given parallel lines shall intercept on it a segment of given length.
3. Through one of the two points of intersection of two equal circles to draw two equal chords, one in each circle, forming a given angle.
4. Through a given point of a circle to draw a chord which shall be twice as long as the distance of this chord from the center of the circle.
5. On a produced diameter of a given circle to find a point such that the tangents drawn from it to this circle shall be equal to the radius of the circle.
6. With a given point as center to describe a circle which shall bisect a given circle, that is, the common chord shall be a diameter of the given circle.
7. Through two given points to draw a circle so that its common chord with a given circle shall be parallel to a given line.
8. Construct a parallelogram so that three of its sides shall have for midpoints three given points.
9. On a given leg of a right triangle to find a point equidistant from the hypotenuse and from the vertex of the right angle.
10. With two given points as centers to draw equal circles so that one of their common tangents shall (i) pass through a (third) given point; (ii) be tangent to a given circle.
11. Through a given point to draw a line so that the two chords intercepted on it by two given equal circles shall be equal.
12. To a given circle, located between two parallel lines, to draw a tangent so that the segment intercepted on it by the given parallels shall have a given length.
13. Construct a right triangle given the hypotenuse and the distance from the middle point of the hypotenuse to one leg.

14. Construct a triangle given an altitude and the circumradii (i.e., the radii of the circumscribed circles) of the two triangles into which this altitude divides the required triangle.

C. GEOMETRIC LOCI

11. Important Loci. In a great many cases the solution of a geometric problem depends upon the finding of a point which satisfies certain conditions. For instance, in order to draw a circle passing through three given points, it is necessary to find a point, the center of the circle, equidistant from the three given points.

The problem of drawing a tangent from a given point to a given circle is solved when we find the point of contact, i.e., the point on the circle at which a right angle is subtended by the segment limited by the given point and the center of the given circle.

If one of the conditions which the required point must satisfy be set aside, the problem may have many solutions. However, the point will not become arbitrary, but will move along a certain path, the geometric locus of the point. Now by taking into consideration the discarded condition and setting aside another, we make the required point describe another geometric locus. A point common to the two loci is the point sought.

The problem of drawing a circle passing through three given points may again serve as an illustration. In order to find a point equidistant from three given points, A, B, C, we disregard one of the given points, say C, and try to find a point equidistant from A and B. Thus stated, the problem has many solutions, the required point being any point of the perpendicular bisector of the segment AB. Now considering the point C and leaving out the point A, the required point describes the perpendicular bisector of the segment BC. The required point lies at the intersection of the two perpendicular bisectors.

The nature of the loci obtained depends upon the condition omitted. In elementary geometry these conditions must be such that the loci shall consist of straight lines and circles. The neatness and simplicity of a solution depend very largely upon the judicious choice of the geometric loci.

Knowledge about a considerable number of geometric loci may often enable one to discover immediately where the required point is to be located.

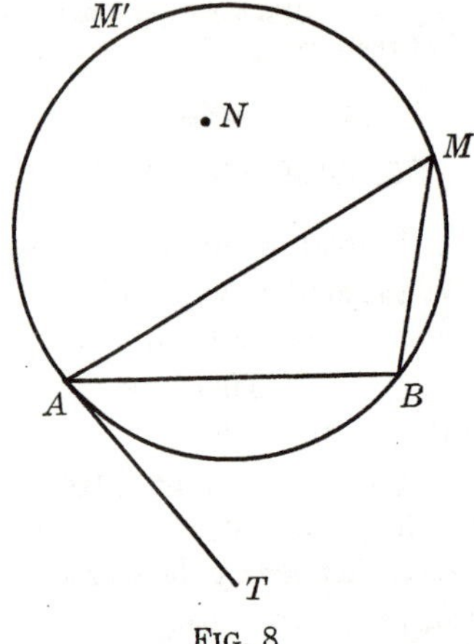

FIG. 8

The following are the most important and the most frequently useful geometric loci:

LOCUS 1. *The locus of a point in a plane at a given distance from a given point is a circle having the given point for center and the given distance for radius.*

LOCUS 2. *The locus of a point from which tangents of given length can be drawn to a given circle is a circle concentric with the given circle.*

LOCUS 3. *The locus of a point at a given distance from a given line consists of two lines parallel to the given line.*

LOCUS 4. *The locus of a point equidistant from two given points is the perpendicular bisector of the segment determined by the two points.*

For the sake of brevity the expression *perpendicular bisector* may be replaced conveniently by the term *mediator.*

LOCUS 5. *The locus of a point equidistant from two intersecting lines indefinitely produced consists of the two bisectors of the angles formed by the two given lines.*

LOCUS 6. *The locus of a point such that the tangents from it to a given circle form a given angle, or, more briefly, at which the circle subtends a given angle, is a circle concentric with the given circle.*

LOCUS 7. *The locus of a point, on one side of a given segment, at which this segment subtends a given angle is an arc of a circle passing through the ends of the segment.*

Let M (Fig. 8) be a point of the locus so that the angle AMB is equal to the given angle. Pass a circle through the three points

A, B, M. At any point M' of the arc AMB the segment AB subtends the same angle as at M; hence every point of the arc AMB belongs to the required locus. On the other hand, any point N not on the arc AMB will lie either inside or outside this arc. In the first case the angle ANB will be larger, and in the second case smaller, than the angle AMB. Hence N does not belong to the locus.

The tangent AT to the circle AMB at the point A makes an angle with AB equal to the angle AMB. Hence AT may be drawn. The construction of the circle AMB is therefore reduced to the problem of drawing a circle tangent to a given line AT at a given point A, and passing through another given point B.

If the condition that the point M is to lie on a given side of the line AB is disregarded, the required locus consists of two arcs of circles congruent to each other and located on opposite sides of AB.

If the given angle is a right angle, the two arcs will be two semicircles of the same radius, and we have: *The locus of a point at which a given segment subtends a right angle is the circle having this segment for diameter.*

LOCUS 8. *The locus of the midpoints of the chords of a given circle which pass through a fixed point is a circle when the point lies inside of or on the circle.*

Let P (Fig. 9) be the given point, and M the midpoint of a chord AB of the given circle (O), the chord passing through the given point P.

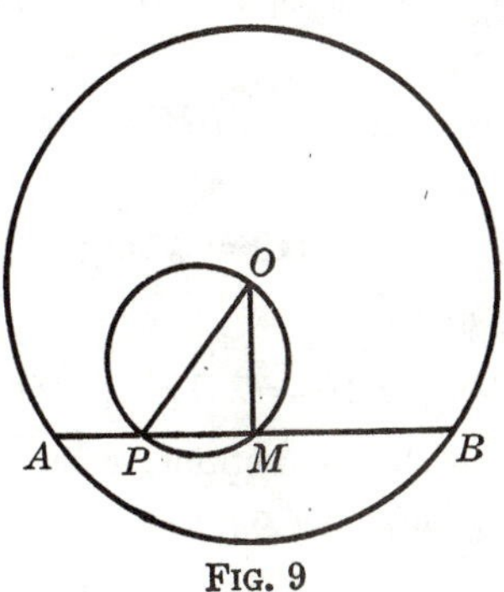

FIG. 9

The segment OP subtends a right angle at M; hence M lies on the circle having OP for diameter (locus 7). On the other hand, any point M of this circle joined to P determines a chord of the circle (O) which is perpendicular to OM at M. Hence M is the midpoint of this chord and therefore belongs to the required locus.

If the point P lies outside of (O), any point M of the locus must lie on the circle having OP for diameter, but not every point of this circle belongs to the locus. The locus consists of the part of the circle (OP) which lies within the given circle (O).

Locus 9. *The locus of the midpoints of all the chords of given length drawn in a given circle is a circle concentric with the given circle.*

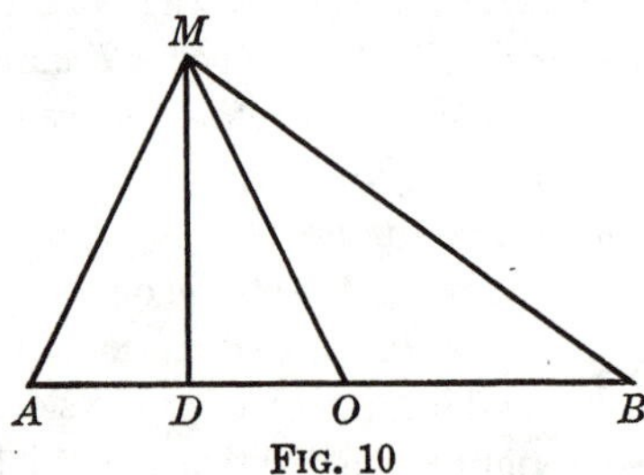

Fig. 10

The chords are all tangent to this circle at their respective midpoints.

Locus 10. *The locus of a point the sum of the squares of whose distances from two given points is equal to a given constant is a circle having for center the midpoint of the given segment.*

Let M (Fig. 10) be a point on the locus. Join M to the midpoint O of the segment AB determined by the two given points A, B and drop the perpendicular MD from M upon AB. From the triangles AOM and BOM we have:

$$AM^2 = OM^2 + OA^2 - 2\,OA \cdot OD. \quad BM^2 = OM^2 + OB^2 + 2\,OB \cdot OD.$$

But:

$$OA = OB = \tfrac{1}{2}\,AB;$$

hence, adding, we have:

$$AM^2 + BM^2 = 2\,OM^2 + 2\,OA^2.$$

Now $AM^2 + BM^2$ is, by assumption, equal to a given constant, say s^2, and $OA^2 = a^2$, where $2\,a$ represents the length of the known segment AB. Hence:

$$OM^2 = \tfrac{1}{2}\,s^2 - a^2. \tag{1}$$

Thus the point M lies at a fixed distance OM from the fixed point O, and therefore the locus of M is a circle with O as center.

The radius OM may be constructed from the formula (1) as follows. OM is one leg of a right triangle which has for its other leg half the length of the given segment AB, and for hypotenuse the side of a square whose diagonal is equal to s.

LOCUS 11. *The locus of a point the ratio of whose distances from two fixed points is constant, is a circle, called the circle of Apollonius, or the Apollonian circle.*

Let A, B be the two given points (Fig. 11) and $p:q$ the given ratio. Let the points C, D divide the segment AB internally and externally in the given ratio, and let M be any point of the locus. We have thus:

$$AC:CB = AD:DB = AM:MB = p:q.$$

The lines MC, MD divide the side AB of the triangle ABM internally and externally in the ratio of the sides MA, MB; hence MC, MD are the internal and external bisectors of the angle AMB, and therefore are perpendicular to each other. Thus at any point M of the required locus the known segment CD subtends a right angle; hence M lies on the circle having CD for diameter (§ 11, locus 7).

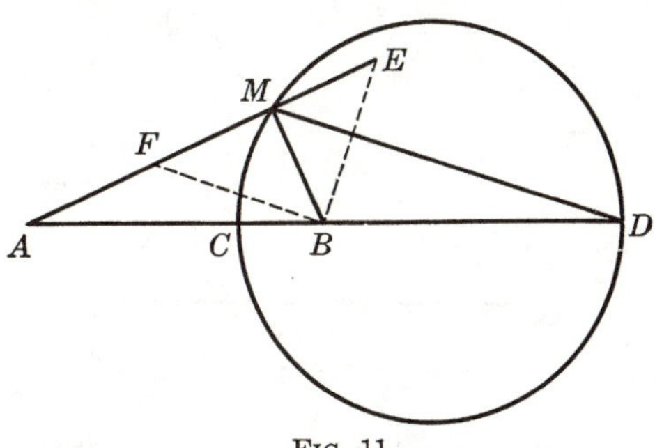

FIG. 11

CONVERSELY. Let us take any point M on this circle and show that this point belongs to the required locus. Through the point B draw the parallels BE, BF to the lines MC, MD meeting AM in E, F, respectively. We thus have:

(1) $$AM:ME = AC:CB, \quad AM:FM = AD:BD.$$

But the second ratios in these proportions are equal, by construction, hence:

$$AM:ME = AM:FM;$$

therefore:

$$ME = MF.$$

Thus M is the midpoint of the segment EF. But EBF is a right angle, for its sides are parallel to the sides of the right angle CMD, and in the right triangle EBF the line MB is equal to half the hypotenuse EF; hence, replacing in the first proportion of (1) the segment ME by the equal segment MB, we obtain:

$$AM:MB = AC:CB = p:q,$$

which shows that M belongs to the locus.

Where, in this converse proof, has the fact been used that M is a point of the circle?

Note. The point C divides the segment AB internally so that $AC:CB = p:q$. But we may also construct the point C' such that $BC':C'A = p:q$. Similarly for the external point of division. Thus the locus actually consists of two Apollonian circles, unless the order in which the two given points are to be taken is specified in the statement of the locus. In actual applications the nature of the problem often indicates the order in which the points are to be considered.

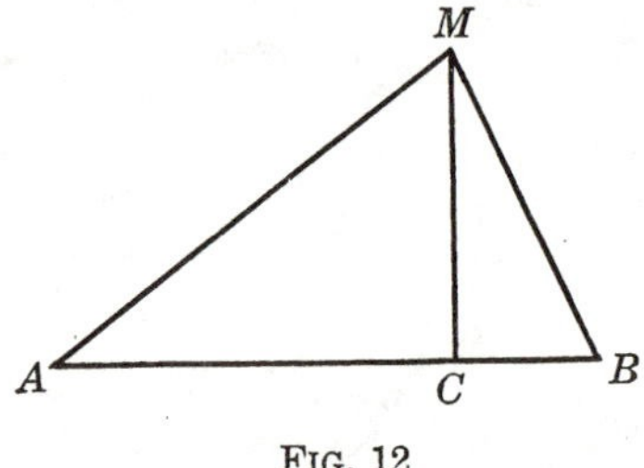

FIG. 12

LOCUS 12. *The locus of a point the difference of the squares of whose distances from two given points is constant, is a straight line perpendicular to the line joining the given points.*

Let A, B (Fig. 12) be the two given points, and let M be a point such that:

$$AM^2 - BM^2 = d^2,$$

where d is given.

If MC is the perpendicular from M to AB, we have from right triangles:

$$AM^2 - AC^2 = MC^2 = MB^2 - BC^2;$$

hence:

$$AM^2 - MB^2 = AC^2 - BC^2 = d^2,$$

or:

$$(AC - BC)(AC + BC) = d^2,$$

and therefore, denoting $AC + BC = AB$ by a:

$$AC - BC = d^2 \div a.$$

This equality gives the difference of the segments AC and BC, while the sum of these segments is equal to a. Thus these segments may be constructed, and the point C located on the line AB. Hence the foot C of the perpendicular dropped upon AB from any point M of the locus is a fixed point of AB. Consequently the point M lies on the perpendicular to the line AB at the point C. It is readily shown that, conversely, every point of this perpendicular belongs to the locus.

Note. We will obtain a different line for the locus of M if we consider that $BM^2 - AM^2 = d^2$. In fact, the locus actually consists of two straight lines, unless the order in which the two given points are to be considered is specified in the statement of the locus.

EXERCISES

1. A variable parallel to the base BC of a triangle ABC meets AB, AC in D, E. Show that the locus of the point $M = (BE, CD)$ is a straight line.
2. On the sides AB, AC of a triangle ABC are laid off two equal segments AB', AC' of variable length. The perpendiculars to AB, AC at B', C' meet in D. Show that the locus of the point D is a straight line. Find the locus of the projection of D upon the line $B'C'$.
3. Find the locus of a point at which two consecutive segments AB, BC of the same straight line subtend equal angles.
4. The base BC of a variable triangle ABC is fixed, and the sum $AB + AC$ is constant. The line DP drawn through the midpoint D of BC parallel to AB meets the parallel CP through C to the internal bisector of the angle A, in P. Show that the locus of P is a circle having D for center.

The following are examples of the use of loci in the solution of problems.

12. Problem. *Draw a circle passing through two given points and subtending a given angle at a third given point.*

Analysis. Let (O) be the required circle passing through the two given points A, B (Fig. 13). The angle formed by the tangents CT, CT' from the given point C to (O) is given; hence in the right triangle OTC we know the acute angle OCT, i.e., we know the shape of this triangle. Thus the ratio $OT:OC$ is known, therefore also the ratios:

$$OA:OC = OB:OC.$$

Consequently we have two loci for the point O (two Apollonian circles), and the point may be found.

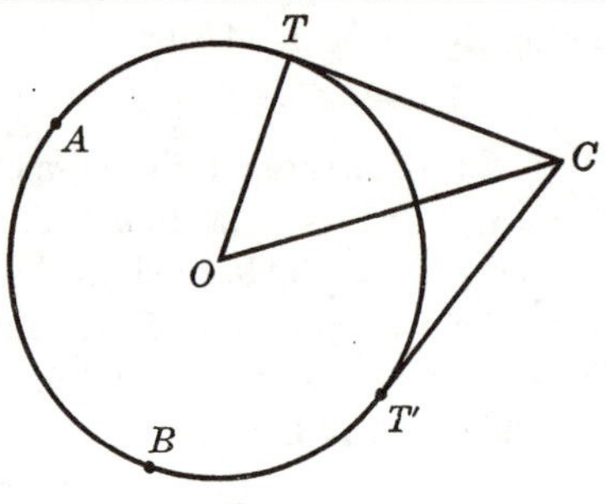

FIG. 13

CONSTRUCTION. On the internal bisector PQ of the given angle P take an arbitrary point Q and drop the perpendicular QR upon one of the sides. Divide each of the two given segments AC, BC internally and externally in the ratio $QR:QP$ in the points E, F and G, H, respectively. A point O common to the two circles having EF and GH for diameters is the center of the required circle.

The proof and the discussion are left to the student.

13. Problem. *Through two given points of a circle to draw two parallel chords whose sum shall have a given length.*

ANALYSIS. Let A, B be the two given points of the circle, center O, and let AC, BD be the two required chords. In the isosceles trapezoid $ABDC$ (Fig. 14) $CD = AB$, and the length AB is known; hence CD is tangent to a known circle having O for center and touching CD at its midpoint F (§ 11, locus 9).

If E is the midpoint of AB we have:

$$2\,EF = AC + BD.$$

Now the point E and the length $AC + BD$ are known; hence we have a second locus for the point F.

CONSTRUCTION. Draw the circle (O, OE). If $2\,s$ is the given length, draw the circle (E, s) meeting (O, OE) in F. The tangent to (O, OE) at F meets the given circle (O) in C and D. The lines AC, BD are the required chords.

PROOF. The two chords AB, CD are equal, for they are equidistant from the center O of the given circle (O). Hence $ABDC$ is an isosceles trapezoid, and therefore:

$$AC + BD = 2\,EF.$$

Now $EF = s$, by construction; hence $AC + BD$ has the required length.

Discussion. The circle (E, s) will not cut the circle (O, OE), if s is greater than $2\,OE$. If $s < 2\,OE$, we obtain two points of intersection F and F', and therefore two solutions.

The tangent to the circle (O, OE) at F determines the two points C, D on the circle (O). These two points and the given points A, B

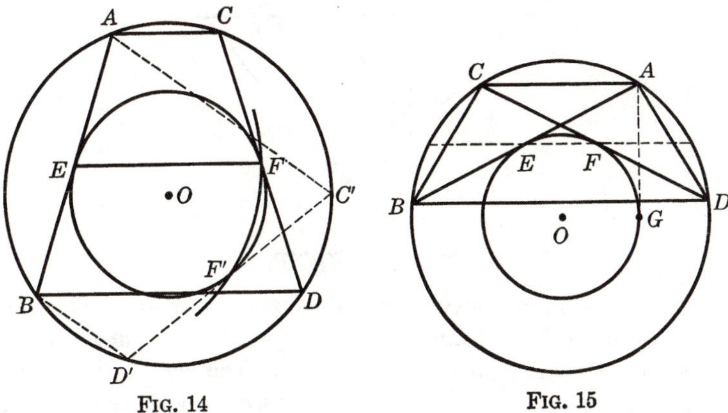

Fig. 14

Fig. 15

determine four lines, namely two sides and the two diagonals of the isosceles trapezoid. The figure will show which two of these four lines are those required.

Let G (Fig. 15) be the point of contact of the second tangent issued from A to the circle (O, OE). If $s < EG$, the tangent to (O, OE) at F will cross the chord AB, and in the resulting trapezoid the line AB will be a diagonal. The line EF is equal to one-half the difference of the two bases.

Consider the case when the two chords are required to have a given difference.

14. Problem. *On a given circle to find a point such that the lines joining it to two given points on this circle shall meet a given line in two points the ratio of whose distances from a given point on this line shall have a given value.*

Let the lines AM, BM joining the given points A, B on the circle to the required point M meet the given line FPQ in the points P, Q (Fig. 16). If F is the given fixed point and the parallel QL to MA through Q meets the line FA in L, we have angle $LQB = AMB$, and the latter angle is known, for the chord AB is given.

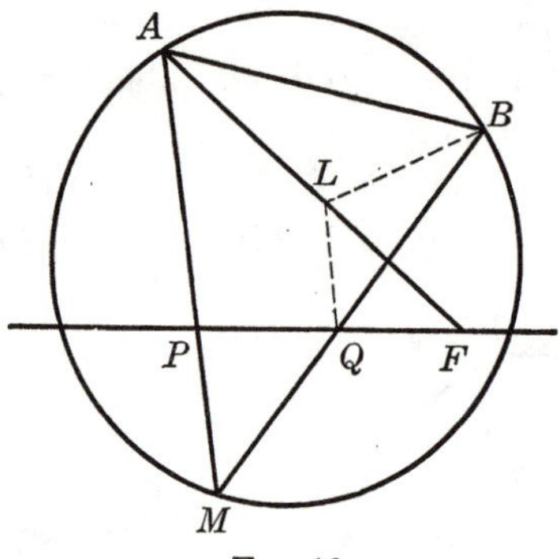

FIG. 16

On the other hand we have:

$$FL:FA = FQ:FP,$$

and the second ratio is given; hence the point L is known. Thus the known segment LB subtends a given angle at the point Q, which gives a locus for this point (§ 11, locus 7). The point Q lies at the intersection of this locus with the given line FPQ. The line BQ meets the circle in the required point M.

The proof and discussion are left to the reader.

EXERCISES

Construct a triangle, given:

1. a, b, A.
2. a, c, h_b.
3. a, h_a, m_a.
4. $a, h_a, b:c$.
5. $a, m_a, b:c$.
6. $a, t_a, b:c$.

Construct a parallelogram, given:

7. An altitude and the two diagonals.
8. The two altitudes and an angle.
9. The two altitudes and a diagonal.
10. A side, an angle, and a diagonal.
11. A side, the corresponding altitude, and the angle between the diagonals.

Construct a quadrilateral $ABCD$, given:

12. The diagonal AC and the angles ABC, ADC, BAC, DAC.
13. The sides AB, BC, the diagonal CA, and the angles ADB, BDC.
14. The sides AB, AD, the angle DAB, and the radius of the inscribed circle.
15. Construct a quadrilateral given three sides and the radius of the circumscribed circle. Give a discussion.
16. Given three points, to find a fourth point, in the same plane, such that its distances to the given points may have given ratios.

17. With a given radius to draw a circle so that it shall touch a given circle and have its center on a given line.

18. With a given radius to draw a circle so that it shall pass through a given point and the tangents drawn to this circle from another given point shall be of given length.

19. In a given circle to inscribe a right triangle so that each leg shall pass through a given point.

20. Construct a triangle given the base, the opposite angle, and the point in which the bisector of this angle meets the base.

21. Construct a triangle given the base and the angles which the median to the base makes with the other two sides.

22. About a given circle to circumscribe a triangle given a side and one of its adjacent angles, so that the vertex of this angle shall lie on a given line.

23. Construct a triangle given the base, an adjacent angle, and the angle which the median issued from the vertex of this angle makes with the side opposite the vertex of this angle.

D. INDIRECT ELEMENTS

Among the conditions which a figure to be constructed may be required to satisfy, elements may be given which do not occur directly in the figure in question. For instance, it may be required that the sum of two sides of a triangle shall have a given length, or that the difference of the base angles shall have a given magnitude, etc. In order to arrive at a method of solution of such a problem, it is necessary to introduce this "indirect element" in the analysis of the problem.

15. Problem. *Construct a triangle given the perimeter, the angle opposite the base, and the altitude to the base* ($2p$, A, h_a).

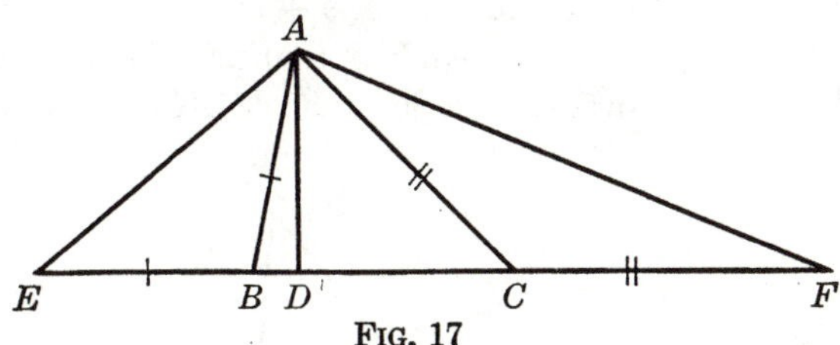

FIG. 17

Let ABC (Fig. 17) be the required triangle. Produce BC on both sides and lay off $BE = BA$, $CF = CA$. Thus $EF = 2p$.

The triangles EAB, FCA are isosceles, hence:

$$\angle E = \angle EAB = \tfrac{1}{2}\angle ABC, \quad \angle F = \angle FAC = \tfrac{1}{2}\angle ACB;$$

hence:

$$\angle EAF = \tfrac{1}{2}B + A + \tfrac{1}{2}C = \tfrac{1}{2}(A + B + C) + \tfrac{1}{2}A = 90^\circ + \tfrac{1}{2}A,$$

and therefore the angle EAF is known. The altitude AD of the triangle ABC is also the altitude of AEF. Thus in the triangle AEF we know the base $EF = 2p$, the opposite angle $EAF = 90° + \frac{1}{2}A$, and the altitude $AD = h_a$; hence this triangle may be constructed. The vertex A of this triangle also belongs to the required triangle ABC. Since $BA = BE$, $CA = CF$, the vertices B, C are the traces on EF of the mediators of AE and AF.

The problem may have two solutions symmetrical with respect to the mediator of EF, or one solution, or none.

Note. In order to construct the required triangle ABC we have constructed another triangle AEF, as an intermediate step. Use of an auxiliary triangle is often very helpful.

16. *Remark.* The angles of the triangle AEF are simply expressed in terms of the angles of ABC, and one of the altitudes of AEF belongs to ABC. These relations afford a simple way of solving the following problems.

Note. Other problems involving the perimeter of a triangle will be discussed later (§§ 157–173).

EXERCISES

Construct a triangle, given:

1. $2p, A, B$.

2. $2p, h_a, B$ (or C).

17. Problem. *Construct a triangle given the base, the opposite angle, and the sum of the other two sides* (a, A, $b + c$).

Let ABC (Fig. 18) be the required triangle. Produce BA and lay off $AD = AC$. In the isosceles triangle ACD we have angle

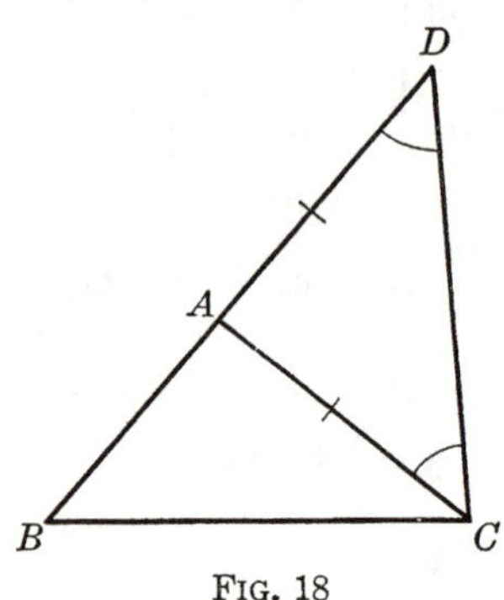

FIG. 18

$D = ACD = \frac{1}{2}\,BAC$. Thus in the triangle BCD we know the base $BC = a$, the side $BD = b + c$, and the angle $D = \frac{1}{2}\,A$; hence this triangle may be constructed. The vertices B, C belong also to the required triangle, and the third vertex A is the intersection of BD with the mediator of the segment CD.

Discussion. The problem is not possible, unless $a < (b + c)$. Assuming that this condition is satisfied, we have given in the auxiliary triangle BCD the angle opposite the smaller side; hence we may have two such triangles, one, or none. From each auxiliary triangle we obtain one and only one required triangle; hence the problem may have two, one, or no solutions.

Note. In the auxiliary triangle BCD we have:

$$\angle D = \tfrac{1}{2}\angle A, B = B,$$

and:

$$\begin{aligned}\angle BCD &= \angle BCA + \angle ACD = C + \tfrac{1}{2}A\\ &= \tfrac{1}{2}C + \tfrac{1}{2}C + \tfrac{1}{2}A + \tfrac{1}{2}B - \tfrac{1}{2}B\\ &= \tfrac{1}{2}(A + B + C) - \tfrac{1}{2}(B - C) = 90^\circ - \tfrac{1}{2}(B - C).\end{aligned}$$

In the triangle BCD the altitude to BD is the altitude h_c of ABC.

Instead of producing the side AB, it is in some cases preferable to produce the side AC.

These relations afford a ready solution of the following problems.

Other problems involving the sum of two sides of a triangle will be considered later (§§ 157–173).

EXERCISES

Construct a triangle, given:

1. $b + c, a, B$ (or C).
2. $b + c, B, h_c$.
3. $b + c, C, a$.
4. $b + c, A, B$.
5. $b + c, a, h_b$ (or h_c).
6. $b + c, a, B - C$.
7. $b + c, h_c, B - C$.
8. $b + c, h_b, B - C$.

9. $b, c, B - C$. *Hint.* Construct $b + c$. In the triangle BCD we have the vertices B, D, and a locus for C. Lay off $BA = c$. The point C lies also on the circle (A, b).

18. Problem. *Construct a triangle given the base, the opposite angle, and the difference of the other two sides* $(a, A, b - c)$.

Let ABC (Fig. 19) be the required triangle. On AC lay off $AD = AB$, so that $CD = b - c$. In the isosceles triangle ADB

$$\angle ADB = \angle ABD = \tfrac{1}{2}(180^\circ - A) = 90^\circ - \tfrac{1}{2}A;$$

hence angle $BDC = 90° + \frac{1}{2}A$. Thus in the triangle BCD we know the base $BC = a$, the opposite angle $BDC = 90° + \frac{1}{2}A$, and the side $CD = b - c$; hence this triangle may be constructed. The vertices

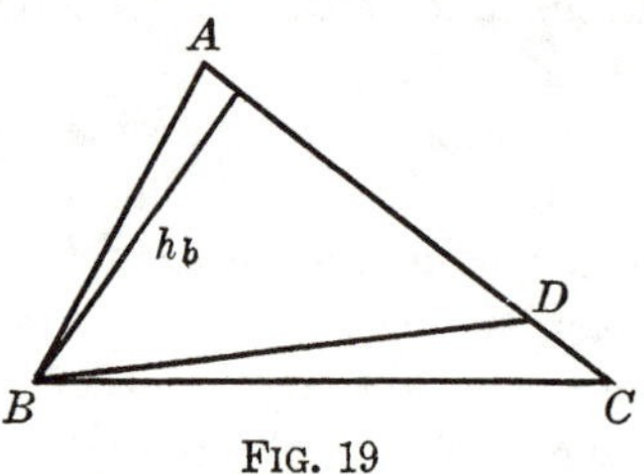

FIG. 19

B, C belong to the required triangle ABC, whose third vertex A lies on CD produced and on the mediator of BD.

The problem is impossible, unless $a > (b - c)$. When this condition is satisfied, the given angle in the triangle BCD lies opposite the larger side, and this triangle may be constructed in one way only. The proposed problem has thus one solution.

Note. The angle BCD of the auxiliary triangle BCD is equal to the angle C of the triangle ABC, angle $BDC = 90° + \frac{1}{2}A$, and:

$$\begin{aligned}\angle CBD &= \angle ABC - \angle ABD = B - (90° - \tfrac{1}{2}A)\\ &= B + \tfrac{1}{2}A + \tfrac{1}{2}C - \tfrac{1}{2}C - 90°\\ &= \tfrac{1}{2}(A + B + C) + \tfrac{1}{2}(B - C) - 90° = \tfrac{1}{2}(B - C).\end{aligned}$$

Also the altitude to CD is the altitude h_b of ABC. These relations may be used to solve the following problems.

EXERCISES

Construct a triangle, given:

1. $b - c, a, C$.
2. $b - c, a, B - C$.
3. $b - c, h_b, C$.
4. $b - c, h_b, B - C$.
5. $b - c, A, B$.
6. $b - c, h_b, A$.

19. Problem. *Construct a triangle given the base, the difference of the other two sides, and the altitude to one of these sides* $(a, b - c, h_c)$.

Let ABC (Fig. 20) be the required triangle. Produce AB and lay off $AE = AC$, so that $BE = b - c$. In the triangle BCE we have $BC = a$, $BE = b - c$, and the altitude to BE equal to h_c; hence this triangle may be constructed, and from this triangle we readily pass to the required triangle ABC.

Note. The angles of BCE are determined in the same way as in the triangle BCD of the preceding problem (§ 18), and these relations may be used to solve the following problems.

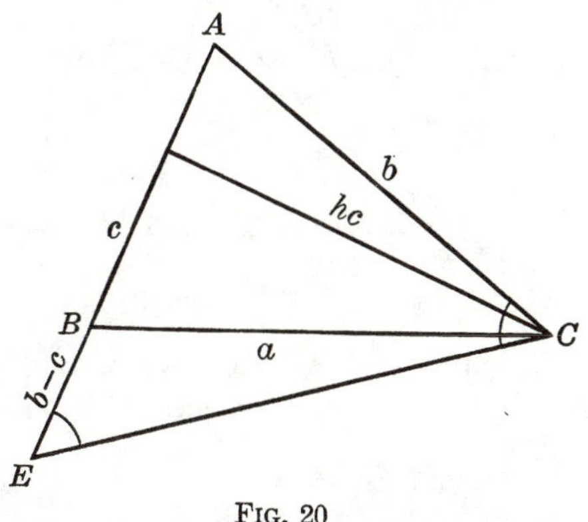

FIG. 20

Other problems involving the difference of two sides of a triangle will be considered later (§§ 157–173).

EXERCISES

Construct a triangle, given:

1. $b - c, h_c, B - C$.

2. $b - c, h_c, A$.

3. $b, c, B - C$. *Hint.* Either BCD or BCE may be used as auxiliary triangle.

20. Problem. *Construct a triangle given the base, the opposite angle, and the sum of the altitudes to the other two sides* $(a, A, h_b + h_c)$.

Let ABC be the required triangle (Fig. 21). Produce the altitude BE and lay off EG equal to the altitude CF. Through G draw the line GH parallel to AC meeting BA, produced, in H. Hence:

$$\angle BHG = \angle BAC = A, \quad \angle BGH = \angle BEA = 90°.$$

Thus in the right triangle BGH we know the leg $BG = h_b + h_c$ and the acute angle $BHG = A$; hence this triangle may be constructed and the length BH determined.

It is readily shown that $BH = b + c$. Draw the line AI parallel

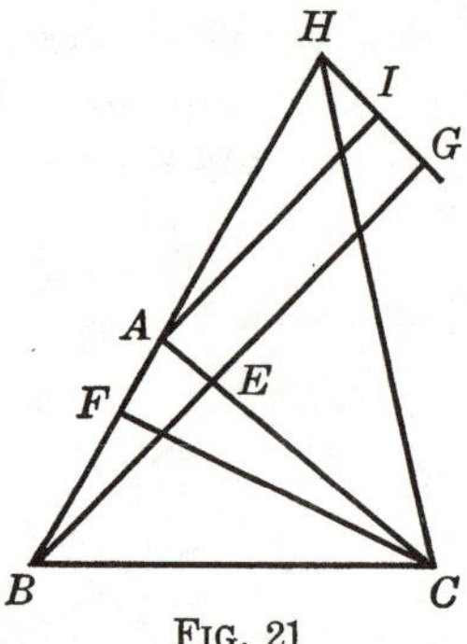

FIG. 21

to BG meeting HG in the point I. $AIGE$ is a rectangle and therefore $AI = EG = h_c$. Now in the right triangles ACF, AHI, we have:

$$AI = CF = h_c, \quad \angle AHI = \angle CAF = A;$$

hence the two triangles are congruent, and $AH = AC = b$; therefore:

$$BH = BA + AH = b + c.$$

Thus, of the required triangle ABC we know now: $a, A, b + c$, and the problem is reduced to a known problem (§ 17).

However, the passage from the auxiliary triangle BGH to the required triangle ABC may be made directly in the figure. The vertex B of BGH belongs to ABC. In order to find C we observe that in the isosceles triangle AHC we have:

$$\angle AHC = \angle ACH = \tfrac{1}{2} A,$$

and since angle $AHG = A$, the line HC is the bisector of the angle H, and this line constitutes a locus for the point C. The circle (B, a) is a second locus for C. The vertex A is then determined on the side BH of BGH by the mediator of CH.

If the given angle A is obtuse, the triangle BGH will include not A but its supplement, and the problem may be solved in the same way.

21. Definition. The right triangle BGH involves the elements:

$$b + c, \quad h_b + h_c, \quad A;$$

hence given any two of these elements the third one is determined. A set of elements of a triangle having this property is sometimes referred to as a *datum*.

EXERCISES

Construct a triangle, given:

1. $h_b + h_c, B, C$.
2. $h_b + h_c, b, c$.
3. $h_b + h_c, b, A$.
4. $h_b + h_c, b + c, a$.
5. $h_b + h_c,\ b + c,\ B - C$. *Hint.* The triangle BGH determines A, and $B + C = 180° - A$ is thus known, hence the angles B and C may be constructed.
6. $h_b + h_c, b - c, A$.

22. Problem. *Construct a triangle given the base, the opposite angle, and the difference of the altitudes to the other two sides* $(a, A, h_c - h_b)$.

Let ABC (Fig. 22) be the required triangle. Draw the altitudes BE, CF, and lay off $FG = BE$, so that $CG = h_c - h_b$. Draw GH parallel to AB. In the right triangle CGH we know the leg $CG = h_c - h_b$ and the angle $CHG = A$; hence this triangle may be constructed, and the length CH is determined.

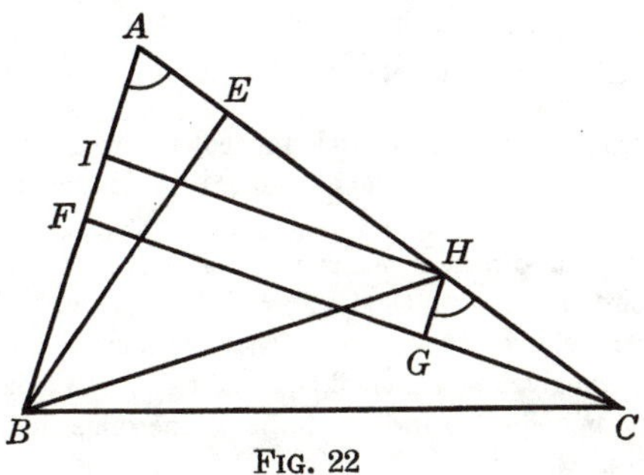

FIG. 22

We now show that $CH = b - c$. Drop the perpendicular HI from H upon AB. We have $HI = FG = BE$, hence the two right triangles ABE, AHI are congruent, having the angle A in common and $BE = HI$; therefore $AH = AB$, so that:

$$CH = CA - AH = CA - AB = b - c.$$

With $b - c$ known, the proposed problem is reduced to a known problem (§ 18).

However, the required triangle may be obtained directly from the triangle CHG. The vertex C belongs to the required triangle. From the isosceles triangle ABH we have angle $AHB = \frac{1}{2}(180° - A)$; but:

$$\angle AHG = 180° - \angle GHC = 180° - A;$$

hence BH is the bisector of the angle AHG and thus furnishes a locus for the vertex B of ABC, the other locus being the circle (C, a). The vertex A is now obtained as the trace on CH, produced, of the mediator of BH.

If the given angle A is obtuse, the triangle CGH will include not A, but its supplement, and the problem may be solved in the same way.

Note. The preceding discussion shows that the elements:

$$b - c, \quad h_c - h_b, \quad A$$

constitute a datum.

EXERCISES

Construct a triangle, given:

1. $h_c - h_b, B, C.$

2. $h_c - h_b, b, c.$

3. $h_c - h_b, b - c, B - C.$

4. $h_c - h_b, A, b + c.$

SUPPLEMENTARY EXERCISES

1. Through a given point to draw a circle tangent to two given parallel lines.

2. Through a given point to draw a line passing through the inaccessible point of intersection of two given lines.

3. Draw a line of given direction meeting the sides AB, AC of a given triangle ABC in the points B', C' such that $BB' = CC'$.

4. Through a given point to draw a line so that the sum (or the difference) of its distances from two given points shall be equal to a given length. Discuss two cases: when the two given points are to lie on the same side, and on opposite sides, of the required line.

5. In a given equilateral triangle to inscribe another equilateral triangle, one of the vertices being given.

6. In a given square to inscribe another square, given one of the vertices.

7. On the side CD of a given parallelogram $ABCD$ to find a point P such that the angles BPA and BPC shall be equal.

8. Construct a parallelogram so that two given points shall constitute one pair of its opposite vertices, and the other pair of vertices shall lie on a given circle.

9. Through a point of intersection of two given circles to draw a line so that the two chords intercepted on it by the two circles shall (i) be equal; (ii) have a given ratio.

10. Through a point of intersection of two circles to draw a line so that the sum of the two chords intercepted on this line by the two circles shall be equal to a given length.

11. Through a point of intersection of two circles to draw a line so that the two chords determined on it by the two circles shall subtend equal angles at the respective centers.

12. Given an angle and a point marked on one side, find a second point on this side which shall be equidistant from the first point and from the other side of the angle.
13. With a given radius to describe a circle having its center on one side of a given angle and intercepting a chord of given length on the other side of this angle.
14. With a given point as center to describe a circle which shall intercept on two given parallel lines two chords whose sum shall be equal to a given length.
15. Draw a line parallel to the base BC of a given triangle ABC and meeting the sides AB, AC in the points B', C' so that the trapezoid $BB'C'C$ shall have a given perimeter.
16. Construct a triangle so that its sides shall pass through three given noncollinear points and shall be divided by these points internally in given ratios.

REVIEW EXERCISES

CONSTRUCTIONS

1. In a given circle to draw a diameter such that it shall subtend a given angle at a given point.
2. Draw a line on which two given circles shall intercept chords of given lengths.
3. Place two given circles so that their common internal (or external) tangents shall form an angle of given magnitude.
4. Construct a right triangle given the altitude to the hypotenuse, two points on the hypotenuse, and a point on each of the two legs.
5. Through a given point of the altitude, extended, of a right triangle to draw a secant so that the segment intercepted on that secant by the sides of the right angle shall have its midpoint on the hypotenuse.
6. Through two given points, collinear with the center of a given circle, to draw two lines so that they shall intersect on the circle, and the chords which the circle intercepts on these lines shall be equal.
7. Through a given point to draw a line so that the segment intercepted on it by two given parallel lines shall subtend a given angle at another given point.
8. Construct a triangle given the base, an adjacent angle, and the trace, on the base, of the circumdiameter (i.e., the diameter of the circumscribed circle) passing through the opposite vertex.
9. Construct a triangle given, in position, the inscribed circle, the midpoint of the base, and a point on the external bisector of a base angle.
10. Construct a triangle ABC given, in position, a line u on which the base BC is to lie, a point D of the circumcircle, a point E of the side AB, the circumradius R, and the length a of the base BC.
11. In a given triangle to inscribe an equilateral triangle of given area.
12. With a given radius to draw a circle so that it shall pass through a given point and intercept on a given line a chord of given length.
13. Draw a circle tangent to two concentric circles and passing through a given point.

14. Construct a right triangle given the altitude to the hypotenuse and the distance of the vertex of the right angle from the trace on a leg of the internal bisector of the opposite acute angle.

15. Construct a rectangle so that one of its vertices shall coincide with a vertex of a given triangle and the remaining three vertices shall lie on the three circles having for diameters the sides of the triangle.

16. Construct a triangle given a median and the circumradii of the two triangles into which this median divides the required triangle.

17. Construct a triangle ABC given $a - b$, $h_b + h_c$, A.

18. On a given line AB to find a point P such that if PT, PT' are the tangents from P to a given circle, we shall have angle $APT = BPT'$.

19. On the sides AB, AC of the triangle ABC to mark two points P, Q so that the line PQ shall have a given direction and that $PQ:(BP + CQ) = k$, where k is a given number (ratio).

20. Draw a line perpendicular to the base of a given triangle dividing the area of the triangle in the given ratio $p:q$.

21. Through a given point to draw a line bisecting the area of a given triangle.

22. Draw a line having a given direction so that the two segments intercepted on it by a given circle and by the sides of a given angle shall have a given ratio.

23. Through a vertex of a triangle to draw a line so that the product of its distances from the other two vertices shall have a given value, k^2.

24. Through two given points A, B to draw two lines AP, BQ meeting a given line PQ in the points P, Q so that $AP = BQ$, and so that the lines AP, BQ form a given angle.

25. Through a given point R to draw a line cutting a given line in D and a given circle in E, F so that $RD = EF$.

26. With a given point as center to draw a circle so that two points determined by it on two given concentric circles shall be collinear with the center of these circles.

27. Inscribe a square in a given quadrilateral.

PROPOSITIONS

28. The circle through the vertices A, B, C of a parallelogram $ABCD$ meets DA, DC in the points A', C'. Prove that $A'D:A'C' = A'C:A'B$.

29. Of the three lines joining the vertices of an equilateral triangle to a point on its circumcircle, one is equal to the sum of the other two.

30. Three parallel lines drawn through the vertices of a triangle ABC meet the respectively opposite sides in the points X, Y, Z. Show that:

$$\text{area } XYZ:\text{area } ABC = 2:1.$$

31. If the distance between two points is equal to the sum (or the difference) of the tangents from these points to a given circle, show that the line joining the two points is tangent to the circle.

32. Two parallel lines AE, BD through the vertices A, B of the triangle ABC meet a line through the vertex C in the points E, D. If the parallel through E to BC meets AB in F, show that DF is parallel to AC.

33. A variable chord AB of a given circle is parallel to a fixed diameter passing through a given point P. Show that the sum of the squares of the distances of P from the ends of AB is constant and equal to twice the square of the distance of P from the midpoint of the arc AB.

34. The points A', B', C' divide the sides BC, CA, AB of the triangle ABC internally in the same ratio, k. Show that the three triangles $AB'C'$, $BC'A'$, $CA'B'$ are equivalent, and find the ratio of the areas ABC, $A'B'C'$.

35. The sides BA, CD of the quadrilateral $ABCD$ meet in O, and the sides DA, CB meet in O'. Along OA, OC, $O'A$, $O'C$ are measured off, respectively, OE, OF, $O'E'$, $O'F'$ equal to AB, DC, AD, BC. Prove that EF is parallel to $E'F'$.

36. The point P of a circle, center O, is projected into N upon a diameter AOB. Along PO lay off $PQ = 2\,AN$. If AQ meets the circle again in R, prove that angle $AOR = 3\,AOP$.

37. If P is any point on a semicircle, diameter AB, and BC, CD are two equal arcs, then if $E = (CA, PB)$, $F = (AD, PC)$, prove that AD is perpendicular to EF.

38. In the triangle ODE the side OD is smaller than OE and O is a right angle. A, B are two points on the hypotenuse DE such that angle $AOD = BOD = 45°$. Show that the line MO joining O to the midpoint M of DE is tangent to the circle OAB.

39. From the point S the two tangents SA, SB and the secant SPQ are drawn to the same circle. Prove that $AP{:}AQ = BP{:}BQ$.

40. On the radius OA, produced, take any point P and draw a tangent PT; produce OP to Q, making $PQ = PT$, and draw a tangent QV; if VR be drawn perpendicular to OA, meeting OA at R, prove that $PR = PQ = PT$.

41. The parallel to the side AC through the vertex B of the triangle ABC meets the tangent to the circumcircle (O) of ABC at C in B', and the parallel through C to AB meets the tangent to (O) at B in C'. Prove that $BC^2 = BC' \cdot B'C$.

42. Two variable transversals PQ, $P'Q'$ determine on two fixed lines OPP', OQQ' two segments PP', QQ' of fixed lengths. If L, M are two points on PQ, $P'Q'$ such that $PL{:}LQ = P'M{:}MQ'$ = a constant ratio, prove that LM is fixed in magnitude and direction.

43. If Q, R are the projections of a point M of the internal bisector AM of the angle A of the triangle ABC upon the sides AC, AB, show that the perpendicular MP from M upon BC meets QR in the point N on the median AA' of ABC.

44. A circle touching AB at B and passing through the incenter I (i.e., the center I of the inscribed circle) of the triangle ABC meets AC in H, K. Prove that IC bisects the angle HIK.

45. AB, CD are two chords of the same circle, and the lines joining A, B to the midpoint of CD make equal angles with CD. Show that the lines joining C, D to the midpoint of AB make equal angles with AB.

46. Three pairs of circles (B), (C); (C), (A); (A), (B) touch each other in D, E, F. The lines DE, DF meet the circle (A) again in the points G, H. Show that GH passes through the center of (A) and is parallel to the line of centers of the circles (B) and (C).

47. The mediators of the sides AC, AB of the triangle ABC meet the sides AB, AC in P, Q. Prove that the points B, C, P, Q lie on a circle which passes through the circumcenter (i.e., the center of the circumscribed circle) of ABC.

48. MNP, $M'N'P'$ are two tangents to the same circle PQP', and AM, BN, AM', BN' are perpendiculars to them respectively from the two given points A, B. If $MP:PN = M'P':P'N'$, prove that the two tangents are parallel.

49. ABC is a triangle inscribed in a circle; DE is the diameter bisecting BC at G; from E a perpendicular EK is drawn to one of the sides, and the perpendicular from the vertex A on DE meets DE in H. Show that EK touches the circle GHK.

50. If the internal bisector of an angle of a triangle is equal to one of the including sides, show that the projection of the other side upon this bisector is equal to half the sum of the sides considered.

51. From two points, one on each of two opposite sides of a parallelogram, lines are drawn to the opposite vertices. Prove that the straight line through the points of intersection of these lines bisects the area of the parallelogram.

52. The point B being the midpoint of the segment AC, the circle (A, AB) is drawn, and upon an arbitrary tangent to this circle the perpendicular CD is dropped from the point C. Show that angle $ABD = 3\,BDC$.

53. Show that the line of centers of the two circles inscribed in the two right triangles into which a given right triangle is divided by the altitude to the hypotenuse is equal to the distance from the incenter of the given triangle to the vertex of its right angle.

54. ABC is an equilateral triangle, D a point on BC such that BD is one-third of BC, and E is a point on AB equidistant from A and D. Show that $CE = EB + BD$.

55. If a line AB is bisected by C and divided by D unequally internally or externally, prove that $AD^2 + DB^2 = 2(AC^2 + CD^2)$.

56. Let M be the midpoint of chord AB of a circle, center O; on OM as diameter draw another circle, and at any point T of this circle draw a tangent to it meeting the outer circle in E. Prove that $AE^2 + BE^2 = 4\,ET^2$.

57. If M, N, P, Q are the midpoints of the sides AB, BC, CD, DA of a square $ABCD$, prove that the intersections of the lines AN, BP, CQ, DM determine a square of area one-fifth that of the given square $ABCD$.

58. If M, N are points on the sides AC, AB of a triangle ABC and the lines BM, CN intersect on the altitude AD, show that AD is the bisector of the angle MDN.

LOCI

59. A, B, C, D are fixed points on a circle (O). The lines joining C, D to a variable point P meet (O) again in Q, R. Find the locus of the second point of intersection S of the two circles PQB, PRA.

60. Find the locus of a point at which two given circles subtend equal angles.

61. Given two points A, B, collinear with the center O of a given circle, and a variable diameter PQ of this circle, find the locus of the second point of intersection of the two circles APO, BQO.

62. On the sides OA, OB of a given angle O two variable points A', B' are marked so that the ratio $AA':BB'$ is constant, and on the segment $A'B'$ the point I is marked so that the ratio $A'I:B'I$ is constant. Prove that the locus of the point I is a straight line.

63. A variable circle, passing through the vertex of a given angle, meets the sides of this angle in the points A, B. Show that the locus of the ends of the diameter parallel to the chord AB consists of two straight lines.

64. AA', BB' are two rectangular diameters of a given circle (O). A variable secant through B meets (O) in M and AA' in N. Show that the point of intersection P of the tangent to (O) at M with the perpendicular to AA' at N describes a straight line.

65. Through the center O and the fixed point A of a given circle (O) a variable circle (C) is drawn meeting (O) again in D. Find the locus of the point of intersection M of the tangents to the circle (C) at the points O and D. Show that the line MC is tangent to a fixed circle concentric with (O).

66. A variable circle touches the sides OB, OD of a fixed angle in B and D; E is the point of contact of this circle with the second tangent to it from a fixed point A of the line OB. Show that the line DE passes through a fixed point.

67. A variable line PAB through the fixed point P meets the sides OA, OB of a given angle O in the points A, B. On the lines OA, OB the points A', B' are constructed so that the ratios $OA':OA$ and $OB':OB$ are constant. Prove that the line $A'B'$ passes through a fixed point.

II

SIMILITUDE AND HOMOTHECY

A. SIMILITUDE

23. Method of Similitude. By disregarding one of the conditions of a problem it is sometimes possible to construct a figure similar to the one required. From the figure thus constructed and the omitted condition it is usually possible to derive an element which enables us to solve the proposed problem. The following examples illustrate this method.

24. Problem. *Construct a square given the sum of its side and its diagonal.*

Since all squares are similar, we begin by constructing a square arbitrarily. Let a', d' denote its side and its diagonal, respectively, and let a, d be the corresponding elements of the required square. From the similitude of the two figures we have:

$$a:a' = d:d', \quad \text{or} \quad (a + d):(a' + d') = a:a'.$$

In the last proportion we know three terms, for $a + d$ is given; hence the segment a may be constructed as a fourth proportional, and the problem is reduced to constructing a square given its side.

25. Problem. *Construct a triangle similar to a given triangle and equivalent to a given square.*

Neglecting the area, construct a triangle $A'B'C'$ similar to the required triangle ABC. If $a' = B'C'$, h' is the altitude to this side, and m' the side of the square equivalent to the area of $A'B'C'$, we have $m'^2 = a' \cdot \frac{1}{2} h'$; hence m' may be constructed as a third proportional.

The side $a = BC$ may now be determined from the proportion

$$a:a' = m:m',$$

where m is the side of the given square, and the required triangle is readily constructed.

On the side $B'C'$ lay off $B'C = a$. The parallel through C to $A'C'$ meets $A'B'$ in the third vertex of the required triangle $AB'C$.

The problem has one and only one solution.

26. Problem. *Construct a triangle given the two lateral sides and the ratio of the base to its altitude* $(b, c, a:h_a = p:q)$.

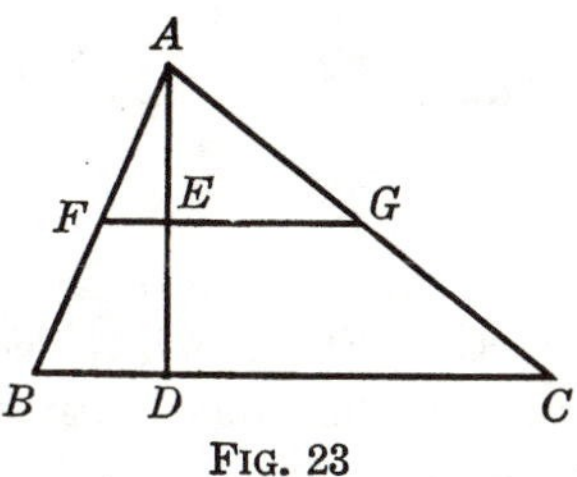

FIG. 23

Let ABC be the required triangle (Fig. 23). On the altitude AD lay off $AE = q$ and through E draw the parallel FEG to BC. From the similar triangles AFG, ABC we have:

$$AF:AG = AB:AC = c:b, \quad FG:AE = BC:AD = p:q;$$

hence $FG = p$.

Thus in the triangle AFG, similar to the required triangle, we know the base $FG = p$, the altitude $AE = q$, and the ratio of the sides $AF:AG = c:b$; hence this triangle may be constructed. Having constructed AFG lay off on AF a segment $AB = c$. The parallel to FG through B will meet AG, produced, in the third vertex C of the required triangle ABC.

The problem may have two, one, or no solutions.

27. Problem. *Construct a trapezoid given the nonparallel sides, the angle between them, and the ratio of the two parallel sides.*

Let $ABCD$ (Fig. 24) be the required trapezoid, and E the point of intersection of the nonparallel sides AD, BC. The triangles ABE, DCE are similar; hence:

$$EC:EB = ED:EA = CD:BA = p:q \text{ (the given ratio)},$$

or:

$$EC:(EB - EC) = ED:(EA - ED) = p:(q - p),$$

or:

$$EC:CB = ED:DA = p:(q - p).$$

Thus the segments EC, ED may be constructed, and therefore also the triangle DCE, of which we have two sides and the included angle.

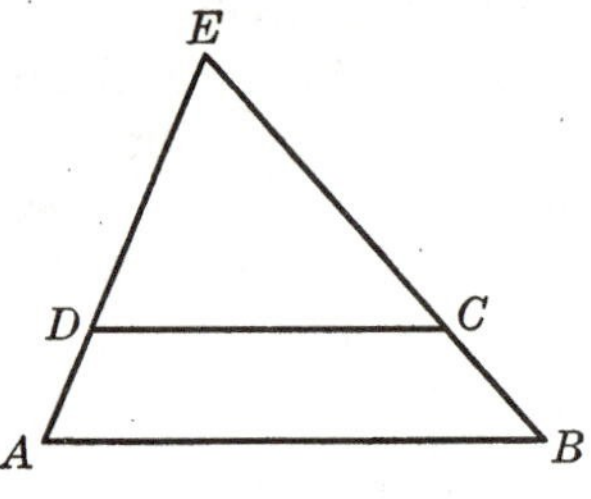

FIG. 24

Having constructed this triangle, produce ED and lay off the given length DA. The parallel to CD through A determines on EC, produced, the fourth vertex B of the required trapezoid $ABCD$.

28. Problem. *About a given circle to circumscribe an isosceles triangle in which the ratio of the lateral side to the base shall have a given value.*

All isosceles triangles in which the value of the ratio of the lateral side to the base is the same are similar, for they are divided by the altitude to the base into similar right triangles.

On the base $B'C'$, of arbitrary length, construct the isosceles triangle $A'B'C'$ so that

$$A'B':B'C' = p:q \text{ (the given ratio).}$$

Let r' and h' be the radius of the inscribed circle and the altitude to the base of this triangle, and let r, h be the analogous elements of the required triangle ABC. From the similitude of the two triangles we have:

$$h:h' = r:r',$$

and in this proportion the last three terms are known; hence h may be constructed.

At an arbitrary point D of the given circle draw the tangent t, and on the line joining D to the center lay off $DA = h$. The tangents from A to the circle and the tangent t form the required triangle ABC.

29. Problem. *Divide a given segment, m, into three parts a, b, c, such that $a:b = p:q$, and $b:c = r:s$, where p, q, r, s are given segments.*

If we determine t from the proportion:

$$q:t = r:s,$$

we have:

$$a:b:c = p:q:t.$$

On one side of an arbitrary angle A lay off $AM = m$, and on the other side lay off $AP = p$, $PQ = q$, $QT = t$. The parallels through P, Q to the line MT meet AM in the points X, Y such that $AX = a$, $XY = b$, $YM = c$.

30. Definition. Given two triangles ABC, $A'B'C'$, angle $A = A'$, $B = B'$, $C = C'$, if the rotation determined by the points A, B, C, taken in that order, is counterclockwise, and the rotation A', B', C' is clockwise, or vice versa, the two triangles are said to be *inversely similar*.

If the senses of rotation ABC, $A'B'C'$ are the same, the triangles are *directly similar*.

EXERCISES

1. If in two triangles two pairs of sides are proportional, and the angle opposite the longer of the two sides considered in one triangle is equal to the corresponding angle in the other triangle, show that the two triangles are similar. Consider the case when the given angle lies opposite the shorter side.
2. If the corresponding sides of two triangles are perpendicular, show that the two triangles are similar.

Construct a triangle, given:

3. $A, B, 2p$.
4. $A, B, b + c$.
5. $A, B, h_a - h_b$.
6. $a:b:c, R$.
7. $A, a:c, h_c$.
8. $A, a:b, 2p$.
9. $B - C, a:(b + c), m_a + m_b$.
10. $a:b, b:c, t_a + t_b - t_c$.

11. Construct a triangle given an angle, the bisector of this angle, and the ratio of the segments into which this bisector divides the opposite side.
12. Construct a right triangle given the perimeter and the ratio of the squares of the two legs.
13. Construct a triangle given the area and the angles which a median makes with the two including sides.
14. Construct a parallelogram given the ratios of one side to the two diagonals, and the area.
15. Given a circle and two radii, produced, draw between them a tangent to the circle which shall be divided by the point of contact in a given ratio.
16. In a given circle to inscribe an isosceles triangle given the sum of its base and altitude.
17. Construct a triangle given a, A, $mb + nc = s$, where m and n are two given constants.
18. If in two triangles two angles are equal and two other angles are supplementary, show that the sides opposite the equal angles are proportional to the sides opposite the supplementary angles.

B. HOMOTHECY

31. Definition. If the corresponding sides of two similar polygons are parallel, the two polygons are said to be *similarly placed*, or *homothetic*.

32. Theorem. *The lines joining corresponding vertices of two homothetic polygons are concurrent (i.e., meet in a point).*

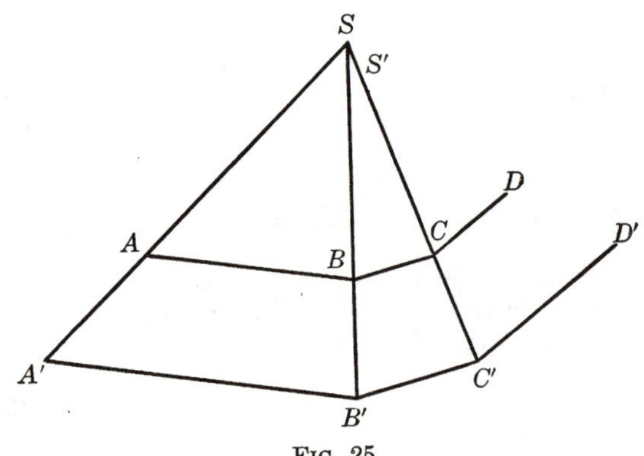

FIG. 25

Let $ABCD\ldots$, $A'B'C'D'\ldots$ be two homothetic polygons, and $S \equiv (AA', BB')$ (Fig. 25). In order to prove the proposition it is necessary to show that a line joining the next pair C, C' of corresponding vertices will also pass through S. If the line CC' does not pass through S, let S' be the point where it meets BB'. From the two pairs of similar triangles SAB and $SA'B'$, $S'BC$ and $S'B'C'$ we have:

$$SB':SB = A'B':AB, \quad S'B':S'B = B'C':BC.$$

But, by assumption:

$$B'C':BC = A'B':AB;$$

hence:

$$SB':SB = S'B':S'B, \quad \text{or} \quad SB':(SB - SB') = S'B':(S'B - S'B'),$$

or:

$$SB':BB' = S'B':BB';$$

therefore S' coincides with S.

33. Definitions. The point S (§ 32) is said to be the *center of similitude*, or the *homothetic center* of the two polygons.

The constant ratio

$$SA' : SA = \cdots = A'B' : AB = \cdots = k$$

is called the *ratio of similitude*, or the *homothetic ratio* of the two figures. This ratio is given either as a number or as the ratio of two given segments, say p, q.

The relation between the two figures is called a *homothecy.*

34. Problem. *Given a polygon $ABCD\ldots$ to construct a second polygon $A'B'C'D'\ldots$ homothetic to the first, so that their homothetic ratio shall have a given value, k, and a given point S shall be their homothetic center.*

On the lines $SA, SB, SC, \ldots$ (Fig. 26) joining the vertices $A, B, C, \ldots$ of the given polygon to the given homothetic center S construct the points $A', B', C', \ldots$ so that:

$$SA' : SA = SB' : SB = SC' : SC = \cdots = k.$$

The polygon $A'B'C'\ldots$ thus constructed satisfies the conditions of the problem.

Indeed, the triangles SAB, $SA'B'$ are similar; hence $A'B'$ is parallel to AB and:

$$A'B' : AB = SA' : SA = k.$$

Likewise for the other pairs of the sides of the two polygons. Corresponding pairs of sides of the two polygons being parallel,

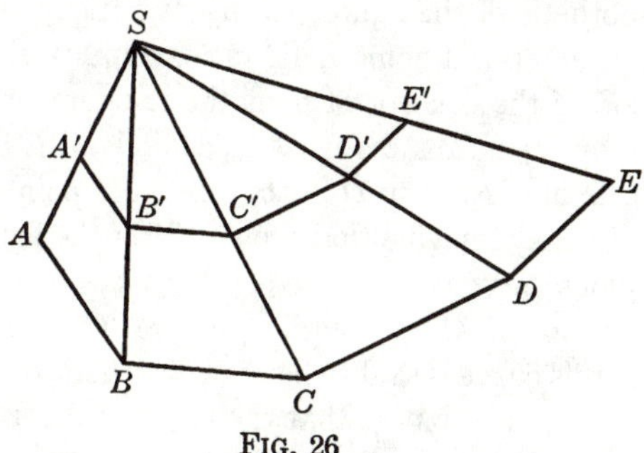

FIG. 26

corresponding angles are equal. Thus the two polygons are similar, they are similarly placed, their ratio of similitude has the given value, k, and the lines joining corresponding vertices obviously meet in S.

35. Definitions. The points A and A', B and B', ... (§ 34) are said to be *corresponding points*, or *homologous points*, or *homothetic points* in the homothecy.

The value of the homothetic ratio, k, may be either positive or negative. In the first case two homologous points lie on the same side of the homothetic center, and the two polygons are *directly homothetic;* in the second case, two homologous points lie on opposite sides of the homothetic center, and the two polygons are *inversely homothetic.*

Of particular interest is the case when $k = -1$. The homothetic center S is then the midpoint of the segment determined by any two corresponding points. The polygons are then said to be *symmetric* with respect to the point S, and S is said to be their *center of symmetry.*

The homothecy having for center S and for ratio k will, for the sake of brevity, be denoted by (S, k).

36. Generalizations. It is clear that given the homothetic center S and the homothetic ratio k, the points $A', B', C', \ldots$ may be constructed, whether the given points $A, B, C, \ldots$ are the vertices of a convex polygon or not. We may therefore extend the concept of homothetic figures in the following way. Given a figure (F) consisting of any number of points $A, B, C, \ldots$ distributed in the plane in any manner, we choose a fixed point S and a fixed ratio k, and construct the points $A', B', C', \ldots$ on the lines $SA, SB, SC, \ldots$ so that:

$$SA':SA = SB':SB = \cdots = k.$$

The figure (F') consisting of the newly constructed points A', B', C', ... is the homothetic of the figure (F), by definition, with S and k as the homothetic center and homothetic ratio, respectively.

The extension of the concept of homothetic figures may be carried further. It is by no means essential to comply with the restriction that the given figure (F) is to consist of isolated points, as we have done so far. We may imagine that a point M of the figure (F) moves along a continuous curve (C). If we construct the homothetic points M' for the positions of M, we obtain a curve (C') belonging to the figure (F'), and this curve is said to be the homothetic of the curve (C).

In particular we may suppose that the point M describes a straight line, or a circle.

37. Theorem. *Given two homothetic figures, if a point of one figure describes a straight line, the homologous point of the second figure also describes a straight line, and the two lines are parallel.*

Let the point M describe the line u of the figure (F). Let P, Q, R be three positions of M and let P', Q', R' be the corresponding points in the homothetic figure (F'). From the pairs of similar triangles SPQ and $SP'Q'$, SQR and $SQ'R'$ it follows that PQ and $P'Q'$, QR and $Q'R'$ are pairs of parallel lines. Now P, Q, R are collinear and the two lines $P'Q'$, $Q'R'$ have the point Q' in common; hence the three points P', Q', R' are collinear. But the two lines PQ, $P'Q'$ are determined by the pairs of points P and Q, P' and Q'; the above argument therefore shows that the corresponding point R' of any point R of the line u lies on the line $P'Q'$, which proves the proposition.

38. *Remark I.* The proposition is often stated more succinctly thus: *The homothetic of a straight line is a straight line parallel to it.*

39. *Remark II.* If the two figures (F), (F') are directly homothetic, the senses PQR, $P'Q'R'$ on the two parallel lines are the same. If the figures (F), (F') are inversely homothetic, the two senses PQR, $P'Q'R'$ are opposite.

40. Corollary. *Corresponding angles in two homothetic figures are equal,* for the two pairs of corresponding sides of the two angles are parallel and are directed either both in the same sense or both in the opposite sense.

41. Problem. *Through a given point to draw a line so that the segment intercepted on it by the sides of a given angle shall be divided by the given point in a given ratio.*

On the line AC (Fig. 27) joining the given point A to an arbitrary point C on the side BC of the given angle CBE lay off the segment AD

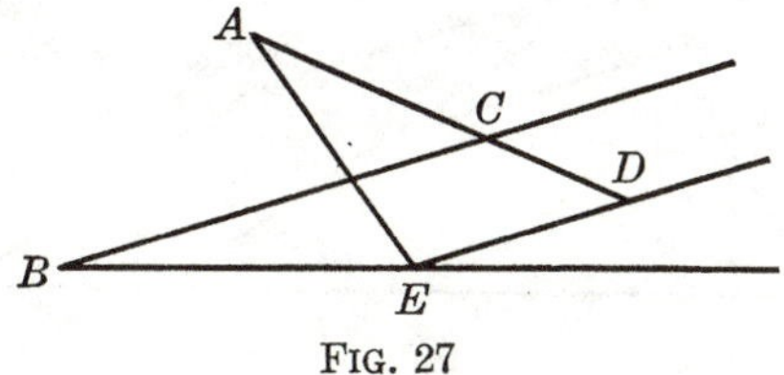

Fig. 27

so that $AC:AD$ shall be equal to the given ratio $p:q$. Thus the points C, D correspond to each other in the homothecy $(A, p:q)$; therefore when the point C describes the given line BC, the point D will describe a line parallel to BC. If E is the point where this parallel meets the other side of the given angle, AE is the required line.

42. Problem. *From a variable point D, within the triangle ABC, the perpendiculars DM, DN are dropped upon AB, AC. If $CN \cdot AC = BM \cdot AB$, find the locus of the point D.*

The points A, M, D, N (Fig. 28) lie on the circle (O) having the midpoint O of AD for center. On the other hand the given equality shows that the squares of the tangents from the points B, C to (O) are

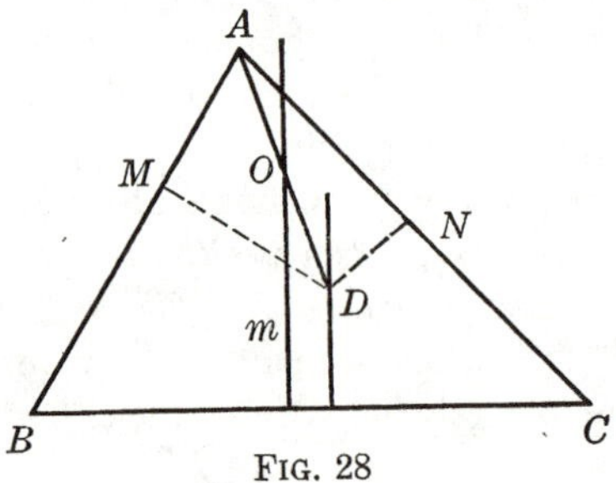

FIG. 28

equal; hence $CO = BO$ (§ 11, locus 2). Thus the locus of O is the mediator m of BC. The locus of D is therefore the straight line which corresponds to the line m in the homothecy $(A, 2)$.

43. Problem. *To draw a secant meeting the sides AB, AC of the given triangle ABC in the points D, E so that $BD = DE = EC$.*

Let $ABCDE$ (Fig. 29) be the required figure. If the parallel to DE through A meets BE in F, and the parallel to AC through F meets

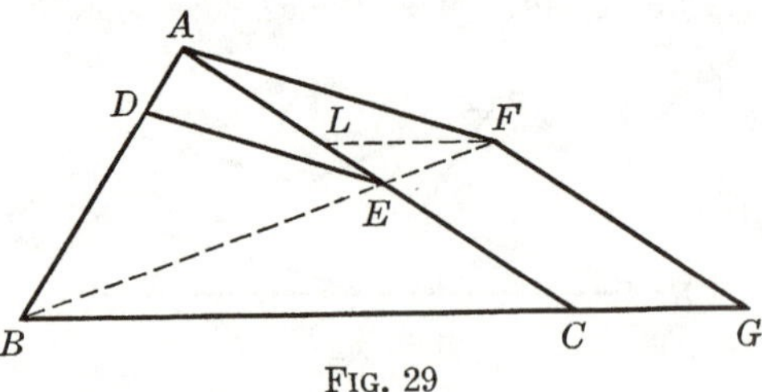

FIG. 29

BC in G, the quadrilaterals $BDEC$, $BAFG$ are clearly homothetic, with B as homothetic center; hence $BA = AF = FG$. Now the quadrilateral $BAFG$ may be constructed in the following way. On CA lay off $CL = AB$, and if the parallel to BC through L meets the circle (A, AB) in F, the parallel to AC through F meets BC in G, the fourth required vertex.

The parallel through the point $E = (AC, BF)$ to the line AF is the required secant.

44. Problem. *Construct a triangle given the vertical angle and the sums of the base with each of the other two sides* $(A, a + b, a + c)$.

Let ABC be the required triangle (Fig. 30). Produce AB and AC. Lay off $BD = BC$, $CE = BC$; hence $AD = a + c$, $AE =$

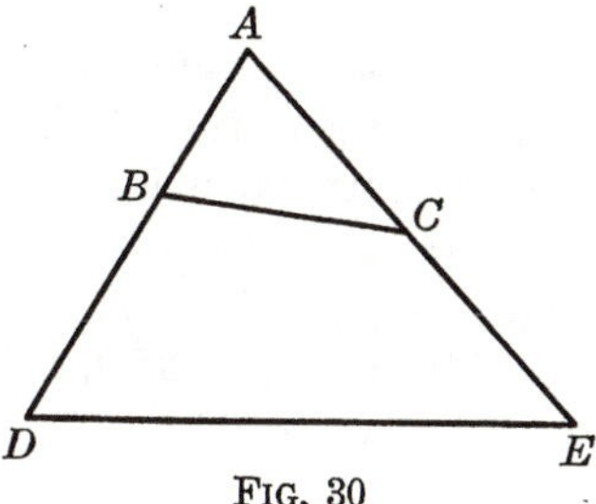

FIG. 30

$a + b$, and the triangle DAE may be constructed, since we know two sides and the included angle. In order to pass from this triangle to the required triangle ABC it is necessary to draw a secant BC so that $DB = BC = CE$, i.e., the problem is reduced to the preceding one (§ 43).

EXERCISES

1. Show that a homothecy is determined, given: (a) the homothetic center and a pair of corresponding points; (b) the homothetic ratio and a pair of corresponding points; (c) two pairs of corresponding points.
2. What figure is the homothetic of a parallelogram? Of a rectangle? Of a square?
3. A, A' and B, B' are two pairs of homologous points in two homothetic figures (F), (F'), and M is a point of the figure (F). The parallels through A', B' to the lines AM, BM, respectively, intersect in M'. Prove that M, M' are two corresponding points in the two figures (F), (F').
4. If two triangles are homothetic, show that their circumcenters, etc., are homologous points, and their altitudes, medians, etc., are homologous lines in the two homothetic figures.
5. Through a given point to draw a line passing through the inaccessible point of intersection of two given lines.
6. Construct a triangle given A, $a - b$, $a - c$.
7. Construct a triangle given A, $a + b$, $a - c$.
8. Draw a line parallel to the base of a given trapezoid so that the segment intercepted on it by the nonparallel sides of the trapezoid shall be trisected by the diagonals.

9. In a given circle to draw a chord which shall be trisected by two radii given in position.

10. Construct a triangle given a median and the two angles which this median makes with the including sides.

11. Through a given point to draw a line so that the two segments determined on it by three given concurrent lines shall have a given ratio.

12. Given three concurrent lines and a fourth line, to draw a secant so that the three segments determined on it by the four lines shall have given ratios.

45. Problem. *In a given triangle to inscribe another triangle whose sides shall be parallel to the sides of a (second) given triangle.*

Let DEF (Fig. 31) be the required triangle inscribed in the given triangle ABC. Draw arbitrarily a line $E'F'$ parallel to EF meeting AC, AB in E', F', and let the parallels through E', F' to ED, FD, respectively, meet in D'. The triangles DEF, $D'E'F'$ are homothetic

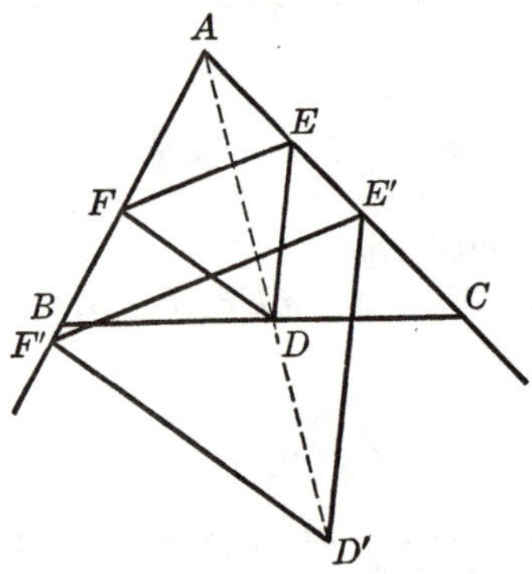

FIG. 31

and the point $A = (EE', FF')$ is their homothetic center; hence the points D, D' are collinear with A. Now the triangle $D'E'F'$ is readily constructed, and the line AD' meets the side BC in the vertex D of the required triangle; the construction is easily completed.

The problem has one and only one solution.

46. Problem. *In a given triangle to inscribe a parallelogram having a given angle and having its adjacent sides in a given ratio.*

Let $DEFG$ (Fig. 32) be the required parallelogram inscribed in the given triangle ABC. Draw an arbitrary parallel $D'E'$ to DE and on the parallel $D'G'$ to DG lay off $D'G'$ so that:

$$D'E':D'G' = DE:DG = k = \text{the given ratio}.$$

Through E', G' draw parallels to EF, GF, respectively.

The parallelograms $DEFG$, $D'E'F'G'$ are homothetic, and the point $A = (DD', EE')$ is their homothetic center; hence the pairs of points F and F', G and G' are collinear with the point A. Now the parallelo-

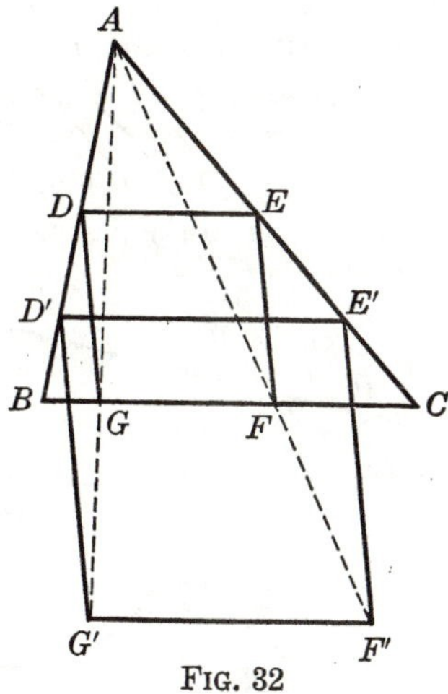

FIG. 32

gram $D'E'F'G'$ is readily constructed, and the lines AF', AG' meet BC in the two vertices F, G of the required parallelogram $DEFG$, which is easily completed.

For convenience the base BC of ABC may be taken for the side $D'E'$ of $D'E'F'G'$.

EXERCISES

1. In a given triangle to inscribe a triangle whose sides shall be parallel to the internal bisectors of the given triangle. Two solutions.
2. In a given triangle to inscribe a triangle whose sides shall be perpendicular to the sides of the given triangle. Two solutions.
3. In a given triangle to inscribe a square.
4. In a given triangle to inscribe a rectangle similar to a given rectangle.
5. In a given triangle to inscribe a parallelogram given the ratio of its sides and the angle between the diagonals.
6. Given a triangle ABC, construct a square so that two vertices shall lie on BA and CA, both produced, and the other two vertices on the side BC.
7. In a given semicircle to inscribe a rectangle similar to a given rectangle.
8. In a given circular sector to inscribe a square. Two cases: (i) one vertex or (ii) two vertices lie on the circumference.
9. Construct a square so that two vertices shall lie on a given line and the other two on a given circle.

10. Draw a parallel to the base of a given triangle so that the segment intercepted on it by the other two sides shall subtend a given angle at a given point of the base.

11. If m is the side of the square inscribed in the triangle ABC so that two vertices lie on the side $BC = a$, and h is the altitude upon BC, show that $m(a + h) = ah$.

47. Theorem. *Given two homothetic figures, if a point of the first figure describes a circle, the corresponding point of the second figure also describes a circle.*

Consider the homothecy (S, k). Let O (Fig. 33) be the center and A a point on the circle belonging to the first figure, and let O', A' be

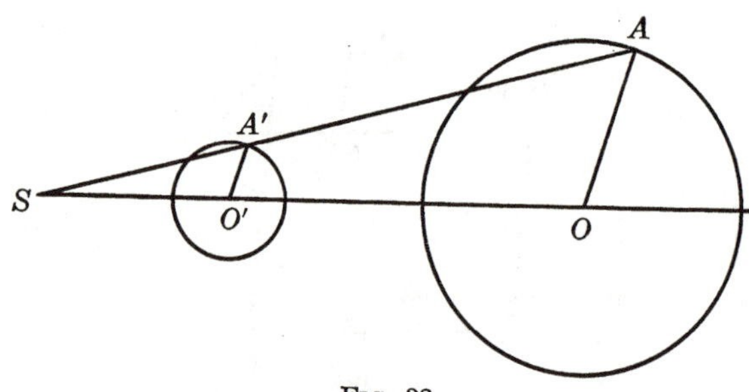

FIG. 33

the homothetic points of the second figure. It follows from the construction that the two triangles SOA, $SO'A'$ are similar; hence:

$$O'A':OA = SA':SA = k; \quad \text{hence} \quad O'A' = k \cdot OA.$$

Thus the length $O'A'$ is constant, i.e., as the point A describes the circle of the first figure, its corresponding point A' moves so as to remain at a fixed distance $k \cdot OA$ from the fixed point O'; hence A' describes a circle having O' for center and $k \cdot OA$ for radius.

The proposition is often stated more succinctly thus: *The homothetic of a circle is a circle.*

Note. It should be carefully observed that (a) the centers O, O' of the two homothetic circles are corresponding points in the two homothetic figures, and that (b) the ratio of the radii of the two circles is equal to the homothetic ratio.

If one of the circles passes through the homothetic center, the other circle will also pass through that center, and the two circles will be tangent at that point.

48. Problem. *If PQ is a variable diameter of a given circle, and A, B two fixed points collinear with the center O of the circle, find the locus of the point $M = (AP, BQ)$.*

If B' (Fig. 34) is the symmetric of B with respect to O, the two triangles OQB, OPB' are congruent, and from the equality of their angles it follows that the lines BM, PB' are parallel; hence:

$$AM:AP = AB:AB'.$$

Now the latter ratio is known; hence the two variable points P, M are collinear with the fixed point A and the ratio of their distances from A is constant, or, in other words, the points P, M correspond to each other in the homothecy $(A, AB':AB)$, and since P describes the given circle, the locus of M is a circle of known center and known radius (§ 47).

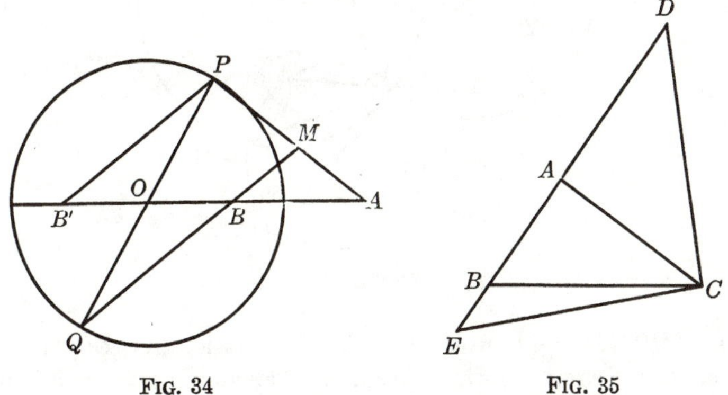

FIG. 34 FIG. 35

49. Problem. *Construct a triangle given the base, the opposite angle, and the ratio of the sum of the other two sides to their difference* $[a, A, (b + c):(b - c) = p:q]$.

Let ABC (Fig. 35) be the required triangle. Lay off $AE = AD = AC$. We have (§§ 17, 18):

$$\angle BDC = \tfrac{1}{2} A, \quad \angle BEC = 90° - \tfrac{1}{2} A;$$

hence we know a geometric locus (§ 11, locus 7) for each of the points E, D. Now the two variable points E, D are collinear with the fixed point B and the ratio $BD:DE$ is given; hence the two points correspond to each other in the homothecy $(B, -p:q)$, i.e., from the known locus of E we may derive a locus for D, and this locus together with the locus for D that we had before determines the position of D. The mediator of DC meets DBE in the third vertex of the required triangle ABC.

As an exercise, draw a complete figure for this problem.

50. Problem. *Through a point of intersection of two circles to draw a line so that the chords intercepted on it by the circles shall have a given ratio.*

Let A (Fig. 36) be a point common to the two given circles (E), (F), and M any point on (E). On the line AM construct the point M' so

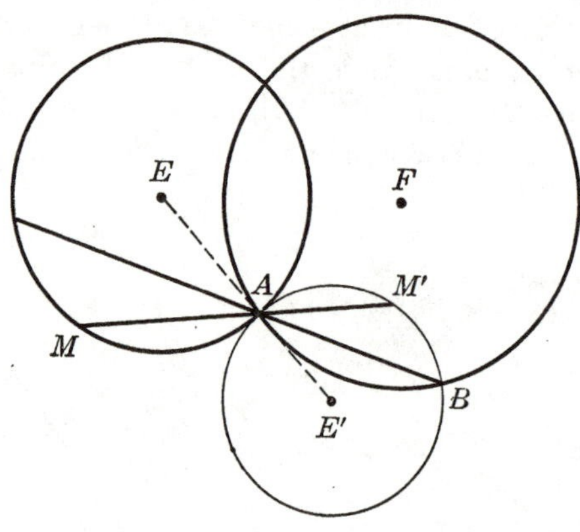

FIG. 36

that $AM':AM = -k$, where k is the given ratio. As the point M describes the circle (E), the point M' will describe the circle (E') corresponding to (E) in the homothecy $(A, -k)$. If B is the second point of intersection of the circles (E'), (F), the line AB is the required secant.

The two circles (E'), (F) have the point A in common and cannot be tangent to each other (§ 47); hence the point B may always be found.

If the sign of the given ratio k is not specified, the point M' may be constructed so that M and M' lie either on the same side or on opposite sides of the point A, and the problem will have two solutions. Furthermore, if the problem does not specify which of the two given circles is to be considered first in forming the given ratio k, the problem will have four solutions.

As an exercise, draw a complete figure for this problem, with all the circles involved.

EXERCISES

1. Through a given point on a circle to draw a chord so that it shall be bisected by a given chord.
2. Through a given point to draw a secant so that the segment intercepted on it by a given line and a given circle shall be divided by the given point in a given ratio.
3. Through a given point to draw a secant so that the ratio of the distances of the given point from the points of intersection of the secant with a given circle shall have a given value.
4. On two given circles to find two points collinear with, and equidistant from, a given point.
5. Construct a triangle given two sides and the bisector of the included angle (b, c, t_a).
6. Draw a line on which two concentric circles shall determine two chords having a given ratio.
7. Given three concentric circles to draw a secant so that the segment between the first and the second circles shall be equal to the segment between the second and the third circles.
8. A variable point P moves on a fixed circle, center C, and A is a fixed point. Find the locus of the point of intersection of the line AP with the internal bisector of the angle ACP.
9. A variable triangle has a fixed base and a fixed circumcircle (i.e., circumscribed circle). Find the locus of the midpoints of the lateral sides, and the locus of the midpoint of the segment joining the midpoints of the lateral sides.
10. From the given points B, C the perpendiculars BB', CC' are dropped upon a variable line $AB'C'$ passing through a fixed point A collinear with B and C. Find the locus of the point $M = (BC', B'C)$.
11. On the base BC of a given triangle ABC find a point P such that $AP^2 : BP \cdot PC$ shall have a given value.

51. Theorem. *If one vertex of a variable triangle is fixed, a second vertex describes a given straight line, and the triangle remains similar to a given triangle, then the third vertex describes a straight line.*

Let A (Fig. 37) be the fixed vertex, ABC the position of the variable triangle when the base BC falls on the given line p, and $AB'C'$ any other position of the variable triangle. The segment AB' subtends equal angles at the points C and C'; hence the quadrilateral $AB'CC'$ is cyclic (§ 252), and angle $ACC' = AB'C'$. Thus:

$$\angle BCC' = BCA + ACC' = BCA + AB'C'.$$

Now the last two angles are given; hence the line CC' makes a fixed angle with the given line p. Moreover, C is a fixed point; hence CC' is a fixed line, which proves the proposition.

52. *Remarks.* It should be observed that the point C is determined on p by the line through A which makes with p an angle equal to the given angle C, and that the line CC' makes with p an angle equal to the given angle A.

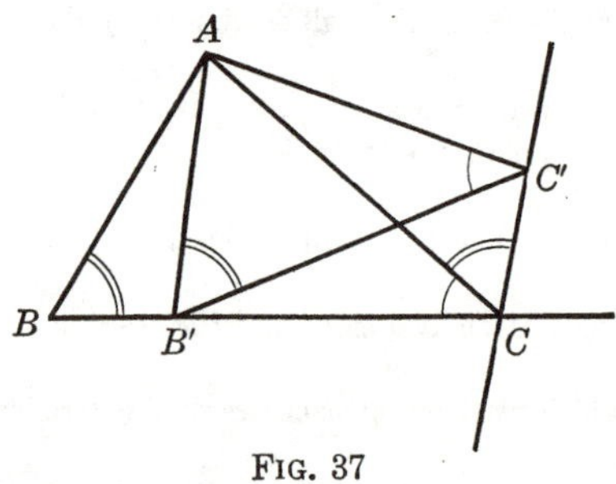

FIG. 37

The locus of the third vertex C was obtained by taking the given angles of the triangle in a certain order, i.e., one of the given angles was placed at the fixed vertex A, and a second at the vertex B describing the given line p. If the conditions of the problem do not assign any specified positions to these angles, the distribution of the angles may be made in six different ways.

Furthermore, after the angles for the vertices A and B have been chosen, the point C on the line p may be chosen on either side of the line AB. Thus the complete locus of the point C consists of twelve lines. They are parallel in pairs.

53. Theorem. *If one vertex of a variable triangle is fixed, a second vertex describes a given circle, and the triangle remains similar to a given triangle, then the third vertex describes a circle.*

Let ABC (Fig. 38) be one position of the variable triangle, where A is the fixed vertex and B lies on the given circle, center O. On AC lay off $AB' = AB$ and on AB' construct a triangle $AO'B'$ congruent to AOB so that the two triangles shall be similarly situated, i.e., so that it shall be possible to bring the triangle AOB into coincidence with $AO'B'$ by rotating AOB about the point A.

Since angle $O'AB' = OAB$, by construction, we have angle $O'AO = B'AB$. Now the last angle is given; hence AO' makes with the fixed line AO a fixed angle; the direction of AO' is therefore fixed. Moreover, AO' is equal to the given length AO; hence the point O' is fixed, and since $O'B' = OB$, the point B' describes a circle (B'). Since the triangle ABC remains similar to a given triangle, we have:

$$AC:AB' = AC:AB;$$

hence the locus of C is a circle (O'') homothetic to the locus of B', which proves the proposition.

As an exercise, show that the triangle AOO'', where O'' is the center of (O''), is directly similar to the given triangle.

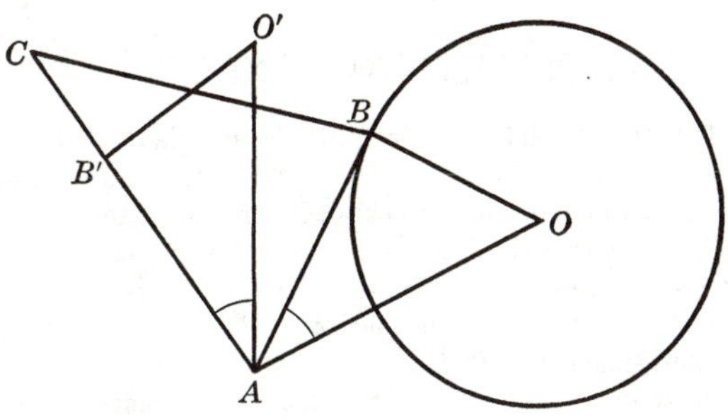

FIG. 38

Note. If the given circle (O), which constitutes the path of the vertex B, is rigidly rotated about the given fixed point A through the given angle BAC, its new position is the locus (B') of the point B'. The locus of C is the circle which corresponds to the circle (B') in the homothecy $(A, AC:AB)$.

An analogous observation may be made with regard to the preceding theorem (§ 51).

Concerning the distributions of the angles of the triangles ABC the remarks made (§ 52) in connection with the preceding theorem may be applied here.

EXERCISES

1. Construct a triangle similar to a given triangle so that one vertex shall coincide with a given point and the other two vertices shall lie on two given lines.

2. In a given triangle to inscribe a triangle similar to a (second) given triangle, one of the vertices being given.

3. Construct a triangle similar to a given triangle so that its vertices shall lie on three given lines.

4. The vertex A of the variable triangle APQ is fixed, and P moves on a fixed line CD; AP meets a fixed line parallel to CD in the point R, and $PQ = AR$; the angle APQ is constant. Prove that the locus of Q is a straight line.

5. A variable regular hexagon has a fixed vertex and its center describes a straight line. Show that the remaining vertices describe straight lines and that these lines are concurrent.
6. Construct a triangle similar to a given triangle so that one vertex shall coincide with a given point and the other two vertices shall lie on two given circles.
7. Construct a triangle similar to a given triangle so that its vertices shall lie on three given circles.
8. Construct a triangle similar to a given triangle so that one vertex shall coincide with a given point, another shall lie on a given circle, and the third on a given line.

SUPPLEMENTARY EXERCISES

1. The lines AL, BL, CL joining the vertices of a triangle ABC to a point L meet the respectively opposite sides in A', B', C'. The parallels through A' to BB', CC' meet AC, AB in P, Q, and the parallels through A' to AC, AB meet BB', CC' in R, S. Show that the four points P, Q, R, S are collinear.
2. $ABCD$ is a rhombus, and P, Q, R, S are the circumcenters of the triangles BCD, CDA, DAB, ABC. Prove that the midpoints of the segments AP, BQ, CR, DS form a rhombus similar to $ABCD$.
3. Construct a triangle ABC given its circumcircle, center O, so that area $OBC{:}OCA{:}OAB = p{:}q{:}r$, where p, q, r are given line segments.
4. Locate two points D, E on the sides AB, AC of a triangle ABC such that $AD{:}DE{:}EC = p{:}q{:}r$, where p, q, r are given line segments.

PROPERTIES OF THE TRIANGLE

A. PRELIMINARIES

54. Problem. *Divide a given segment, AB, internally and externally in a given ratio $p:q$.*

Through the ends A, B (Fig. 39) of the given segment draw any pair of parallel lines AF, GBH, and lay off $AF = p$, $BG = BH = q$.

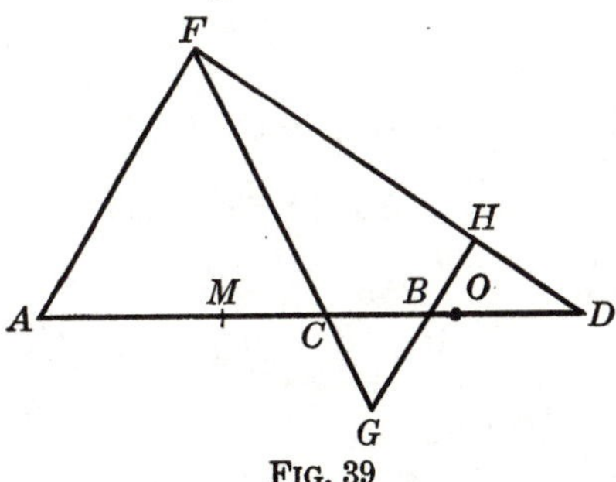

FIG. 39

The lines FG, FH meet the line AB in the required points C, D, as is readily seen from the two pairs of similar triangles CAF and CBG, DAF and DBH.

55. *Remark I.* To avoid ambiguity the statement of the problem should indicate whether the segments of AB proportional to p and q shall be adjacent to A and B, respectively, as is the case of the figure, or to B and A. Otherwise there is another pair of points, C', D', the symmetric of C, D with respect to the midpoint M of AB, which solves the problem. However, in actual applications the nature of the problem usually distinguishes between the points A, B and thus leads to only one pair of points of division.

56. *Remark II.* If $p > q$, we have $AC > CB$ and $AD > DB$, so that the midpoint M of AB lies outside the segment CD. The same relative position of the points M, C, D prevails, if $p < q$.

If $p = q$, the point C coincides with M, and there is no external point of division, since FH is then parallel to AB.

57. Theorem. *If the points C, D divide the segment AB internally and externally in the ratio $p:q$, the points B, A divide the segment DC internally and externally in the ratio $(p + q):(p - q)$.*

We have (Fig. 39):

$$AC:CB = p:q, \quad AD:DB = p:q;$$

hence:

$$(1) \quad (AC + CB):CB = (p + q):q, \quad (AD - DB):DB = (p - q):q,$$
$$(2) \quad AC:(AC + CB) = p:(p + q), \quad AD:(AD - DB) = p:(p - q).$$

Substituting AB for $AC + CB$ and for $AD - DB$ in (1) and (2), and combining the two proportions in each of these two lines we obtain:

$$DB:BC = (p + q):(p - q),$$
$$DA:AC = (p + q):(p - q).$$

58. Corollary. *If $AB = a$ and $CD = b$, we have $b = 2apq:(p^2 - q^2)$.*

Indeed, we have:

$$AD:AB = p:(p - q), \quad AC:AB = p:(p + q), \quad CD = AD - AC;$$

hence the announced result.

59. Definitions. Instead of saying that the points C, D (§ 57) divide the segment AB internally and externally in the same ratio, we shall sometimes say that the points C, D *divide the segment AB harmonically*, or that the points C, D *separate the points A, B harmonically*, or that the points C, D are *harmonic conjugates* with respect to the points A, B.

The preceding proposition (§ 57) states that the relation between the two pairs of points is mutual, so that the points A, B, in turn, divide the segment CD harmonically, and the points A, B separate the points C, D harmonically. We may thus refer to the two pairs of points A, B and C, D as *two pairs of harmonic points*, and to the two segments AB, CD as *two harmonic segments*.

60. Problem. *Given three collinear points A, B, C, construct the harmonic conjugate D of C with respect to the points A, B.*

Through the points A, B draw any pair of parallel lines AF, GBH (Fig. 39) and through C any transversal meeting AF, BG in F, G. On GB lay off $BH = BG$. The line FH meets AB in the required point D.

61. Theorem. *The feet of the two pairs of perpendiculars dropped upon a given line from two pairs of harmonic points are also two pairs of harmonic points.*

Indeed, the perpendiculars are four parallel lines; hence the segments determined by their feet are to each other as the corresponding segments determined by the two given pairs of harmonic points.

62. Theorem. *If the points C, D divide the segment AB harmonically in the ratio $p:q$, the midpoint O of the segment CD divides the segment AB externally in the ratio $p^2:q^2$.*

The point O lies outside the segment AB (§ 56). Now:

$$AO = \tfrac{1}{2}\,CD + AC, \quad BO = \tfrac{1}{2}\,CD - BC.$$

Substituting for CD, AC, BC their values (§§ 57, 58), we obtain:

$$AO = ap^2:(p^2 - q^2), \quad BO = aq^2:(p^2 - q^2);$$

hence the announced relation.

63. Corollary. *We have* (§ 58):

$$OA \cdot OB = OC^2.$$

64. *Remark I.* If M, O are the midpoints of the segments AB, CD, we have:

$$MO = AO - AM = AO - \tfrac{1}{2}\,AB;$$

hence:

$$MO = a(p^2 + q^2):2(p^2 - q^2).$$

65. *Remark II.* The point M lies outside the segment CD (§ 56).

66. Theorem. *The sum of the squares of two harmonic segments is equal to four times the square of the distance between the midpoints of these segments.*

We have (§ 58):

$$AB^2 + CD^2 = a^2 + 4\,a^2p^2q^2:(p^2 - q^2)^2 = a^2(p^2 + q^2)^2:(p^2 - q^2)^2;$$

hence (§ 64):

$$AB^2 + CD^2 = 4\,MO^2.$$

67. Problem. *Construct two segments given their product, t^2, and their sum, or their difference, a.*

SOLUTION. Construct two segments u, v having the segment t for mean proportional. On the perpendicular erected at one end, say B,

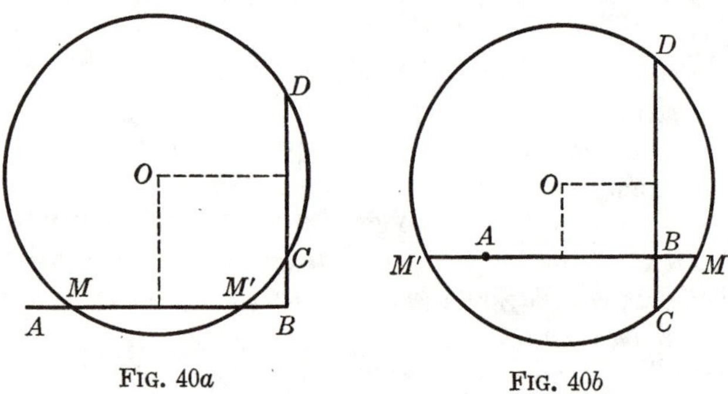

FIG. 40*a* FIG. 40*b*

of the segment $AB = a$ lay off the segment $BC = u$, $BD = v$, on the same side of AB (Fig. 40 *a*), or on opposite sides of AB (Fig. 40 *b*), depending on whether a is the sum or the difference of the required segments. With the point of intersection O of the mediators of the segments AB, CD, as center and radius $OC = OD$ draw a circle meeting AB in the points M, M'; then MA, MB (or $M'A, M'B$) are the required segments.

Indeed, we have:

$$MA \cdot MB = BM' \cdot BM = BC \cdot BD = uv = t^2.$$

The construction above is a graphical solution of the quadratic equations:

$$x^2 - ax + t^2 = 0, \quad x^2 - ax - t^2 = 0.$$

The roots of the quadratic equations

$$x^2 + ax + t^2 = 0, \quad x^2 + ax - t^2 = 0$$

differ from the roots of the first equations in their signs only; hence these roots may be found by this construction. Thus the construction gives a graphical solution of any quadratic equation whose first coefficient is unity.

EXERCISES

1. Place two segments, of given lengths, on the same line so that they shall be harmonic. *Hint.* Use § 66.

2. Show that the pairs of bisectors of the angles of a triangle determine on the respectively opposite sides three segments such that the reciprocal of one is equal to the sum of the reciprocals of the other two. *Hint.* Use § 58.

3. Show that the internal (or external) bisector of an angle of a triangle is divided harmonically by the feet of the perpendiculars dropped upon it from the two other vertices of the triangle.

B. THE CIRCUMCIRCLE

68. Theorem. *If a triangle inscribed in a given circle has a constant angle, then the opposite side is tangent to a fixed circle concentric with the given circle.*

The proposition may be considered as the converse of locus 7 (§ 11). The proof is left to the student.

69. Corollary. *We have the datum:*

$$a, A, R.$$

70. Problem. *Construct a triangle given the base, the vertical angle, and the ratio of the distances of the midpoint of the base from the traces, on the base, of the altitude and of the bisector of the vertical angle.*

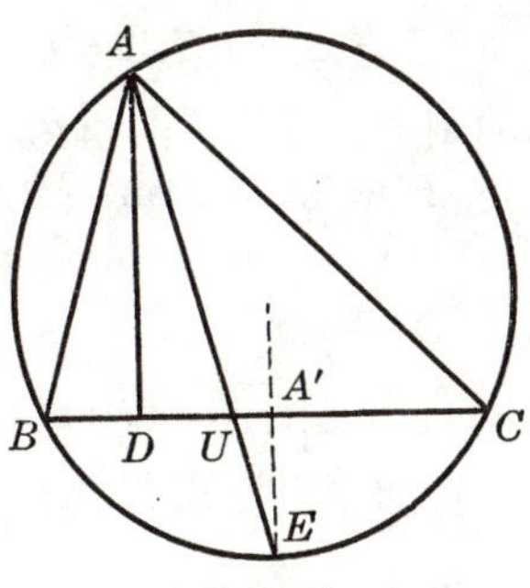

Fig. 41

Let ABC (Fig. 41) be the required triangle. Let A' be the midpoint of BC, and D, U the traces on BC of the altitude AD and the bisector AU of the angle A. The bisector AU, produced, and the mediator of BC both bisect the arc BC, say, in the point E.

From the similar right triangles ADU, $EA'U$ we have:

$$A'U : UD = A'E : AD.$$

Now the base BC and the angle A being given, the circumcircle ABC is determined; hence the segment $A'E$ is known; also the first ratio in the above proportion is known; hence the altitude AD may be constructed. The solution of the problem is now readily completed.

71. Problem. *Construct a triangle given the circumradius, the sum of the lateral sides, and the difference of the base angles* (R, $b + c$, $B - C$).

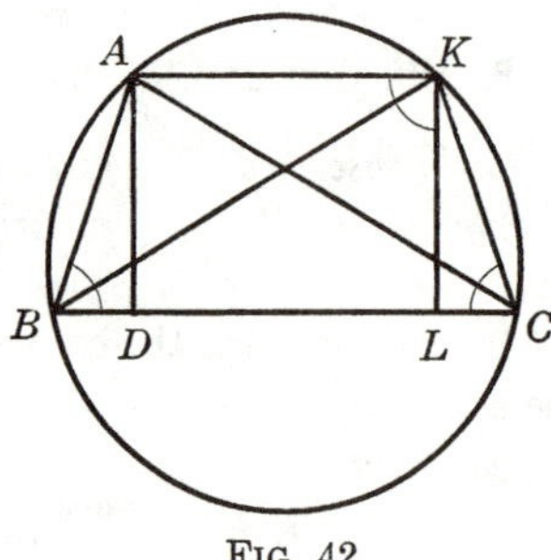

FIG. 42

Let ABC be the required triangle (Fig. 42). Let the parallel to BC through A meet the circumcircle of ABC in K. In the isosceles trapezoid $ABCK$ we have:

$$\angle ABK = ABC - KBC = B - BKA = B - BCA = B - C.$$

The given radius R and the known angle $ABK = B - C$ determine the segment AK (§ 69). Thus in the triangle ABK we know the base AK, the opposite angle $B - C$, and the sum of the sides:

$$AB + BK = AB + AC = b + c;$$

hence this triangle may be constructed (§ 17). The parallel to AK through B will meet the circumcircle of ABK in the third vertex C of the required triangle ABC.

72. Problem. *Construct a triangle given the circumradius, the base, and the product of the sum of the other two sides by one of them* [R, a, $(b + c)b$].

Let ABC be the required triangle inscribed in the given circle, center O (Fig. 43). Produce BA and lay off $AD = AC$. We have angle $BDC = \frac{1}{2}$ BAC, and the latter angle is determined by R

and a (§ 69). Thus if we place the side $BC = a$ in the given circle, radius R, we have a locus for D (§ 11, locus 7).

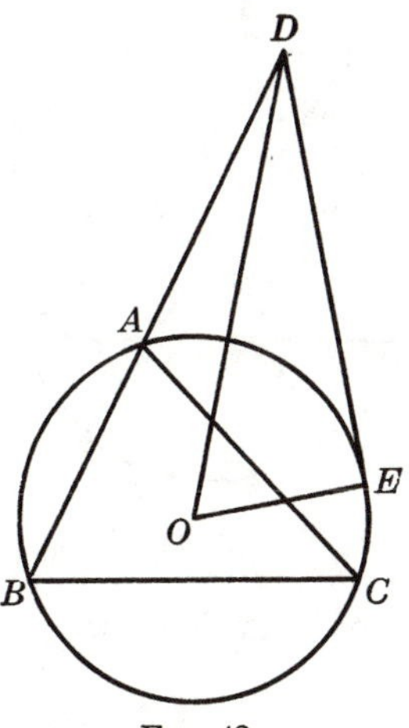

FIG. 43

Let DE be the tangent from D to the circle. We have:

$$DE^2 = DA \cdot DB = b(b + c),$$

i.e., we know the length of the tangent from D to the circle (O); hence we have a second locus for D (§ 11, locus 2), and the auxiliary triangle DBC may thus be constructed. The mediator of CD meets BD in the third vertex A of the required triangle ABC.

EXERCISES

Construct a triangle, given:

1. $R, A, b^2 + c^2$.
2. $R, A, b:c$.
3. R, a, h_b.
4. $R, a, b + c$.
5. $R, b, B - C$.
6. R, m_b, C.
7. $R, A, 2p$.
8. $R, A, a - b$.
9. $R, b + c, h_b + h_c$.
10. $R, b - c, h_c - h_b$.
11. $R, A, h_b + h_c$.
12. $R, a, B - C$.
13. $R, b, a:h_a$.
14. $a, A, b(b + c)$.
15. $R, b - c, B - C$.

73. Theorem. *The angle between the circumdiameter and the altitude issued from the same vertex of a triangle (a) is equal to the difference of the other two angles of the triangle, and (b) is bisected by the bisector of the angle of the triangle at the vertex considered.*

(a) Let AD be the altitude and AK the circumdiameter issued from the vertex A (Fig. 44). The angles B and K are equal since they sub-

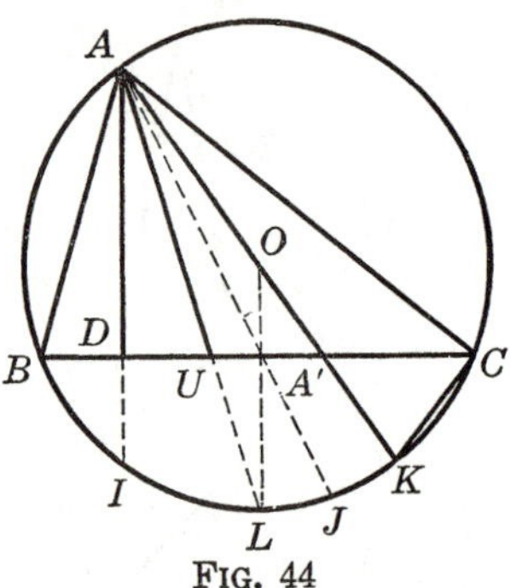

FIG. 44

tend the same chord of the circumcircle; hence we have from the right triangles ABD, ACK:

$$\angle BAD = \angle KAC = 90° - B,$$

and:

$$\angle DAK = A - 2(90° - B) = A + 2B - 180° = A + 2B - A - B - C = B - C.$$

The student may verify that the proposition holds when one of the angles B, C is obtuse.

(b) Let AU be the bisector of the angle A. Since $\angle BAD = CAO$, we have angle $DAU = OAU$.

74. COROLLARY. *In the right triangle ADU we have $AD = h_a$, $AU = t_a$, and angle $DAU = \frac{1}{2}(B - C)$, i.e., we have the datum:*

$$h_a, \quad t_a, \quad B - C.$$

75. Theorem. *The difference of the two angles which an internal bisector of an angle of a triangle makes with the opposite side is equal to the difference of the angles adjacent to this side.*

From the two triangles AUB, AUC (Fig. 44) we have:

$$\angle AUC = B + \tfrac{1}{2}A, \quad \angle AUB = C + \tfrac{1}{2}A;$$

hence:

$$\angle AUC - \angle AUB = B - C.$$

76. Problem. *Construct a triangle given the altitude, the median, and the bisector issued from the same vertex (h_a, m_a, t_a).*

The two right triangles having for their common leg $AD = h_a$ (Fig. 44) and for hypotenuses $AA' = m_a$ and $AU = t_a$ may be constructed. The circumcenter O of the required triangle ABC lies on the perpendicular to DA' at A' and on the line which makes with AU an angle equal to the angle DAU (§ 73 b); hence O may be determined. The circle (O, OA) meets the line DA' in the two vertices B, C of the required triangle ABC.

77. Problem. *Construct a triangle given the base, the altitude to the base, and the bisector of the vertical angle* (a, h_a, t_a).

Let ABC be the required triangle. Let $AD = h_a$, $AU = t_a$. If F is the symmetric of B with respect to D, we have:

$$\angle FAC = \angle AFD - \angle C = B - C,$$

and this angle is known from the right triangle ADU (§ 73).

Let the perpendicular to BC at B meet AF in G. The right triangle GBC may be constructed, for BC is given and $GB = 2\,h_a$. We have thus for the vertex A two loci: the arc of a circle at the points of which GC subtends an angle of $180° - (B - C)$, and the mediator of BG.

78. Problem. *Construct a triangle given the base, the median to the base, and the difference of the base angles* $(a, m_a, B - C)$.

Let ABC (Fig. 44) be the required triangle. We have angle $A'OK = DAO = B - C$; hence angle $AOA' = 180° - (B - C)$.

Let AA' produced meet the circumcircle in J. We have then:

$$A'A \cdot A'J = A'B \cdot A'C; \quad \text{hence} \quad A'J = a^2 : 4\,m_a.$$

Thus the segment $A'J$ may be constructed as the third proportional of the segments a, $4\,m_a$, and the circumcenter O lies on the mediator of AJ. Hence the following construction.

Place $AA' = m_a$ and on it construct the arc of a circle at which AA' subtends the angle of $180° - (B - C)$. Produce AA' by $A'J = a^2 : 4\,m_a$. The arc is met by the mediator of AJ in the circumcenter O of the required triangle. The circle (O, OA) meets the perpendicular to OA' at A' in the vertices B, C of the required triangle ABC.

The angle subtended by AA' at O is equal to $B - C$, if the points A, O lie on opposite sides of BC, or, in other words, when A is an obtuse angle.

EXERCISES

Construct a triangle, given:

1. $R, h_a, B - C$.
2. $R, t_a, B - C$.
3. A, h_a, t_a.
4. $a, A, h_a : t_a$.
5. R, h_a, t_a.
6. $R, m_a, B - C$.

79. Theorem. *The internal and external bisectors of an angle of a triangle pass through the ends of the circumdiameter which is perpendicular to the side opposite the vertex considered.*

The internal bisector AU of the angle A (Fig. 45) passes through the midpoint L of the arc BLC of the circumcircle which does not contain the point A. The circumdiameter LL' passing through L is perpendicular to BC, and the line AL' joining A to the other end L' of this diameter is perpendicular to AL, for $L'AL$ is a right angle; hence AL' is the external bisector of the angle A, which proves the proposition.

80. Theorem. *The angle which the external bisector makes with the opposite side is equal to half the difference of the two angles adjacent to this side.*

The angle which the external bisector AU' (Fig. 45) makes with the base BC has its sides respectively perpendicular to the internal bisector AU and the altitude AD; hence the proposition (§ 73).

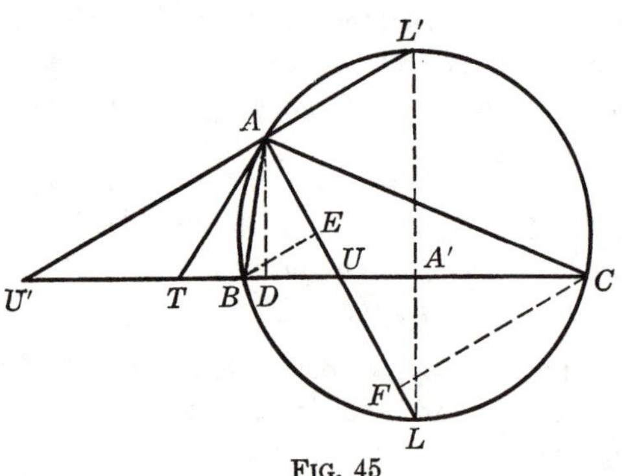

FIG. 45

81. COROLLARY I. *If the tangent at A to the circumcircle meets BC in T, we have* $TA = TU = TU'$

Indeed, we have (Fig. 45):

$$\angle TAU = TAB + BAU, \quad \angle TUA = UCA + UAC.$$

Now angle $TAB = UCA$, and angle $BAU = UAC$; hence the triangle TAU is isosceles, and $TA = TU$.

The circle (T, TA) will pass through U, for $TA = TU$, and since UAU' is a right angle, the circle will also pass through U'; hence $TA = TU'$.

82. Corollary II. *The altitude AD of ABC is also an altitude of the triangle UAU′*; hence:

$$t_a, \quad t_a', \quad h_a, \quad B - C$$

constitute a datum such that any two of these four magnitudes being given, the remaining two are determined.

83. Problem. *Through the midpoint E of the arc BEC of a circle to draw a secant meeting the chord BC in U and the circle again in A, so that AU shall have a given length, t.*

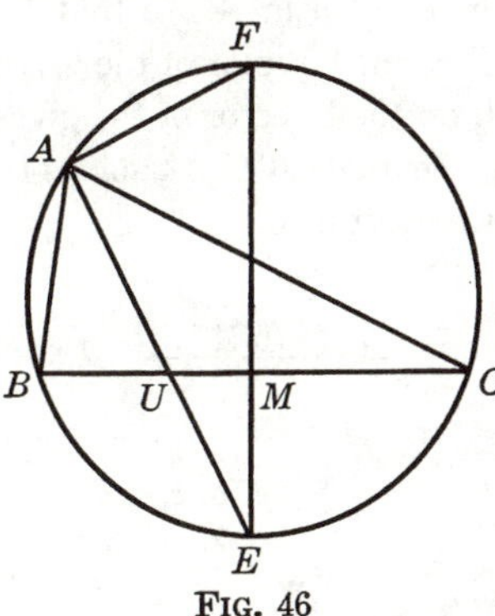

Fig. 46

Let the diameter EMF (Fig. 46) passing through E meet the chord BC in M. From the two equiangular right triangles EUM, EFA we have:

$$EF:EU = EA:EM.$$

Now if we put:

$$EF = 2R, \quad EM = m, \quad EU = x, \quad EA = x + t,$$

the proportion may be written:

$$2R \cdot m = x(x + t),$$

and the segment x may be constructed (§ 67).

The known length of the segment EU determines the point U; hence the required secant EUA.

84. The Problem of Pappus. *Through a given point on the bisector of a given angle to draw a secant on which the given angle shall intercept a segment of given length.*

Let B, C (Fig. 46) be the traces of the required secant $UBC = a$ on the sides of the given angle BAC; the circumradius of the triangle ABC is determined (§ 69), and in the circumcircle the bisector AU of the angle A passes through the midpoint E of the arc BEC subtended by $BC = a$. We are thus led to the following construction.

On an arbitrary line lay off $B'C'$ equal to the given length a, and on $B'C'$ as chord construct the locus of the points at which $B'C'$ subtends an angle equal to the given angle A (§ 11, locus 7). Through the midpoint E' of the other arc of this circle draw a secant $E'U'A'$ meeting $B'C'$ in U' and the circle again in A', so that $U'A'$ shall be equal to the given segment AU of the bisector of the given angle (§ 83).

The line through U, on the bisector of the given angle, making with AU an angle equal to the angle $B'U'A'$, solves the proposed problem.

The problem has two solutions.

EXERCISES

Construct a triangle, given:

1. $h_a, t_a, b:c$. *Hint.* Use § 57.
2. $t_a, B - C, b:c$.
3. $h_a, B - C, b:c$.
4. $h_a, t_a', b:c$.
5. $t_a', B - C, b:c$.
6. b, c, t_a.
7. $t_a, m_b, b:c$.

8. If the line joining a vertex of a triangle to the circumcenter is parallel to the opposite side, show that the two bisectors of the corresponding angle are equal, and conversely.

9. Construct a triangle given the base and an adjacent angle, so that the two bisectors of this angle shall be equal.

10. **(a)** If the two bisectors of the angle A of the triangle ABC are equal, and the circle having BC for diameter cuts the sides AB, AC in the points P, Q, show that $CP = CQ$. **(b)** Construct a triangle given a side and the foot of the altitude on this side so that the two bisectors of an angle adjacent to this side shall be equal.

85. Theorem. *The product of two sides of a triangle is equal to the altitude to the third side multiplied by the circumdiameter.*

The angle $ABD = AKC$ (Fig. 44); hence, from the similar right triangles:

$$AB:AK = AD:AC, \quad \text{or} \quad AB \cdot AC = AK \cdot AD,$$

i.e.,

$$bc = 2R \cdot h_a.$$

86. Corollary. *The area of a triangle is equal to the product of its three sides divided by the double circumdiameter of the triangle.*

Indeed, multiplying both sides of the above equality (§ 85) by a and observing that $ah_a = 2S$, where S is the area of the triangle, we have:

$$abc = 4RS.$$

EXERCISES

Construct a triangle, given:

1. a, A, bc.

2. h_a, t_a, bc.

3. Through a fixed point A of a given circle two variable chords AB, AC are drawn so that their product is constant. Show that the chord BC is tangent to a fixed circle.

4. Show that the product of the distances of a point of the circumcircle of a triangle from the sides of the triangle is equal to the product of the distances of the same point from the sides of the tangential triangle (i.e., the triangle formed by the tangents to the circumcircle at the vertices) of the given triangle.

C. MEDIANS

87. Theorem. (*a*) *The line joining the midpoints of two sides of a triangle is parallel to the third side and is equal to one-half its length.* (*b*) *The line drawn parallel to a side of a triangle through the midpoint of another side passes through the midpoint of the third side.*

88. Theorem. *The three medians of a triangle meet in a point, and each median is trisected by this point.*

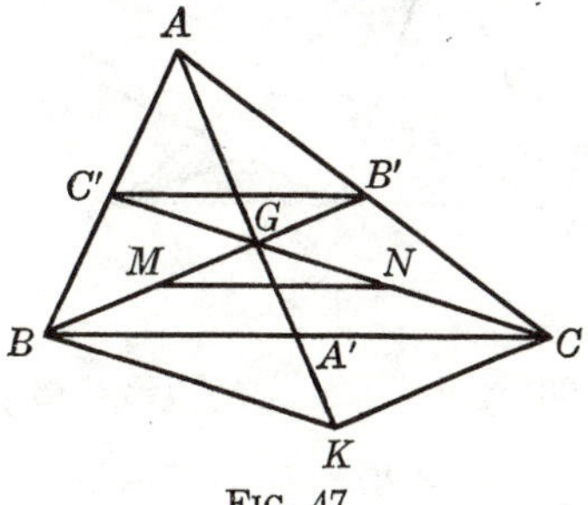

FIG. 47

Let G be the point of intersection of the two medians BB', CC' (Fig. 47).

From the two similar triangles GBC, $GB'C'$ we have:

$$BG:B'G = CG:C'G = BC:B'C' = 2:1.$$

Thus any two medians trisect each other, which proves the proposition.

89. Definition. The point G is called the *centroid* of the triangle ABC.

90. Problem. *Construct a triangle given its medians* (m_a, m_b, m_c).

Produce $C'B'$ (Fig. 48) and lay off $B'K = B'C'$. The diagonals of $AKCC'$ bisect each other; hence $AKCC'$ is a parallelogram, and $AK = CC' = m_c$.

Again, $A'BB'K$ is a parallelogram, for $A'B$ is equal and parallel to $B'K$; hence:

$$A'K = BB' = m_b.$$

Thus the triangle $AA'K$ has for its sides the given medians and may be constructed.

To pass to the required triangle we observe that AA' is bisected by $C'B'K$, in L; hence KL is a median of $AA'K$ and is therefore known. Moreover:

$$C'L = LB' = \tfrac{1}{3}\,KL;$$

hence the points B', C' may be determined and the triangle ABC readily completed.

91. Theorem. *With the medians of a triangle a new triangle is constructed. The medians of the second triangle are equal to three-fourths of the respective sides of the given triangle.*

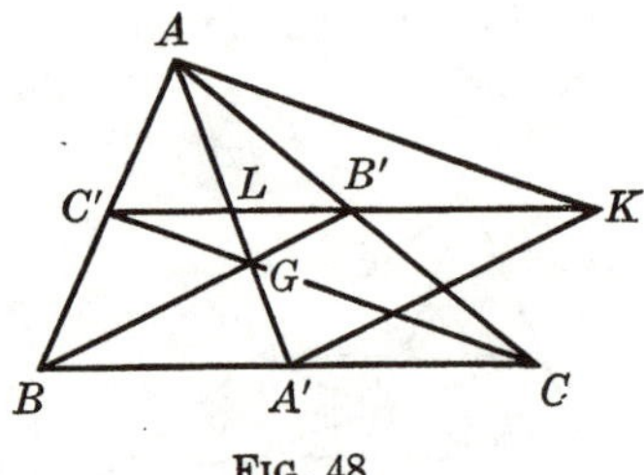

FIG. 48

We have seen (§ 90) that $C'L = \frac{1}{3}\,KL$ (Fig. 48); hence:

$$KL = \tfrac{3}{4}\,KC' = \tfrac{3}{4}\,BC = \tfrac{3}{4}\,a,$$

and since KL is any one of the medians of the triangle $AA'K$, the proportion is proved.

92. Consequence. Given a triangle, 1; with the medians of 1 as sides a triangle, 2, is constructed; with the medians of 2 a new triangle, 3, is constructed, etc. The triangles of odd rank, $1, 3, 5, \ldots,$ are similar, and the triangles of even rank. $2, 4, 6, \ldots$ are similar. In each group the sides of the triangle are equal to three-fourths of the sides of the preceding triangle.

93. Theorem. *The area of the triangle having for sides the medians of a given triangle is equal to three-fourths of the area of the given triangle.*

The area of the triangle $AA'K$ (Fig. 48) is equal to the sum of the areas of the triangles AKL and $A'KL$. Now these two triangles have the same base KL and the altitude of each is equal to one-half the corresponding altitude of the triangle ABC; hence:

$$\text{area } AA'K : \text{area } ABC = KL:BC = \tfrac{3}{4} \text{ (§ 91).}$$

94. Theorem. *In a given circle an infinite number of triangles may be inscribed having for their centroid a given point, within the circle.*

On the given circle (O) take an arbitrary point A and join it to the given centroid G. On AG, produced, lay off $GA' = \frac{1}{2}GA$ and join A' to the center O of (O). The perpendicular to OA' at A' will meet the circumcircle (O) in the other two vertices of the triangle ABC.

The point A yields a solution, if A' falls within the circle (O). Now the locus of A' is a circle (N) which corresponds to (O) in the homothecy $(G, -\frac{1}{2})$. Hence if (N) lies entirely within (O), every point of (O) yields a solution. Otherwise (O) contains an arc on which A cannot be taken.

95. Problem. *Construct a triangle given an altitude and a median, issued from the same vertex, and the angle at this vertex* (h_a, m_a, A).

Let A', B' be the midpoints of the sides BC, AC of the required triangle ABC, and AD the altitude. The right triangle ADA' may be constructed. The median AA' subtends at the point B' the known angle of $180° - A$, which gives a locus for B', and on the other hand B' lies on the line joining the midpoints of AA' and AD. The point B' is thus determined, and the triangle ABC is readily completed.

EXERCISES

1. Show that the line joining the midpoint of a median to a vertex of the triangle trisects the side opposite the vertex considered.
2. Construct a triangle given, in position, two vertices and the centroid.
3. Construct a triangle so that its medians shall lie on three given concurrent lines. Show that one of the vertices may be taken arbitrarily on one of the given lines.
4. Show that a parallel to a side of a triangle through the centroid divides the area of the triangle into two parts, in the ratio 4:5.
5. Show that the lines joining the midpoints of the sides of a triangle divide the triangle into four congruent triangles.

6. Show that the lines joining the centroid of a triangle to its vertices divide the triangle into three equivalent triangles, and that the centroid is the only point having this property.
7. Show that (a) in a triangle a smaller median corresponds to the greater of two sides; (b) if two medians of a triangle are equal, the triangle is isosceles; (c) if two medians of a triangle are proportional to the sides to which they are drawn, the triangle is isosceles.
8. Show that the distances of a point on a median of a triangle from the sides including the median are inversely proportional to these sides.
9. A line is drawn through the centroid of a triangle. Show that the sum of the distances of the line from the two vertices of the triangle situated on the same side of the line is equal to the distance of the line from the third vertex.
10. Show that the (algebraic) sum of the distances of the vertices of a triangle from any line in the plane is equal to the sum of the distances of the midpoints of the sides of the triangle from this line.
11. Show that the sum of the medians of a triangle is smaller than the perimeter of the triangle and greater than three-fourths of that perimeter.

Construct a triangle, given:

12. b, c, m_a. 13. m_b, m_c, b. 14. a, m_b, m_c.
15. m_b, m_c, h_a. *Hint.* The distance of G from $BC = \frac{1}{3} h_a$.

96. Definition. The triangle having for its vertices the midpoints of the sides of a given triangle is called the *medial*, or *complementary, triangle* of the given triangle.

97. Theorem. *A triangle and its medial triangle have the same centroid.*

The median AA' of the triangle ABC bisects the segment $B'C'$ (Fig. 48); hence AA' is also a median of the triangle $A'B'C'$. Similarly for the other medians.

98. Consequence. The complementary triangle $A'B'C'$ corresponds to the given triangle ABC in the homothecy $(G, 2{:}-1)$.

The point P' which corresponds in this homothecy to a given point P is called the *complementary point* of the point P.

99. Definition. The triangle formed by the parallels to the sides of a given triangle through the respectively opposite vertices is called the *anticomplementary triangle* of the given triangle.

100. Theorem. *A triangle and its anticomplementary triangle have the same centroid.*

Indeed, if we start with $A'B'C'$ (Fig. 48) as the given triangle, then ABC is its anticomplementary triangle; hence the proposition (§ 97).

101. Consequence. The anticomplementary triangle $A''B''C''$ corresponds to the given triangle ABC in the homothecy $(G, 1{:}-2)$.

A point P'' which corresponds in this homothecy to a given point P is called the *anticomplementary* point of the point P.

102. Problem. *Construct a triangle given an angle and the medians to the sides of this angle* (A, m_b, m_c).

Having placed $BB' = m_b$, we have a locus for the vertex A (§ 11, locus 7), from which we derived a locus for the vertex C by the homothecy $(B', -1)$. Now, having located the centroid G on BB', the circle $(G, \frac{1}{3}2m_c)$ is a second locus for the vertex C.

103. Problem. *Construct a triangle given the base, the ratio of the medians to the lateral sides, and the difference of the squares of these sides* $(a, m_b:m_c, b^2 - c^2)$.

Having placed $BC = a$, we have (Fig. 47):

$$BG:CG = \tfrac{1}{3}2m_b:\tfrac{1}{3}2m_c = m_b:m_c;$$

hence G lies on a known circle (§ 11, locus 11), from which we derive a locus for the vertex A by the homothecy $(A', 1:3)$. A second locus for A is given by the condition $b^2 - c^2$ (§ 11, locus 12).

104. Problem. *The right angle ACB revolves about its vertex C and its sides meet the two fixed perpendicular lines OAX, OBY in A and B. Find the locus of the centroid of the triangle ABC.*

The line AB is the diameter of the circle $OACB$; hence the midpoint M of AB lies on the mediator, d, of the chord OC, which is fixed. The locus of M is therefore the line d, and the locus of the centroid G of ABC is the line corresponding to d in the homothecy $(C, 2:3)$.

105. Definitions. Two points on the side of a triangle are said to be *isotomic points*, if they are equidistant from the midpoint of this side.

The lines joining two isotomic points to the vertex opposite the side considered are said to be *isotomic lines*.

EXERCISES

Construct a triangle, given:

1. m_b, m_c, h_b.
2. b, c, m_b.
3. a, A, m_b.
4. $a, h_a, m_b:m_c$.
5. If L is the harmonic conjugate of the centroid G of a triangle ABC with respect to the ends A, A' of the median AA', show that $LA' = A'A$.
6. Construct a triangle given, in position, the centroid, one vertex, and the directions of the sides through this vertex.
7. Find the locus of the centroid of a variable triangle having a fixed base and a fixed circumcircle.
8. The complementary triangle and the anticomplementary triangle of a given triangle are homothetic. Find the center and the ratio of this homothecy.

9. The centroids of the four triangles determined by the vertices of a quadrilateral taken three at a time form a quadrilateral homothetic to the given quadrilateral. Find the center and the ratio of this homothecy.

10. From the midpoints A'', B'', C'' of the lines $B'C'$, $C'A'$, $A'B'$ joining the projections A', B', C' of a point P upon the sides of the triangle ABC, perpendiculars are dropped upon the sides BC, CA, AB of ABC. Show that these perpendiculars are concurrent. *Hint.* The two sets of perpendiculars are corresponding lines in the two homothetic figures $A'B'C'$ and $A''B''C''$.

11. If K, K' are two isotomic points on the side BC of the triangle ABC, and the line AK meets the side $B'C'$ of the complementary triangle in K'', show that the line $K'K''$ passes through the centroid G of ABC and that $K'G:GK'' = 2:1$. *Hint.* Consider the triangle AKK'.

12. Let L, L' and M, M' be two pairs of isotomic points on the two sides AC, AB of the triangle ABC, and L'', M'' the traces of the lines BL, CM on the sides $A'C'$, $A'B'$ of the complementary triangle $A'B'C'$ of ABC. Show that the triangles $AL'M'$, $A'L''M''$ are inversely homothetic. Find in the figure other pairs of homothetic triangles.

13. Construct a triangle given an angle, the sum of the two sides forming that angle, and the median to one of these sides $(A, b + c, m_c)$.

106. Theorem. *Twice the square of a median of a triangle is equal to the sum of the squares of the two including sides diminished by half the square of the third side.*

The proof was given in connection with the derivation of locus 10, § 11.

We have thus, in the usual notation:

$$2\,m_a^2 = b^2 + c^2 - \tfrac{1}{2}\,a^2, \quad 2\,m_b^2 = c^2 + a^2 - \tfrac{1}{2}\,b^2,$$
$$2\,m_c^2 = a^2 + b^2 - \tfrac{1}{2}\,c^2.$$

107. Theorem. *The sum of the squares of the medians of a triangle is equal to three-fourths the sum of the squares of the sides* (§ 106).

108. Corollary. *The sum of the squares of the distances of the centroid of a triangle from the vertices is equal to one-third the sum of the squares of the sides.*

109. Theorem. *If M is any point in the plane of the triangle ABC, centroid G, we have:*

$$MA^2 + MB^2 + MC^2 = GA^2 + GB^2 + GC^2 + 3\,MG^2.$$

If D is the midpoint of AG, we have, applying the formula for the median (§ 106) to the triangles MBC, MDA', MAG:

$$MB^2 + MC^2 = 2\,MA'^2 + \tfrac{1}{2}\,BC^2,$$
$$MD^2 + MA'^2 = 2\,MG^2 + \tfrac{1}{2}\,DA'^2,$$
$$MA^2 + MG^2 = 2\,MD^2 + \tfrac{1}{2}\,AG^2.$$

Now $DA' = AG$; hence, multiplying the second equality by 2 and adding, we obtain, after simplification:

$$MA^2 + MB^2 + MC^2 - 3\,MG^2 = \tfrac{1}{2}\,BC^2 + \tfrac{3}{2}\,AG^2.$$

Considering, in turn, the medians BB' and CC', we obtain two analogous relations, and adding the three, we have:

$$3(MA^2 + MB^2 + MC^2 - 3\,MG^2) = \tfrac{1}{2}(BC^2 + CA^2 + AB^2) + \tfrac{3}{2}(GA^2 + GB^2 + GC^2).$$

Observing that (§ 108):

$$BC^2 + CA^2 + AB^2 = 3(GA^2 + GB^2 + GC^2),$$

we arrive readily at the announced result.

110. Corollary. *If the point M (§ 109) coincides with the circumcenter O of ABC, we have:*

$$3\,R^2 = GA^2 + GB^2 + GC^2 + 3\,OG^2,$$

or:

$$OG^2 = R^2 - \tfrac{1}{9}(a^2 + b^2 + c^2).$$

EXERCISES

1. If two points are equidistant from the centroid of a triangle, show that the sums of the squares of their distances from the vertices of a triangle are equal, and conversely.
2. Construct a triangle, given a, m_a, $b + c$.
3. Find the locus of a point which moves so that the sum of the squares of its distances from two vertices of a given triangle is equal to the square of its distance from the third vertex of the same triangle.
4. A variable chord of a fixed circle subtends a right angle at a fixed point. Prove that the locus of the midpoint of the chord is a circle.
5. If a, b, c are the sides, and S the area, of a given triangle, show that the area of the triangle formed by the feet of the perpendiculars from the centroid of the triangle upon its sides is equal to:

$$4\,S^3(a^2 + b^2 + c^2)/9\,a^2b^2c^2.$$

6. Construct a triangle, given the ratios $a:b$, $b:c$, and the sum of the medians.
7. If $A''B''C''$ is the anticomplementary triangle of the triangle ABC, and A' the midpoint of BC, show that:

$$A'B''^2 - A'C''^2 = 2(AB^2 - AC^2).$$

D. TRITANGENT CIRCLES

a. BISECTORS

111. Theorem. *The angle formed by two internal (external) bisectors equals a right angle increased (diminished) by half the third angle of the triangle.*

If I is the point of intersection of the internal bisectors BI, CI of the triangle ABC, we have

$$\angle BIC = 180° - \tfrac{1}{2}B - \tfrac{1}{2}C = 90° + \tfrac{1}{2}A.$$

The proof for the external bisectors is analogous.

112. Theorem. *The foot of the perpendicular from a vertex of a triangle upon a bisector issued from a second vertex lies on the side of the medial triangle opposite the first vertex considered.*

If P is the foot of the perpendicular from the vertex A upon the bisector BP issued from the vertex B, and Q is the trace of AP on the side BC, the two right triangles BPA, BPQ are congruent, and the point P bisects the segment AQ; hence the proposition.

113. COROLLARY. *The feet of the four perpendiculars dropped from a vertex of a triangle upon the four bisectors of the other two angles are collinear.*

114. Theorem. *In a triangle the larger of two angles has the shorter internal bisector.*

In a triangle ABC let angle B be larger than angle C (Fig. 49), and let BD, CE be the internal bisectors of those two angles. On the

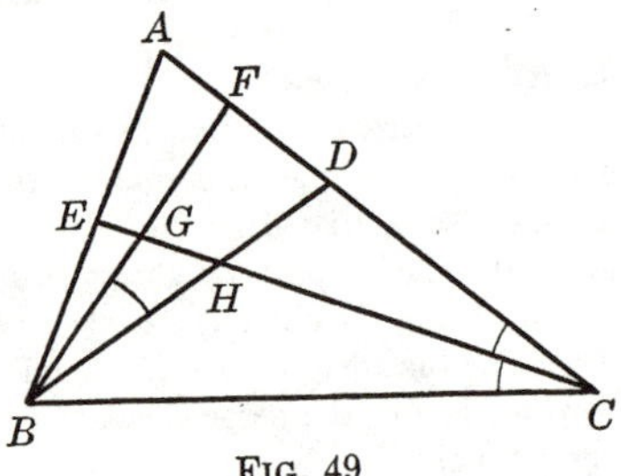

FIG. 49

segment AD take the point F such that angle $FBD = ACE = ECB$. Let H, G be the traces of the lines BD, BF on the bisector CE.

The two triangles FBD, FGC are similar, for they are equiangular; hence:

(a) $$BF:CF = BD:CG.$$

Now the triangle BFC has a smaller angle at the vertex C than at B; hence $BF < CF$, and therefore, by (a), $BD < CG < CE$, which proves the proposition.

115. COROLLARY. *If two internal bisectors of a triangle are equal, the triangle is isosceles.*

Note. This proposition is not valid for two external bisectors.

EXERCISES

1. One vertex of a variable triangle is fixed and the opposite side, of variable length, lies on a fixed straight line. Show that the locus of the projections of the fixed vertex upon the bisectors of the other two angles of the variable triangle is a straight line.
2. Show that the feet of the perpendiculars dropped from two vertices of a triangle upon the internal (external) bisector of the third angle, and the midpoint of the side joining the first two vertices, determine an isosceles triangle whose equal sides are parallel to the sides of the given triangle which include the bisector considered.
3. If the line joining the feet of two internal bisectors of a triangle is parallel to the third side, show that the triangle is isosceles.
4. Show that the sum of the reciprocals of the internal bisectors of a triangle is greater than the sum of the reciprocals of the sides of the triangle.

b. TRITANGENT CENTERS

116. Theorem. *The three internal bisectors of the angles of a triangle meet in a point, the incenter I of the triangle.*

117. Theorem. *The external bisectors of two angles of a triangle meet on the internal bisector of the third angle.*

118. Definitions. The point I_a of the internal bisector AII_a (Fig. 50) of the triangle ABC is called the *excenter* of the triangle relative to the vertex A, or the excenter of A.

The point I_a is equidistant from the three sides of the triangle and is therefore the center of a circle (I_a) touching the three sides of ABC, the points of contact with AB, AC lying on these sides, produced, and the point of contact with the side BC lying on this side not produced. On that account the circle (I_a) and its center I_a are sometimes said to be relative to the side BC, or to the vertex A, or to the angle A.

The two analogous points I_b, I_c and two analogous circles (I_b), (I_c) are relative to the other two vertices B, C, or to the other two sides CA, AB (Fig. 50).

The *inscribed circle, or incircle* (I) and the three *escribed circles, or excircles* (I_a), (I_b), (I_c) are sometimes referred to as the four *tritangent*

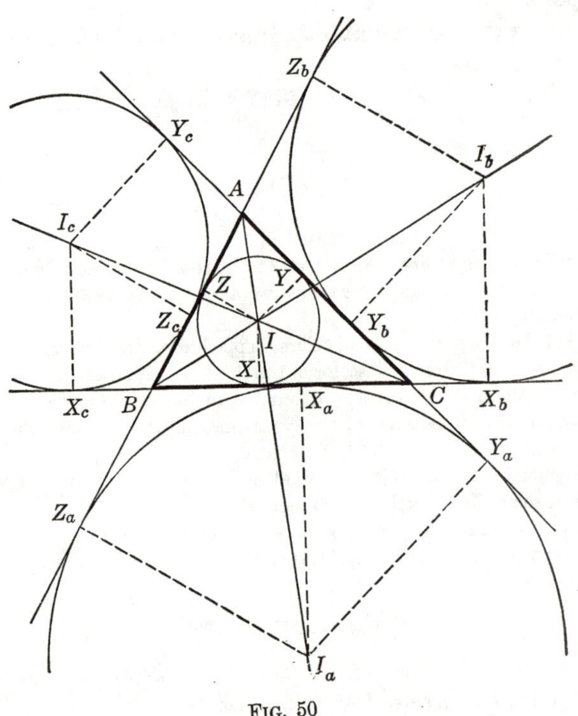

FIG. 50

circles of the triangle ABC, and their centers as the four *tritangent centers* of the triangle.

119. Corollaries. From the two preceding theorems (§§ 116, 117) we have:

Corollary I. *The four tritangent centers of a triangle lie on six lines, the bisectors of the angles of the triangle.*

Each tritangent center lies on three lines, and on each line lie two tritangent centers.

Corollary II. *Given the six midpoints of the arcs subtended by the sides of a given triangle on its circumcircle, the tritangent centers of the triangle may be constructed with the ruler alone.*

120. Theorem. *Two tritangent centers divide the bisector on which they are located, harmonically.*

Let U be the trace of AII_a on BC (Fig. 51). The traces I, I_a of the bisectors BI, BI_a of the angle B on the base AU of the triangle BAU divide the base AU harmonically.

Similarly for the other pairs of tritangent centers.

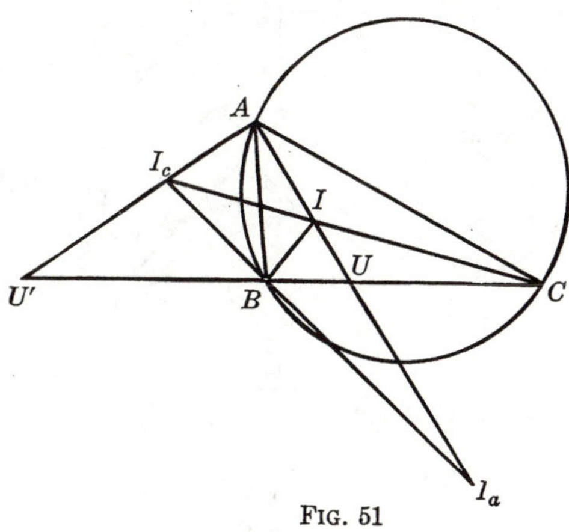

FIG. 51

121. Theorem. *In a triangle, a tritangent center lying on an internal (external) bisector divides this bisector in the ratio of the sum (difference) of the sides forming the angle considered to the side opposite that angle.*

If U, U' are the traces on BC of AI, AI_c (Fig. 51) we have, in the triangle ABC:

$$BU:CU = c:b, \qquad BU':CU' = c:b,$$
$$BU:(BU + CU) = c:(b + c), \qquad BU':(CU' - BU') = c:(b - c),$$
$$BU:a = c:(b + c), \qquad BU':a = c:(b - c).$$

Now in the triangles ABU, ABU' we have:

$$AI:IU = c:BU, \quad AI_c:U'I_c = c:BU',$$

or, substituting for BU, BU' the values obtained above:

$$AI:IU = (b + c):a, \quad AI_c:I_cU' = (b - c):a.$$

Similarly for the other bisectors.

122. Theorem. *Two tritangent centers of a triangle are the ends of a diameter of a circle passing through the two vertices of the triangle which are not collinear with the centers considered.*

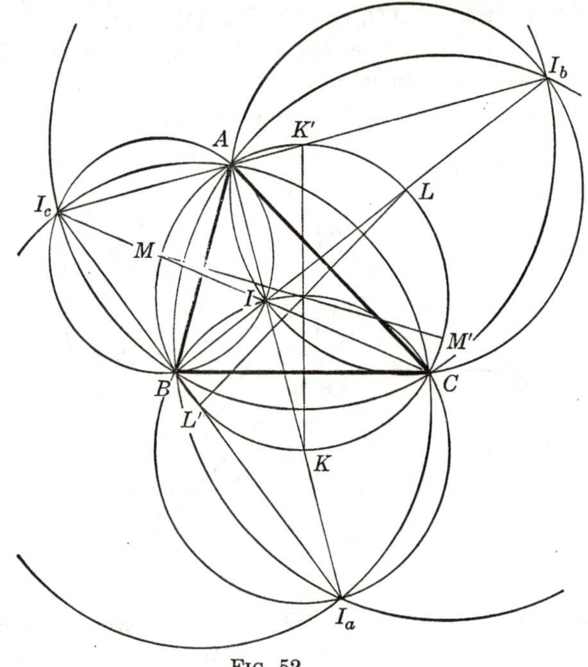

FIG. 52

(a) Consider the incenter I and the excenter I_a. The bisectors BI and BI_a, CI and CI_a (Fig. 52) are perpendicular to each other; hence the proposition.

Moreover, the center of this circle lies on the intersection of the mediator of BC with II_a, and these two lines both pass through the midpoint K of the arc BC of the circumcircle (O) of ABC which does not contain the vertex A; hence the center of the circle (II_a) considered coincides with K.

(b) Consider the two excenters I_b, I_c. The bisectors BI_b and BI_c, CI_b and CI_c are perpendicular to each other; hence the proposition.

Moreover, it may be shown, in the same way as in (a), that the center of this circle (I_bI_c) is the midpoint K' of the arc BC of the circumcircle (O) of ABC which contains the vertex A.

The points K, K' (§ 121) are the ends of the diameter of the circumcircle (O) of ABC perpendicular to the side BC. The diameters LL', MM' of (O) perpendicular to AC, AB may each play the same role as KK'; hence the proposition.

123. Theorem. *The four tritangent centers of a triangle lie on six circles which pass through the pairs of vertices of the triangle and have for their centers the midpoints of the arcs subtended by the respective sides of the triangle on its circumcircle* (Fig. 52).

124. COROLLARY. *Given the midpoints of the arcs subtented by the sides of a given triangle on its circumcircle, the tritangent centers of the triangle may be constructed with compass alone.*

125. Theorem. *If a variable triangle has a fixed base and a fixed circumcircle, the tritangent centers of the triangle describe two circles passing through the two fixed vertices and having for centers the extremities of the circumdiameter perpendicular to the fixed side.*

If the circumcircle (O) of the triangle ABC and the side BC are given, the circles (II_a), (I_bI_c) (§ 122) are determined, regardless of the position of the vertex A; hence the proposition.

126. Theorem. *The midpoints of the six segments determined by the four tritangent centers of a triangle lie on a circle, namely, the circumcircle of the triangle* (Fig. 52).

127. Theorem. *The product of the distances of two tritangent centers of a triangle from the vertex of the triangle collinear with them is equal to the product of the two sides of the triangle passing through the vertex considered.*

If the side AB meets the circle (II_a) again in B', we have:

$$AI \cdot AI_a = AB \cdot AB'.$$

Now the lines AB, AC are equally inclined on the diameter AII_a of the circle; therefore $AB' = AC$; hence:

$$AI \cdot AI_a = AB \cdot AC.$$

Similarly for the other pairs of tritangent centers.

128. Observation. The two tritangent centers on an internal bisector of an angle of a triangle lie on the same side of this vertex, but the opposite is true of two tritangent centers located on an external bisector.

129. Problem. *Construct a triangle given the base, the internal bisector of the opposite angle, and the sum of the lateral sides* $(a, t_a, b + c = s)$.

The two segments s, a being given, their ratio $s:a$ is known. Divide the given segment $AU = t_a$ internally and externally in the known ratio $s:a$, in the points I, I_a (§ 121); in the circle (II_a) having II_a for diameter draw a chord BC, of given length a, through the point U (§ 11, locus 8), and ABC is the required triangle. The problem has two solutions.

130. Problem. *Construct a triangle given in position the three points where the internal bisectors produced meet the circumcircle.*

The three given points K, L, M (Fig. 52) determine the circumcircle (O) of the required triangle ABC and are the midpoints of the arcs subtended by the sides of ABC on (O).

The points L, M are the centers of two circles passing through A and through the incenter I of ABC (§ 122); hence AIK is perpendicular to LM, i.e., A is the second point of intersection of (O) with the perpendicular from the given point K to the known line LM. The perpendiculars from L, M to the lines MK, KL, respectively, determine on (O) the other two vertices B, C of the required triangle ABC.

EXERCISES

1. Show that an external bisector of an angle of a triangle is parallel to the line joining the points where the circumcircle is met by the external (internal) bisectors of the other two angles of the triangle.
2. Construct a triangle given the points where the external bisectors of the angles meet the circumcircle.

Construct a triangle given, in position:

3. I_a, I_b, I_c.
4. I, I_b, I_c.
5. O, I, I_a.

6. Construct a triangle given the base, the external bisector of the opposite angle, and the difference of the other two sides $(a, t_a', b - c)$.
7. Construct a triangle given two sides and the distance of their common vertex from the incenter (b, c, AI).

c. TRITANGENT RADII

131. Notation. Unless otherwise stated, the radius of the inscribed circle or, more briefly, the *inradius* of a triangle will be denoted by r, and the radii of the escribed circles or, more briefly, the *exradii* relative to the vertices A, B, C of the triangle ABC by r_a, r_b, r_c, respectively.

The four radii will be referred to as the *tritangent* radii of the triangle ABC.

132. Theorem. *The inradius of a triangle is equal to the area divided by half the perimeter.*

If S is the area of the triangle ABC, we have (Fig. 53):

$$\begin{aligned} S &= \text{area } ABI + \text{area } BCI + \text{area } CAI \\ &= \tfrac{1}{2}\,rc + \tfrac{1}{2}\,ra + \tfrac{1}{2}\,rb = \tfrac{1}{2}\,r(a + b + c) = pr. \end{aligned}$$

133. Theorem. *An exradius of a triangle is equal to the area divided by the difference between half the perimeter and the side of the triangle relative to the exradius considered.*

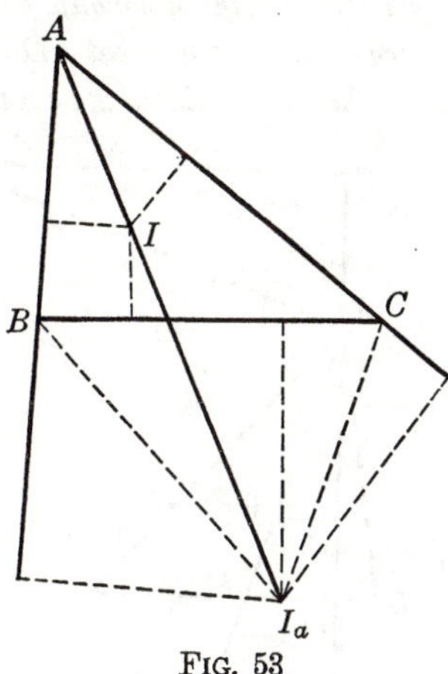

FIG. 53

We have (Fig. 53):

$$S = \text{area } ABI_a + \text{area } ACI_a - \text{area } BCI_a$$
$$= \tfrac{1}{2}\, cr_a + \tfrac{1}{2}\, br_a - \tfrac{1}{2}\, ar_a = \tfrac{1}{2}\, r_a(b + c - a) = r_a(p - a).$$

134. COROLLARY I. *The product of the four tritangent radii of a triangle is equal to the square of its area.*

$$rr_ar_br_c = S^4 : p(p - a)(p - b)(p - c) = S^4 : S^2 = S^2.$$

135. COROLLARY II. *The reciprocal of the inradius is equal to the sum of the reciprocals of the three exradii of the triangle.*

$$\frac{1}{r_a} + \frac{1}{r_b} + \frac{1}{r_c} = \frac{(p - a) + (p - b) + (p - c)}{S} = \frac{p}{S} = \frac{1}{r}.$$

136. Theorem. *The reciprocal of the inradius of a triangle is equal to the sum of the reciprocals of the altitudes of the triangle.*

We have:

$$2\,S = 2\,pr = ah_a = bh_b = ch_c,$$
$$2\,p : \frac{1}{r} = a : \frac{1}{h_a} = b : \frac{1}{h_b} = c : \frac{1}{h_c}$$
$$= (a + b + c) : \left(\frac{1}{h_a} + \frac{1}{h_b} + \frac{1}{h_c}\right),$$

and since $2\,p = a + b + c$, the proposition is proved.

137. Corollary. *The sum of the reciprocals of the exradii of a triangle is equal to the sum of the reciprocals of the altitudes.*

138. Theorem. *If on an arbitrary line the segments $DP = r$, $DQ = r_a$ are laid off in opposite senses, and the harmonic conjugate A of D with respect to P, Q is constructed* (§ 60), *the segment AD is equal to the altitude h_a of the triangle relative to the same side as the exradius considered.*

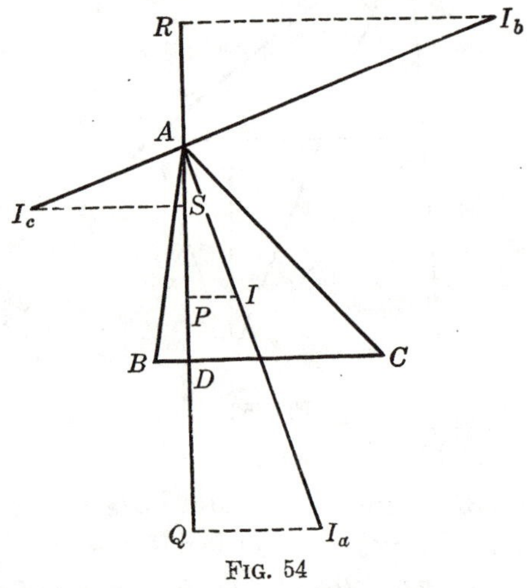

Fig. 54

The feet P, Q of the perpendiculars dropped from the points I, I_a (Fig. 54) upon the altitude AD of the triangle ABC divide the segment AD harmonically (§ 61), and $DP = r$, $DQ = r_a$; hence the proposition.

139. Corollary I. *Given any two of the three segments in one of the groups:*

$$r, r_a, h_a; \quad r, r_b, h_b; \quad r, r_c, h_c,$$

the third segment of the group may be constructed.

For, given any three of the four harmonic points A, P, D, Q, the fourth point is determined (§ 60).

140. Corollary II. *The point D obviously divides the segment QP internally in the ratio $QD{:}DP = r_a{:}r$; hence* (§ 58):

$$\begin{aligned} h_a &= 2(r_a + r)r_a r:(r_a^2 - r^2) \\ &= 2\,rr_a:(r_a - r). \end{aligned}$$

We have analogous formulas for h_b and h_c.

OTHERWISE. We have (§§ 132, 133):

$$2S = ah_a = 2pr, \quad \text{and} \quad 2S = ah_a = 2(p-a)r_a;$$

hence:

$$a(h_a - r) = (b+c)r, \quad a(h_a + r_a) = (b+c)r_a,$$

or:

$$(h_a - r):(h_a + r_a) = r:r_a.$$

Solving for h_a we obtain the same value as above.

Note. The points P, Q divide the altitude AD internally and externally in the ratio (§ 57):

$$(r_a - r):(r_a + r).$$

141. Theorem. *If on an arbitrary line the segments* $DR = r_b$, $DS = r_c$ *are laid off, in the same sense, and the harmonic conjugate* A *of* D *with respect to* R, S *is constructed* (§ 60), *the segment* AD *is equal to the altitude to the side joining the two vertices of the triangle relative to the exradii considered.*

The feet R, S of the perpendiculars from the points I_b, I_c upon AD (Fig. 54) divide the segment AD harmonically (§ 61), and $DR = r_b$, $DS = r_c$; hence the proposition.

142. COROLLARY I. *Given any two of the segments in one of the groups:*

$$h_a, r_b, r_c; \quad h_b, r_c, r_a; \quad h_c, r_a, r_b,$$

the third segment of the group may be constructed.

For given any three of the four harmonic points D, S, A, R, the fourth point is determined.

143. COROLLARY II. *The point* D *obviously divides the segment* $RS = r_b - r_c$ *externally in the ratio* $RD:RS = r_b:r_c$; *hence* (§ 58):

$$\begin{aligned} h_a &= 2(r_b - r_c)r_br_c:(r_b^2 - r_c^2) \\ &= 2\,r_br_c:(r_b + r_c). \end{aligned}$$

We have two analogous formulas for h_b and h_c.

OTHERWISE. We have (§ 133):

$$2S = ah_a = (a+c-b)r_b, \quad 2S = ah_a = (a+b-c)r_c;$$

hence:

$$a(h_a - r_b) = (c-b)r_b, \quad a(r_c - h_a) = (c-b)r_c,$$

or:

$$(h_a - r_b):(r_c - h_a) = r_b:r_c.$$

Solving for h_a, we obtain the same value as above.

Note. The points S, R divide the altitude AD internally and externally in the ratio (§ 57)

$$(r_b - r_c):(r_b + r_c).$$

144. Data. Let A' be the midpoint of the side BC (Fig. 55), and let X, X_a, X_b, X_c be the feet of the perpendiculars from the tritangent

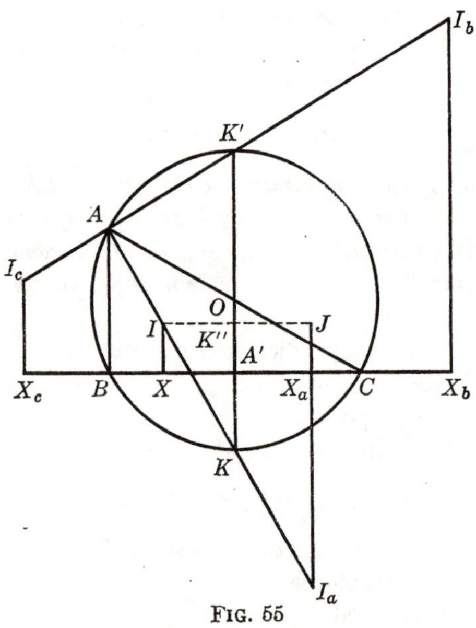

FIG. 55

centers I, I_a, I_b, I_c upon the side BC. Let the parallel to BC through I intersect the diameter KK' in K'' and I_aX_a, produced, in J.

(a) The point K being the midpoint of the side II_a of the triangle IJI_a, we have:

$$KA' + A'K'' = KK'' = \tfrac{1}{2} I_aJ = \tfrac{1}{2}(I_aX_a + X_aJ),$$

and:

$$A'K'' = X_aJ = r;$$

hence:

$$A'K = \tfrac{1}{2}(r_a - r).$$

Now when R and a are given, the segment $A'K$ is determined; hence we have the datum:

$$R, \quad a, \quad r_a - r.$$

(b) The point K' being the midpoint of the side I_bI_c of the trapezoid $I_bI_cX_cX_b$, we have:

$$A'K' = \tfrac{1}{2}(I_bX_b + I_cX_c) = \tfrac{1}{2}(r_b + r_c).$$

Thus we have the datum:

$$R, \quad a, \quad r_b + r_c.$$

Analogous results are obtained considering the sides CA, AB of the triangle and the circumdiameters perpendicular to these sides.

145. COROLLARY. *The sum of the exradii of a triangle is equal to the inradius increased by four times the circumradius.*

Indeed, $KK' = KA' + A'K'$.

146. Carnot's Theorem. *The sum of the distances of the circumcenter of a triangle from the three sides of the triangle is equal to the circumradius increased by the inradius.*

Indeed, we have (Fig. 55):

$$OA' = OK - A'K = R - \tfrac{1}{2}(r_a - r),$$

and similarly:

$$OB' = R - \tfrac{1}{2}(r_b - r), \quad OC' = R - \tfrac{1}{2}(r_c - r).$$

Adding these three segments and taking into account the preceding proposition (§ 145), we obtain the stated result.

Note. In the case of an obtuse-angled triangle the distance from the circumcenter to the side opposite the obtuse angle is to be taken negatively.

147. Problem. *Construct a triangle given one side, the circumradius, and the inradius* (a, R, r).

In the circle of radius R construct a chord $BC = a$. The angle A is now determined (§ 69), and at the point I the side BC subtends an angle of $90° + \tfrac{1}{2} A$ (§ 111); hence we have a locus for I (§ 11, locus 7). A second locus for I is the parallel to BC at a distance r from it. The problem may have two solutions.

148. Problem. *Construct a triangle given one base angle, the sum of the two lateral sides, and the inradius* $(B, b + c, r)$.

Produce BA (Fig. 56) and lay off $AD = AC$. The triangle BID may be constructed for we know the base $BD = b + c$, the angle $IBD = \tfrac{1}{2} B$, and the altitude to $BD = r$. Now:

$$\angle DCI = DCA + ACI = \tfrac{1}{2} A + \tfrac{1}{2} C = 90° - \tfrac{1}{2} B;$$

hence C lies on a circle described on DI as a chord (§ 11, locus 7). This circle is met by the second tangent, besides BD, from B to the circle (I, r) in the vertex C of the required triangle ABC. The third vertex A of ABC is the intersection of BD with the second tangent from C to the incircle (I, r).

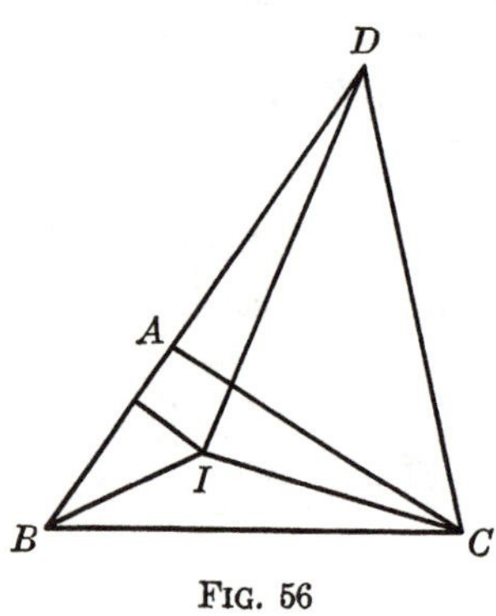

FIG. 56

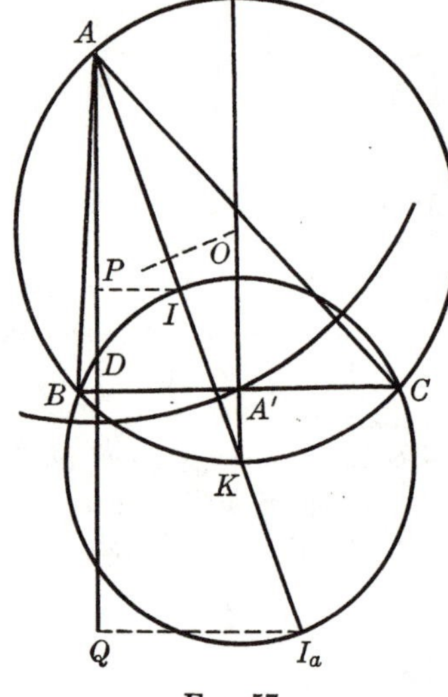

FIG. 57

149. Problem. *Construct a triangle given the inradius, and exradius, and a median relative to the same side of the triangle* (r, r_a, m_a).

On an arbitrary line lay off $DP = r$ (Fig. 57), and in the opposite sense lay off $DQ = r_a$. If A is the harmonic conjugate of D with respect to P, Q, the segment AD is the altitude relative to the side BC of the required triangle ABC (§ 138).

Let the circle (A, m_a) meet the perpendicular to AD at D in A'; draw the perpendicular to DA' at A' and lay off, on the same side of DA' as DQ, the segment $A'K = \frac{1}{2}(r_a - r)$ (§ 142 a). The mediator of AK meets $A'K$, produced, in the circumcenter O of ABC. The circle (O, OA) meets DA' in the vertices B, C of ABC.

150. Problem. *Construct a triangle given, in position,* I, I_a, *and the length of the altitude* h_a, *as well as the ratio* $r_a : r$.

The points which divide the segment II_a externally and internally in the given ratio are, obviously, the vertex A and the trace U on the side BC of the bisector AI. A tangent from U to the circle (A, h_a) meets the circle on II_a as diameter in the remaining two vertices of the required triangle ABC. The problem may have two solutions, one solution, or none.

151. Problem. *Construct a triangle given the base, the altitude to the base, and the sum of the other two sides* (a, h_a, $b + c$).

The relation (§ 132):

$$ah_a = (a + b + c)r$$

determines r as a fourth proportional, and h_a, r determine r_a (§ 139). Again, a and $r_a - r$ determine the circumdiameter (§ 146). The triangle is now readily constructed, since we have a, h_a, and R.

EXERCISES

Construct a triangle, given:

1. a, $2p$, S.
2. R, r, r_a.
3. a, R, r_a.
4. h_a, $B - C$, r.
5. h_a, $B - C$, r_a.
6. h_a, t_a, r (or r_a).
7. h_a, r, $b + c$.
8. h_a, r_a, $b + c$.
9. h_a, $B - C$, r_a.
10. r_a, r, $2p$.
11. A, r_b, r_c. *Hint.* The centers I_b, I_c are readily determined on the external bisector of A by parallels to the sides of this angle.
12. a, r_b, r_c.
13. Show that in a variable triangle having a fixed incircle the sum of the reciprocals of the exradii is constant.

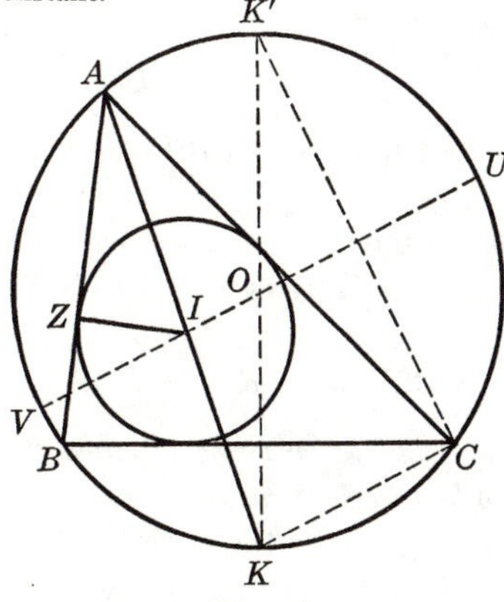

FIG. 58

152. Euler's Theorem. *The distance, d, between the circumcenter and the incenter of a triangle is given by the relation:*

$$d^2 = R(R - 2r),$$

where R and r are the circumradius and the inradius, respectively.

We have (Fig. 58):

$$AI \cdot IK = IU \cdot IV = (R + d)(R - d),$$

and (§ 122 a):

$$IK = KC = BK;$$

hence:

(a) $$R^2 - d^2 = AI \cdot KC.$$

On the other hand, the two right triangles AIZ, $KK'C$ are similar, for angle $IAZ = K'$; hence:

(b) $$AI \cdot CK = IZ \cdot KK' = 2Rr,$$

and the announced relation follows from (a) and (b).

153. Theorem. *The distances between the circumcenter and the excenters I_a, I_b, I_c of the triangle ABC are given by the relations:*

$$OI_a^2 = R(R + 2r_a), \quad OI_b^2 = R(R + 2r_b), \quad OI_c^2 = R(R + 2r_c).$$

The proofs are analogous to the preceding one (§ 152).

154. Corollary.

$$II_a^2 = 4R(r_a - r), \quad I_bI_c^2 = 4R(r_b + r_c).$$

Indeed, applying the formula for the median (§ 106) to the two triangles OII_a, OI_bI_c, we have:

$$OI^2 + OI_a^2 = 2OK^2 + \tfrac{1}{2}II_a^2, \quad OI_b^2 + OI_c^2 = 2OK'^2 + \tfrac{1}{2}I_bI_c^2,$$

and using the preceding formulas (§ 153) we obtain the announced relations.

We have analogous formulas for the segments II_b, II_c, I_aI_b, I_aI_c.

155. Theorem. *If the line of centers d of two given circles (O, R), (I, r) satisfies the relation:*

$$OI^2 = d^2 = R(R - 2r),$$

an infinite number of triangles may be circumscribed about the circle (I, r) which shall also be inscribed in the circle (O, R).

From any point A (Fig. 58) of the circle (O, R) draw two tangents AB, AC to the circle (I, r). The line AI bisects the angle BAC, and it follows from the given relation that:

$$AI \cdot IK = 2Rr;$$

from the similar triangles AIZ, $KK'C$ we have:

$$AI \cdot CK = 2Rr,$$

and therefore $IK = KC$. Thus I is the incenter of the triangle ABC (§ 122 a), which proves the proposition.

EXERCISES

1. If the inradius of a triangle is equal to half the circumradius, show that the triangle is equilateral.
2. Construct a triangle given the inradius, the circumradius, and the difference of two angles $(r, R, B - C)$. *Hint.* Consider the triangle AIO.
3. In a variable triangle having a fixed circumcircle and a fixed incircle the sum of the exradii is constant.
4. Prove the formulas:

(a) $$OI^2 + OI_a^2 + OI_b^2 + OI_c^2 = 12\,R^2.$$
(b) $$II_a^2 + II_b^2 + II_c^2 = 8\,R(2\,R - r).$$
(c) $$I_aI_b^2 + I_bI_c^2 + I_cI_a^2 = 8\,R(4\,R + r).$$

5. If JJ' is the diameter of the incircle of a triangle perpendicular to the diameter of that circle passing through the circumcenter O of the triangle, show that the perimeter of the triangle OJJ' is equal to the circumdiameter of the given triangle.

d. POINTS OF CONTACT

156. Notation. Unless otherwise stated, the points of contact of the four tritangent circles (I), (I_a), (I_b), (I_c) of the triangle ABC with the side BC will be denoted by X, X_a, X_b, X_c. The letters Y and Z will be used for the sides CA and AB, respectively.

157. Theorem. *The distance of a vertex of a triangle from the point of contact of the inscribed circle with a side issued from this vertex is equal to half the perimeter decreased by the length of the side opposite the vertex considered.*

We have (Fig. 59):

$$AZ = AY, \quad BZ = BX, \quad CX = CY.$$

Now:

$$\begin{aligned} AZ + AY &= AB + AC - BZ - CY \\ &= AB + AC - BX - CX \\ &= AB + AC - BC = 2\,p - 2\,a; \end{aligned}$$

hence:

$$AZ = AY = p - a.$$

Similarly:

$$BZ = BX = p - b, \quad CX = CY = p - c.$$

158. Theorem. *The distance of a vertex of a triangle from the point of contact of the excircle relative to this vertex with a side issued from the vertex considered equals half the perimeter of the triangle.*

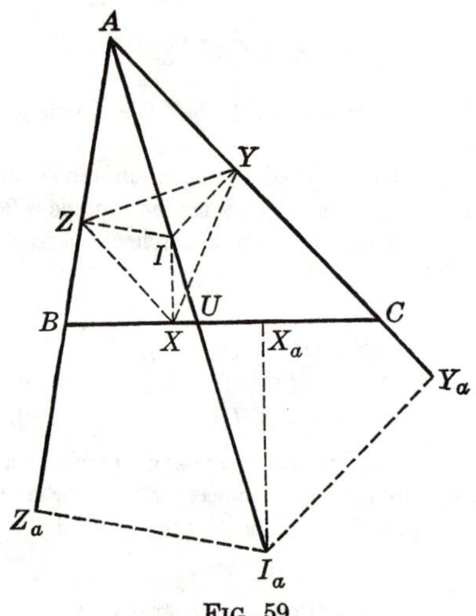

FIG. 59

We have (Fig. 59):

$$AZ_a = AY_a, \quad BX_a = BZ_a, \quad CX_a = CY_a.$$

Now:

$$\begin{aligned} AZ_a + AY_a &= AB + AC + BZ_a + CY_a \\ &= AB + AC + BX_a + CX_a \\ &= AB + AC + BC = 2p; \end{aligned}$$

hence:

$$AZ_a = AY_a = p.$$

Similarly:

$$BX_b = BZ_b = CX_c = CY_c = p.$$

159. COROLLARY. *We have* (Fig. 59):

$$\begin{aligned} BX_a &= BZ_a = AZ_a - AB = p - c, \\ CX_a &= CY_a = AY_a - AC = p - b. \end{aligned}$$

Similarly for the other two sides of the triangle.

160. Theorem. *The points of contact of a side of a triangle with the incircle and the excircle relative to this side are two isotomic points* (§ 105).

Indeed, we have (§§ 157, 159):

$$BX = p - b, \quad CX_a = p - b.$$

161. Corollary. *We have:*

$$XX_a = BC - BX - CX_a = a - 2(p - b) = b - c.$$

Similarly:

$$YY_b = a - c, \quad ZZ_c = a - b.$$

162. Theorem. $ZZ_a = YY_a = a$. (Fig. 59.)

We have:

$$ZZ_a = BZ + BZ_a = p - b + p - c = 2p - (b + c);$$

hence the proposition.

Similarly for the other sides of the triangle.

163. Theorem. *The two points of contact of a side of a triangle with the two excircles relative to the other two sides are two isotomic points.*

The distance between these two points of contact is equal to the sum of the other two sides.

We have (Fig. 60):

$$BX_c = CX_c - CB = p - a, \quad CX_b = BX_b - BC = p - a;$$

hence the points X_b, X_c are isotomic.

$$X_bX_c = X_bC + BC + BX_c = (p - a) + a + (p - a) = b + c,$$

which proves the second part of the proposition.

Similarly for the other sides of ABC.

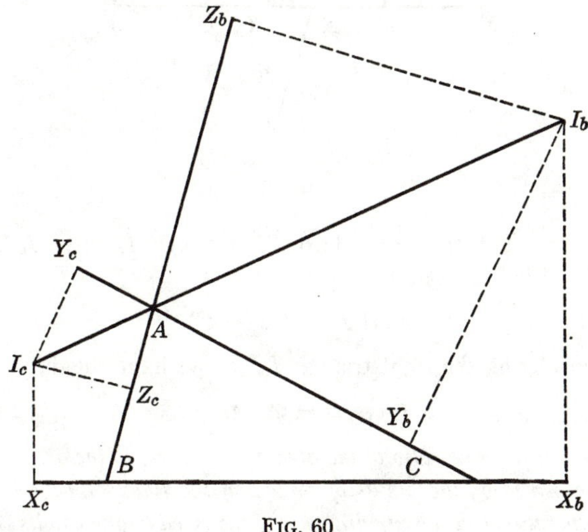

Fig. 60

164. Theorem. $Y_bY_c = Z_bZ_c = a$. (Fig. 60.)

We have:

$$Y_bY_c = CY_c - CY_b = p - CX_b = p - (p - a);$$

hence the proposition.

165. Suggestion. The six points B, C, X, X_a, X_b, X_c on the side BC determine $6 \cdot \frac{5}{2} = 15$ segments. The student is advised to list these segments, with their values, in terms of the sides of the triangle. The same thing may be done for the segments on the other sides of ABC.

166. Datum. Let the parallel to BC through I meet I_aX_a, produced, in J (Fig. 61). We have (§ 73):

$$\angle JI_aI = \angle XII_a = \tfrac{1}{2}(B - C).$$

Thus, considering the right triangle JII_a, we have the datum:

$$b - c, \quad B - C, \quad r_a + r.$$

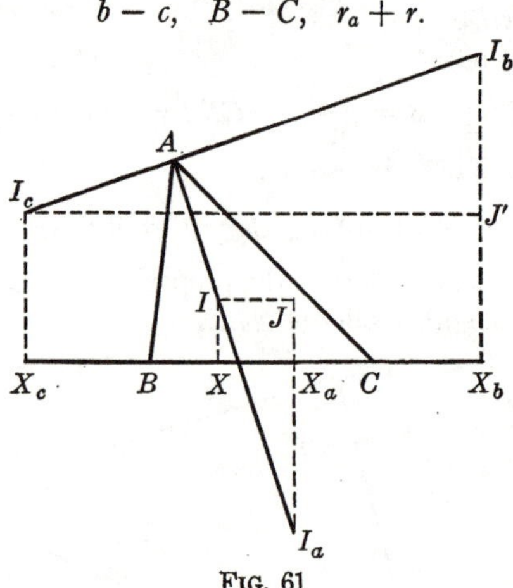

FIG. 61

167. Datum. Let the parallel to BC through I_c meet I_bX_b in J' (Fig. 61). We have (§ 80):

$$\angle I_bI_cJ' = \tfrac{1}{2}(B - C).$$

Thus, considering the right triangle I_bI_cJ', we have the datum:

$$b + c, \quad B - C, \quad r_b - r_c.$$

168. Theorem. *The ratio of the area of a triangle to the area of the triangle determined by the points of contact of the sides with the incircle is equal to the ratio of the circumdiameter of the given triangle to its inradius.*

In the two triangles ABC, IYZ (Fig. 59) the angles at A and I are supplementary; hence:

$$\text{area } IYZ : \text{area } ABC = r^2 : bc = r^2 a : abc,$$

and we have analogous relations for the triangles IZX and IXY. Adding the three relations, we have:

$$\text{area } XYZ : \text{area } ABC = r^2(a + b + c) : abc;$$

hence (§§ 132, 86) the stated result.

As an exercise, state and prove analogous propositions relating to the points of contact with excircles.

169. Problem. *Construct a triangle given the perimeter, an angle, and the internal bisector of that angle* $(2\,p, A, t_a)$.

The triangle AI_aZ_a (Fig. 59) may be constructed. On AI_a lay off $AU = t_a$ and draw the circle (I_a, I_aZ_a). The line AZ_a, the second tangent from A to this circle, and a tangent drawn to it from the point U are the three sides of the required triangle.

The problem may have two, one, or no solutions.

170. Problem. *Construct a triangle given an angle, the sum of the two including sides, and the inradius* $(A, b + c = s, r)$.

The triangle AIZ (Fig. 59) may be constructed. We have thus $AZ = p - a = \frac{1}{2}(b + c - a)$, and since $b + c$ is given, we have a, and the problem is reduced to (a, A, r).

171. Problem. *Construct a triangle given the difference of the two lateral sides, a base angle, and the inradius* $(b - c, B, r)$.

The triangle BIX (Fig. 59) may be constructed. Producing BX by the length $\frac{1}{2}(b - c)$ we obtain the midpoint A' of BC (§§ 160, 161); hence the vertex C. The two tangents from the points B, C to the circle (I, r) intersect in the third vertex of the required triangle ABC.

172. Problem. *Construct a triangle given the difference of two sides, the altitude to one of these sides, and the inradius* $(b - c, h_b, r)$.

The triangle IXX_a (Fig. 62) may be constructed. Since the midpoint A' of XX_a is also the midpoint of BC, the perpendicular from A' to AC is equal to $\frac{1}{2}\,h_b$. Draw the circles (I, r) and $(A', \frac{1}{2}\,h_b)$. A common external tangent to these two circles meets $XA'X_a$ in the vertex C of the required triangle, and the symmetric B of C with respect to A' is a second vertex of the triangle; the third vertex A is the trace of the tangent from B to (I, r) on the common tangent considered.

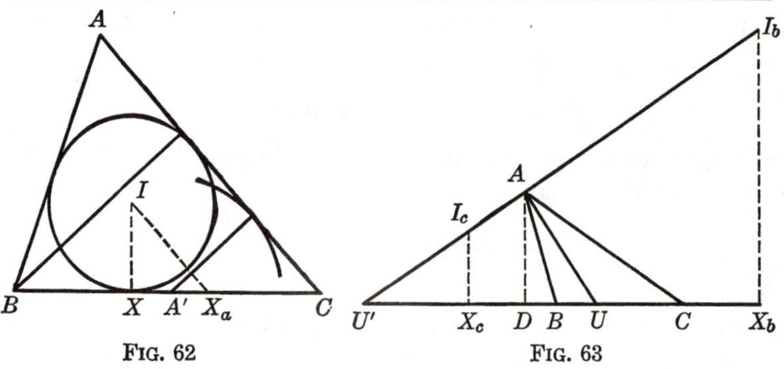

FIG. 62 FIG. 63

173. Problem. *Construct a triangle given the internal and external bisectors of an angle, and the sum of the sides including that angle* ($t_a, t_a', b + c$).

Let ABC be the required triangle (Fig. 63). The right triangle AUU' may be constructed, and the altitude AD of ABC is also the altitude of AUU'. Now the tritangent centers I_b, I_c are separated harmonically by the points A, U', hence the points X_b, X_c are separated harmonically by the points D and U' (§ 61). But $X_bX_c = b + c$ (§ 163); hence the points X_b, X_c are determined by placing on the line UDU' the segment of given length $b + c$ so that it shall be harmonic to the segment DU' (§ 66).

Let I_b, I_c be the points of intersection of the line AU' with the perpendiculars to UU' at X_b, X_c; draw the circles (I_b, I_bX_b), (I_c, I_cX_c). The common internal tangents of these two circles will pass through A and form with UU' the required triangle ABC.

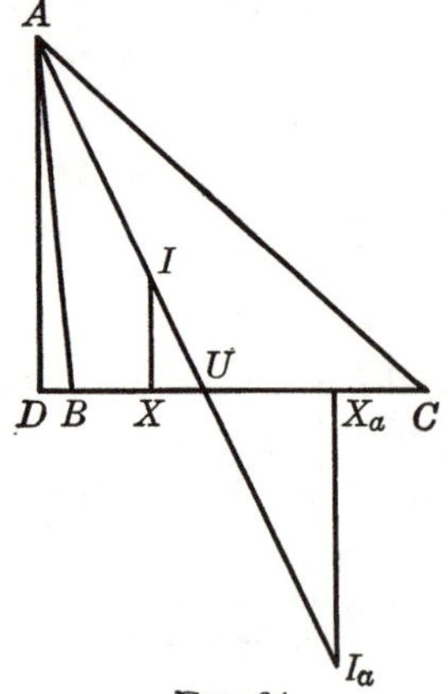

FIG. 64

174. Problem. *Construct a triangle given, in position, a vertex and the points of contact of the opposite side with the incircle and the excircle relative to that side* (A, X, X_a).

Let ABC be the required triangle (Fig. 64). The tritangent centers I, I_a are separated harmonically by A and U (§ 120); hence the foot D of the perpendicular AD from A to the base XX_a is separated harmonically from U by the points X, X_a, and therefore the point U may be constructed (§ 60).

Let the perpendiculars at X, X_a to XX_a meet the line AU in I, I_a. The common external tangents of the circles (I, IX), (I_a, I_aX_a) will pass through A and will form with XX_a the required triangle.

EXERCISES

Construct a triangle, given:

1. $2p, A, r$.
2. $2p, R, r_a$.
3. $b - c, r, r_a$.
4. $b + c, r_b, r_c$.
5. $b - c, h_a, r_a$.
6. $b + c, A, r_a$.
7. $b + c, B - C, r_c$.
8. $b + c, h_b + h_c, r$. *Hint.* $b + c$ and $h_b + h_c$ determine the angle A (§ 21).
9. $b - c, h_c - h_b, r$.
10. $b + c, r_c, h_c$.

11. Construct a triangle given, in position, a vertex and the two points of contact of the opposite side with the two excircles not relative to that side.

12. Show that the line joining a vertex of a triangle to the point of contact of the opposite side with the excircle relative to that side bisects the perimeter of the triangle.

13. Show that the three circles having for their centers the vertices of a triangle, and passing through the points of contact of the incircle with the sides of the triangle, are tangent to each other.

14. Show that a parallel through a tritangent center to a side of a triangle is equal to the sum, or the difference, of the two segments on the other two sides of the triangle between the two parallel lines considered.

15. Show that the lines tangent to the incircle of a triangle and parallel to the sides cut off three small triangles the sum of whose perimeters is equal to the perimeter of the given triangle.

16. Through a given point to draw a secant so that it shall form with the sides of a given angle a triangle of given perimeter.

17. Prove the formula: $AZ \cdot BX \cdot CY = rS$.

18. Construct a triangle ABC given, in position, the excenter I_a, the indefinite line s, containing the corresponding side BC, and the lengths of the inradius r and the circumradius R. *Hint.* The distance from I_a to s is equal to r_a; hence II_a is known (§ 154), and the midpoint of II_a lies at a distance $\frac{1}{2}(r_a - r)$ from s (§ 144 a).

19. Show that (a) the sum of the legs of a right triangle diminished by the hypotenuse is equal to the diameter of the inscribed circle; (b) the altitude upon the hypotenuse of a right triangle is equal to the inradius of the triangle increased by the sum of the inradii of the two triangles into which the altitude divides the given triangle.

20. Show that the area of a right triangle is equal to the product of the two segments into which the hypotenuse is divided by its point of contact with the incircle.

E. ALTITUDES

a. THE ORTHOCENTER

175. Theorem. *The three altitudes of a triangle are concurrent.*

Let H be the point of intersection of the altitudes BE, CF of the triangle ABC (Fig. 65), and let the line AH meet BC in D. The

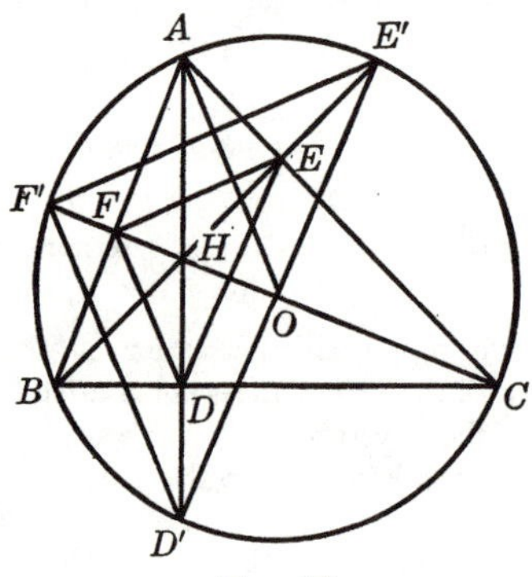

FIG. 65

points E, F lie on the circle having BC for diameter, and in the cyclic quadrilateral $BCEF$ we have angle $FCB = FEB$. Similarly, in the cyclic quadrilateral $AEHF$, we have angle $FEH = FAH$; therefore:

$$\angle BAD = FAH = FEH = FEB = FCB.$$

Thus the two triangles ABD, BCF have the angle B in common and two other angles equal; hence:

$$\angle ADB = BFC = 90°,$$

i.e., AHD is the third altitude of the triangle ABC.

176. Definition. The common point of the three altitudes is called the *orthocenter* of the triangle. It will usually be denoted by the letter H.

177. Theorem. *In a given triangle the three products of the segments into which the orthocenter divides the altitudes are equal.*

Indeed, in the two similar triangles BHF, CHE we have (Fig. 65):

$$HF:HE = BH:CH, \quad \text{or} \quad CH \cdot HF = BH \cdot HE,$$

and we may prove, in the same way, that $BH \cdot HE = AH \cdot HD$.

OTHERWISE. In the circle $BCEF$ (Fig. 65) we have:

$$BH \cdot HE = CH \cdot HF.$$

178. Theorem. *The segment of the altitude extended between the orthocenter and the second point of intersection with the circumcircle is bisected by the corresponding side of the triangle.*

Angle $BD'D = C$ (Fig. 65), for they intercept the same arc AB on the circumcircle, and angle $BHD = C$, because the sides of the two angles are respectively perpendicular; hence the triangle BHD' is isosceles, and since BD is perpendicular to HD', we have $HD = DD'$.

179. Corollary. *The product of the segments into which a side of a triangle is divided by the foot of the altitude is equal to this altitude multiplied by the distance of the side from the orthocenter.*

Indeed, in the circumcircle of ABC (Fig. 65) we have $DB \cdot DC = DA \cdot DD'$, and $DD' = DH$ (§ 178).

180. Theorem. *The circumcircle of the triangle formed by two vertices and the orthocenter of a given triangle is equal to the circumcircle of the given triangle.*

The two triangles HBC, $D'BC$ (Fig. 65) are congruent (§ 178), and the circumcircle of $D'BC$ coincides with the circumcircle of ABC; hence the proposition.

181. Theorem. *Any three equal circles having for centers the vertices of a given triangle cut the respective sides of the medial triangle in six points equidistant from the orthocenter of the given triangle.*

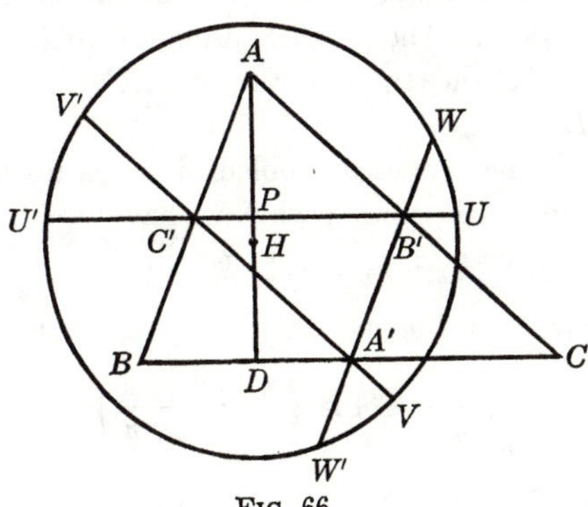

Fig. 66

Let U, U'; V, V'; W, W' be the points of intersection of the sides $B'C', C'A', A'B'$ of the medial triangle $A'B'C'$ of the triangle ABC with the three circles, of equal radii, having for centers the vertices A, B, C (Fig. 66).

If H is the orthocenter of ABC and P the trace of the altitude AD on the side $B'C'$, we have:

$$PU^2 = PU'^2 = AU^2 - AP^2 = HU^2 - HP^2;$$

hence:

$$AU^2 - HU^2 = AP^2 - HP^2 = (AP + PH)(AP - PH)$$
$$= AH(PD - PH) = AH \cdot HD,$$

or:

$$HU^2 = AU^2 - AH \cdot HD.$$

Now, by assumption:

$$AU = BV = CW;$$

hence (§ 177):

$$HU = HV = HW.$$

182. Problem. *Construct a triangle, given its altitudes* (h_a, h_b, h_c).

We have:

$$ah_a = bh_b = ch_c \, (=2S).$$

Dividing by $h_a h_b$, we obtain:

$$a : h_b = b : h_a = c : m,$$

where $m = h_a h_b : h_c$ and is therefore readily constructed as a fourth proportional to the three given altitudes.

The required triangle is thus similar to the triangle DEF in which $EF = h_b$, $FD = h_a$, and $DE = m$. On the altitude DK of this triangle lay off $DL = h_a$. The parallel through L to EF meets DE, DF in the vertices B, C of the required triangle ABC, whose third vertex A coincides with D.

The solution of the proposed problem depends upon the construction of the triangle DEF, so that we must have:

$$h_a + h_b > m > h_a - h_b,$$

or, replacing m by its value and dividing by $h_a h_b$:

$$\left(\frac{1}{h_b} + \frac{1}{h_a}\right) > \frac{1}{h_c} > \left(\frac{1}{h_b} - \frac{1}{h_a}\right).$$

EXERCISES

1. Prove that the circumcenter of a triangle is the orthocenter of its medial triangle.
2. Construct a triangle, given, in position, the orthocenter and a vertex, and the directions of the sides passing through that vertex.
3. Construct a triangle, given, in position, the orthocenter and two vertices.

4. Construct a triangle, given, in position, the feet of the altitudes to the two lateral sides, and the line on which the base is located.
5. If p, q, r are the distances of a point inside a triangle ABC from the sides of the triangle, show that:

$$(p:h_a) + (q:h_b) + (r:h_c) = 1.$$

How should the statement of the problem be modified to make it applicable to points outside the triangle?
6. Show that the three perpendiculars to the sides of a triangle at the points isotomic to the feet of the respective altitudes are concurrent.
7. Show that with the altitudes of a given triangle as sides it is possible to construct a second triangle, if, and only if, the reciprocal of one side of the given triangle is smaller than the sum and greater than the difference of the reciprocals of the other two sides of the given triangle.
8. The perpendicular at the orthocenter H to the altitude HC of the triangle ABC meets the circumcircle of HBC in P. Show that $ABPH$ is a parallelogram.

Construct a triangle, given:

9. h_b, h_c, m_a.
10. h_a, m_a, h_b.
11. Find the locus of the orthocenter of a variable triangle having a fixed base and a constant vertical angle.
12. In order to draw a line through a given point M and the inaccessible point of intersection of two given lines p, q, drop the perpendiculars u, v from M upon p, q, respectively. The required line is the perpendicular from M to the line joining the points (uq) and (pv). Justify this construction.

b. THE ORTHIC TRIANGLE

183. Definition. The triangle having for its vertices the feet of the altitudes of a given triangle is called the *orthic triangle* of the given triangle.

184. Theorem. *The three triangles cut off from a given triangle by the sides of its orthic triangle are similar to the given triangle.*

The angles B and AEF (Fig. 65) are equal, for each is supplementary to the angle FEC of the cyclic quadrilateral $BCEF$, and angle $AFE=C$, for a similar reason; hence the triangles AEF, ABC are similar.

Likewise for the triangles BDF, CDE.

185. Definition. Two opposite sides of a cyclic quadrilateral are said to be *antiparallel* with respect to the other two sides of the quadrilateral.

Thus EF, BC (Fig. 65) are antiparallel with respect to AB, AC; and FD, AC are antiparallel with respect to AB, BC.

186. Theorem. *A vertex of a triangle is the midpoint of the arc determined on its circumcircle by the two altitudes, produced, issued from the other two vertices.*

The quadrilateral $BCEF$ being cyclic (Fig. 65), we have angle $ABE = ACF$; hence arc $AE' =$ arc AF'.

Similarly for the vertices B and C.

187. *Remark.* The line EF joins the midpoints of two sides of the triangle $HE'F'$; hence EF is parallel to the side EF', and $EF = \frac{1}{2} E'F'$. Similarly for the other sides of the orthic triangle DEF. Hence the two triangles DEF, $D'E'F'$ correspond to each other in the homothecy $(H, \frac{1}{2})$.

188. Theorem. *The radii of the circumcircle passing through the vertices of a triangle are perpendicular to the corresponding sides of the orthic triangle.*

Indeed, the radius OA is perpendicular to the chord $E'F'$ (§ 186), and therefore to EF (§ 187).

189. Corollary. *The angle which a side of a triangle makes with the corresponding side of the orthic triangle is equal to the difference of the angles of the given triangle adjacent to the side considered* (§ 73).

190. Definition. The tangents to the circumcircle at the vertices of a given triangle form a triangle called the *tangential triangle* of the given triangle.

191. Theorem. *The tangential and the orthic triangles of a given triangle are homothetic.*

Indeed, corresponding sides of these two triangles are perpendicular to the same circumradius of the basic triangle (§ 188).

192. Theorem. *The altitudes of a triangle bisect the internal angles of its orthic triangle.*

The line $D'A$ (Fig. 65) bisects the angle $F'D'E'$ (§ 186), and the lines DF, DE are respectively parallel to $D'F', D'E'$ (§ 187); hence $D'DA$ also bisects the angle FDE. Similarly for the other altitudes.

Otherwise. From the three cyclic quadrilaterals $BDHF, BCEF, CDHE$ (Fig. 65), we have:

$$\angle HDF = HBF = EBF = ECF = ECH = EDH.$$

193. Corollary. *The sides of a triangle bisect the external angles of its orthic triangle.*

The vertices and the orthocenter of a given triangle are the tritangent centers of the orthic triangle of the given triangle.

194. Problem. *Construct a triangle given the points where the altitudes produced meet the circumcircle.*

The three given points D', E', F' determine the circumcircle (O) of the required triangle ABC whose vertices are the midpoints of the arcs $D'E'$, $E'F'$, $F'E'$ (§ 186).

How many solutions may the problem have?

EXERCISES

1. If O is the circumcenter and H the orthocenter of a triangle ABC, and AH, BH, CH meet the circumcircle in D', E', F', prove that the parallels through D', E', F' to OA, OB, OC, respectively, meet in a point.

2. Show that (a) the product of the segments into which a side of a triangle is divided by the corresponding vertex of the orthic triangle is equal to the product of the sides of the orthic triangle passing through the vertex considered; (b) the product of the six segments into which the sides of a triangle are divided by the feet of the altitudes is equal to the square of the product of the three sides of the orthic triangle.

3. If P, Q are the feet of the perpendiculars from the vertices B, C of the triangle ABC upon the sides DF, DE, respectively, of the orthic triangle DEF, show that $EQ = FP$. *Hint.* Use § 193.

4. DP, DQ are the perpendiculars from the foot D of the altitude AD of the triangle ABC upon the sides AC, AB. Prove that the points B, C, P, Q are concyclic, and that angle $DPB = CQD$.

5. Show that the four projections of the foot of the altitude on a side of a triangle upon the other two sides and the other two altitudes are collinear. *Hint.* Use §§ 193, 135.

6. Show that the perimeter of the orthic triangle of an acute-angled triangle ABC is smaller than twice any altitude of ABC. *Hint.* Use § 192.

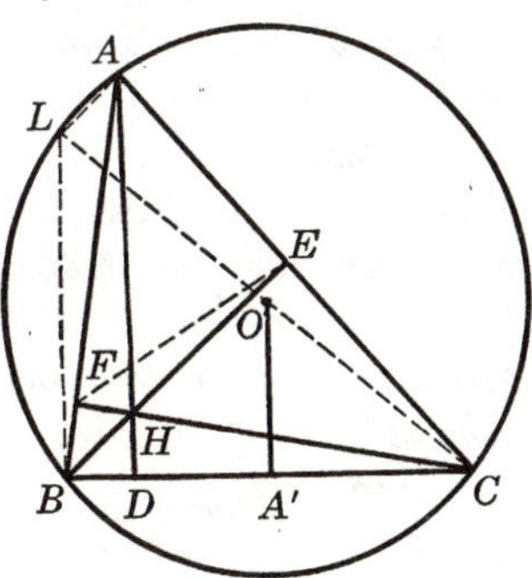

FIG. 67

195. Theorem. *The distance of a side of a triangle from the circumcenter is equal to half the distance of the opposite vertex from the orthocenter.*

Let L be the diametric opposite of the vertex C on the circumcircle (Fig. 67); the segment OA' joins the midpoints of two sides of the right triangle BCL; hence $OA' = \frac{1}{2} BL$.

On the other hand, the pairs of opposite sides of the quadrilateral $ALBH$ are respectively perpendicular to the lines AC, BC; hence $ALBH$ is a parallelogram, and $BL = AH$, which proves the proposition.

196. Theorem. *In a triangle the distances of the vertices from the orthocenter, increased by the corresponding exradii, are equal* (Fig. 55).

We have (§ 144):

$$r_a - r = 2\,KA' = 2(KO - OA');$$

hence (§ 195):

$$2\,OA' + r_a = AH + r_a = 2\,KO + r = 2\,R + r,$$

and similarly:

$$BH + r_b = CH + r_c = 2\,R + r.$$

197. Theorem. *The ratio of a side of a triangle to the corresponding side of the orthic triangle is equal to the ratio of the circumradius to the distance of the side considered from the circumcenter.*

The segment AH (Fig. 67) is the circumdiameter of the triangle AEF, and the lines BC, EF are corresponding sides in the two similar triangles AEF, ABC (§ 184); hence (§ 195):

$$BC:EF = 2\,R:AH = R:OA'.$$

198. Corollary. *The sum of the ratios of the sides of the orthic triangle to the corresponding sides of a given acute-angled triangle is equal to the ratio of the sum of the circumradius and the inradius of the given triangle to the circumradius of the same triangle.*

Indeed, the sum of these ratios is equal to the sum of the distance of the sides of the given triangle from the circumcenter divided by the circumradius; hence (§ 146) the proposition.

199. Theorem. *The perimeter of the orthic triangle of an acute-angled triangle is equal to the doubled area of the given triangle divided by the circumradius of the given triangle.*

If e, f, g are the sides of the orthic triangle of the triangle ABC, we have (§ 197):

$$e + f + g = (a\cdot OA' + b\cdot OB' + c\cdot OC'):R,$$

and the sum in the parentheses represents the double area of ABC; hence the proposition.

200. Problem. *Construct a triangle given, in position, the traces D, U, A' on the base of the triangle of the altitude, the internal bisector, and the* *median issued from the opposite vertex, and the distance, d, of that vertex from the orthocenter.*

On the perpendicular to DUA' at A' lay off $A'O = \frac{1}{2} d$, and O is the circumcenter of the required triangle ABC (§ 193). The point A lies on the perpendicular to DUA' at D. Now the bisector of the angle A is also the bisector of the angle DAO (§ 73); hence the point A lies on a tangent from O to the circle (U, UD). Thus A may be determined.

The circle (O, OA) meets the line DUA' in the two vertices B, C of ABC.

EXERCISES

1. In the triangle ABC show that: $AH^2 + BC^2 = 4\,OA^2$.
2. Construct a triangle given, in position, a vertex, the midpoint of the opposite side, and the orthocenter.
3. Construct a triangle given, in position, the circumcircle and the orthocenter, and the length of one side.
4. Construct a triangle given the circumradius, the distance of a vertex from the orthocenter, and the median issued from that vertex.
5. Construct a triangle given the base, the distance of the opposite vertex from the orthocenter, and the radius of the inscribed circle.
6. A variable line of constant length has its extremities on two fixed intersecting lines. Prove that the locus of the orthocenter of the variable triangle formed by the three lines is a circle.
7. The base BC and the circumcircle (O) of a variable triangle ABC are fixed. Find the locus of the orthocenter of the triangle having for vertices the traces on (O) of the internal bisectors of the angles of ABC.

c. THE EULER LINE

201. Theorem. *The circumcenter, the orthocenter, and the centroid of a triangle are collinear, and the distance from the centroid to the orthocenter is equal to twice the distance from the centroid to the circumcenter.*

Let Q be the point where the altitude AD of the triangle ABC is met by the line OG joining the circumcenter O to the centroid G of the triangle. The two triangles AGQ, OGA' are similar; hence:

$$GQ:OG = AG:GA' = 2:1.$$

Now if we take another altitude, its point of intersection with OG will be the same point Q, since $GQ = 2\,OG$. In other words, the point Q coincides with the orthocenter H of ABC, and we have $GH = 2\,GO$.

It should be noticed that we have here a new proof that the altitudes of a triangle meet in a point (§ 173), and also that $AH = 2OA'$ (§ 195).

202. Definition. The straight line containing the circumcenter, the orthocenter, and the centroid of a triangle is called the *Euler line* of the triangle.

203. Theorem. *We have:*

$$OH^2 = 9R^2 - (a^2 + b^2 + c^2),$$
$$GH^2 = 4R^2 - 4(a^2 + b^2 + c^2).$$

Indeed, we have $OH = 3OG, GH = 2OG$ (§ 201); hence the formulas (§ 110).

204. Theorem. *The sum of the squares of the distances of the vertices of a triangle from the orthocenter is equal to twelve times the square of the circumradius diminished by the sum of the squares of the sides of the triangle.*

$$HA^2 + HB^2 + HC^2 = 12R^2 - (a^2 + b^2 + c^2).$$

The formula follows readily from previously established relations (§§ 108, 109, 203).

205. Theorem. *The homothetic center of the orthic and the tangential triangles of a given triangle* (§ 191) *lies on the Euler line of the given triangle.*

The tangential triangle (T) and the orthic triangle DEF of the given triangle ABC have for their respective incenters the circumcenter O and the orthocenter H (§ 193) of ABC; hence the points O, H are corresponding points in two homothetic figures, and are therefore collinear with the homothetic center of these figures, which proves the proposition.

If ABC is obtuse-angled, the points O, H are corresponding excenters of the triangles (T) and DEF.

EXERCISES

1. Construct a triangle given, in position, the points O and A, and the lengths HA and GA.
2. Construct a triangle given, in position, the circumcircle and the centroid, and the difference of two angles.
3. Two rectangular chords AB, CD of a circle revolve about a fixed point P. Show that the orthocenters of the variable triangles ABC, ABD describe the same circle. Find the loci of their centroids.

4. The Euler line of a triangle passes through a vertex of the triangle only if the triangle is either isosceles, or right-angled.
5. The line joining the centroid of a triangle to a point P on the circumcircle bisects the line joining the diametric opposite of P to the orthocenter.

F. THE NINE-POINT CIRCLE

206. Definition. The midpoints of the segments joining the orthocenter of a triangle to its vertices are called the *Euler points* of the triangle.

The three Euler points determine the *Euler triangle* of the given triangle.

207. Theorem. *In a triangle the midpoints of the sides, the feet of the altitudes, and the Euler points lie on the same circle.*

Let P, Q, R be the Euler points of the triangle ABC (Fig. 68). The segments QC', RB' are equal and parallel, for they join the midpoints

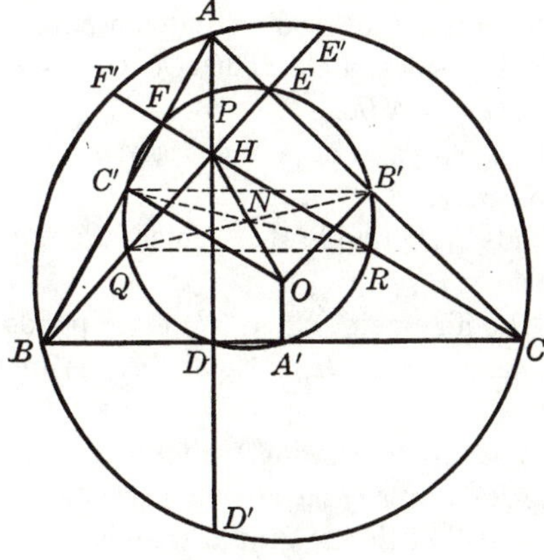

Fig. 68

of the lateral sides of the two triangles AHB, AHC having AH for common base; moreover, these segments make right angles with the line $B'C'$, for AH is perpendicular to $B'C'$. Thus $B'C'QR$ is a rectangle; therefore QB', RC' are equal and bisect each other, say, in the point N. In the same manner it may be shown that PA' is equal to QB' and has N for its midpoint.

Thus N is the center of a circle (N) passing through A', B', C', P, Q, R, and also through the points D, E, F, for the diameters PA', QB', RC' of (N) subtend right angles at these three points, respectively.

208. Definition. The circle (N) (§ 207) is often referred to as the *nine-point circle*, and the point N as the *nine-point center* of the given triangle.

209. Theorem. (*a*) *The radius of the nine-point circle is equal to half the circumradius of the triangle.*

(*b*) *The nine-point center lies on the Euler line, midway between the circumcenter and the orthocenter.*

We have (§ 195):

$$OA' = \tfrac{1}{2}\,AH = AP = PH;$$

hence $APA'O$ and $PHA'O$ are parallelograms. From the first it follows that the diameter PA' of the circle (N) is equal to the circumradius AO of ABC, and from the second that the diagonal HO passes through the midpoint N of the diagonal PA' and is bisected by N.

OTHERWISE. The triangles ABC, $A'B'C'$ correspond to each other in the homothecy $(G, -2)$ (§ 98); hence the circumradius of ABC is equal to twice that of $A'B'C'$ and the circumcenters O, N of ABC, $A'B'C'$ lie on opposite sides of G. Thus $OG{:}GN = 2$. Now (§ 201) $2\,OG = GH$; hence $ON = NH$.

210. *Remark.* The two pairs of points O and N, G and H separate each other harmonically.

Indeed, the points N, O divide the segment HG internally and externally in the ratio $2{:}1$.

Consequently, the nine-point circle (N) corresponds to the circumcircle (O) in the homothecy $(H, 2{:}1)$, as well as in the homothecy $(G, -2{:}1)$.

211. Theorem. *The circumcenter of the tangential triangle of a given triangle lies on the Euler line of the given triangle.*

The circumcenter O'' of the tangential triangle (T) of ABC and the circumcenter N of the orthic triangle DEF of ABC are corresponding points in the homothecy in which (T) and DEF correspond to each other (§ 191). Now the center of this homothecy lies on the Euler line (§ 205), and so does the point N (§ 209 b); hence O'' also lies on this line.

212. COROLLARY. Let p be the inradius of DEF, and R, q, the circumradii of the triangles ABC, (T). The homothetic ratio of the triangles (T), DEF is equal to the ratio $R{:}P$ of their inradii, and also to the ratio $q{:}\frac{1}{2}R$ of their circumradii; hence:

$$R{:}p = q{:}\tfrac{1}{2}\,R, \quad \text{or} \quad R^2 = 2\,pq.$$

As an exercise, state this result verbally.

213. Theorem. *The circumcenter of a triangle lies on the Euler line of the triangle determined by the points of contact of the sides of the given triangle with its inscribed circle.*

The given triangle is the tangential triangle of the second triangle considered; hence the proposition (§ 211).

Note. The proposition remains valid for the escribed circles.

214. Theorem. *The projections of the orthocenter of a triangle upon the two bisectors of an angle of the triangle lie on the line joining the midpoint of the side opposite the vertex considered to the nine-point center of the triangle.*

The feet J, J' of the perpendiculars from the orthocenter H of the triangle ABC (Fig. 69) upon the bisectors AJ, AJ' of the angle

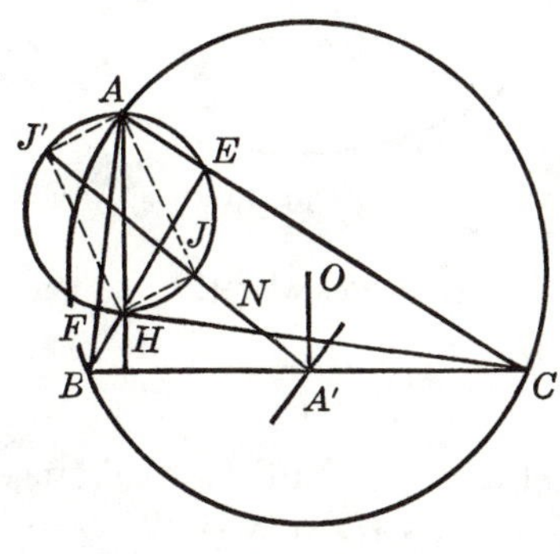

FIG. 69

A lie on the circle (AH) having AH for diameter. Those two points are thus the ends of the circumdiameter of the triangle AEF inscribed in (AH) perpendicular to the side EF (§ 79).

The circle having BC for diameter passes through E, F (§ 175), and so does the nine-point circle (§ 207); hence the centers A', N of these circles lie on the mediator JJ' of EF, which proves the proposition.

215. Feuerbach's Theorem. *The nine-point circle of a triangle touches each of the four tritangent circles of the triangle.*

In a triangle ABC (Fig. 70) the incenter I and the excenter I_a relative to the vertex A separate harmonically the point A and the trace U of the bisector AI on the side BC (§ 120); hence the points of contact X, X_a of the tritangent circles (I), (I_a) with the side BC separate harmonically U and the foot D of the altitude AD of ABC (§ 61).

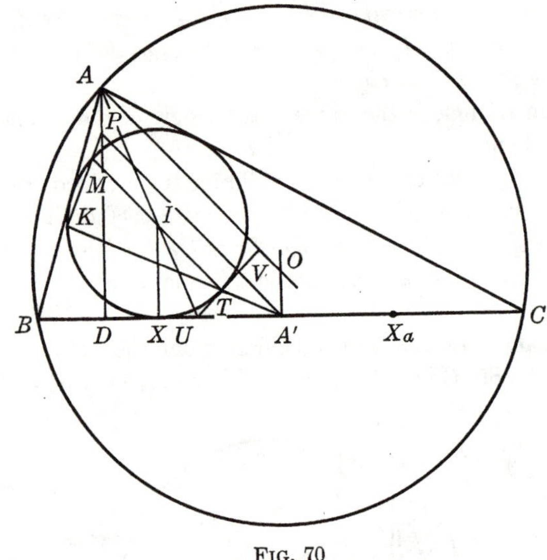

FIG. 70

The midpoint of XX_a coincides with the midpoint A' of BC (§ 160); therefore (§ 63):

$$A'U \cdot A'D = A'X^2. \tag{1}$$

Let T be the point of contact of the second tangent UT from U to the incircle (I). The lines $UT, UX = BC$ are symmetric with respect to AU, and so are the circumdiameter AO of ABC and the altitude AD (§ 73 b); hence UT is perpendicular to AO and therefore also to the diameter $A'P$ of the nine-point circle (N) of ABC, where P is the Euler point of the altitude AD (§ 207).

Let K be the second point of intersection of (I) with $A'T$ and $V = (A'P, UT)$. From the two similar right triangles $A'DP, A'VU$ we have, using the relation (1):

$$A'V \cdot A'P = A'D \cdot A'U = A'X^2 = A'T \cdot A'K;$$

hence the two pairs of points V, P; T, K are concyclic; thus angle $PKT = TVA'$, i.e., PK is perpendicular to KTA', and therefore K lies on the nine-point circle (N), since PA' is a diameter of (N) (§ 207). Moreover, the second point of intersection M of PK with (I) is the diametric opposite of T on (I).

The lines $TM, A'VP$ are parallel, for both are perpendicular to UTV; hence the point $K = (PM, A'T)$ is collinear with the midpoints of the segments $TM, A'P$. Now $TM, A'P$ are diameters of the circles (I), (N), respectively, and K is a point common to these two circles; hence (I), (N) touch each other at the point K.

216. *Remark I.* The point K is called the *Feuerbach point* of the circle (I).

From the above proof it follows that if T is the symmetric of X with respect to the bisector AU, the foot of the perpendicular from the Euler point P upon the line $A'T$ is the Feuerbach point of (I).

217. *Remark II.* It may be shown in a similar manner that if T_a is the symmetric of X_a with respect to AU, the foot K_a of the perpendicular PK_a from P upon the line $A'T_a$ is the Feuerbach point of the excircle (I_a).

Similarly for the other two excircles.

218. Problem. *Construct a triangle, given, in position, the nine-point center N and one vertex, A, and the directions of the internal bisector t and of the altitude h passing through the given vertex A.*

The circumcenter O lies on the symmetric d of the line h with respect to t (§ 73 b), and also on the symmetric h' of h with respect to N (§ 209 b); hence O is determined, and also H, on the line h.

If D' is the second point of intersection of AH with the circle (O, OA), the mediator of HD' meets this circle in the other two vertices B, C of the required triangle ABC.

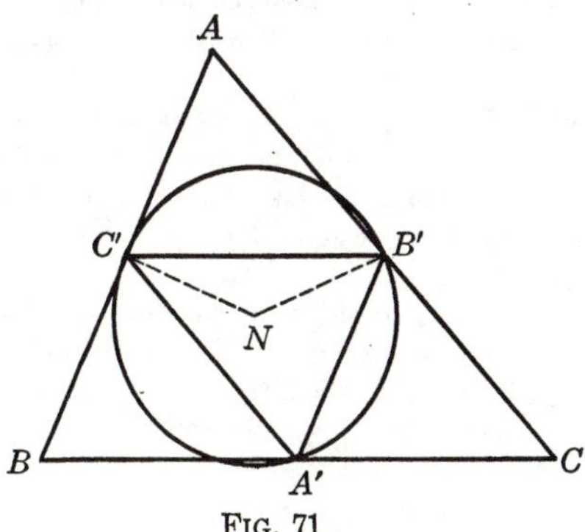

FIG. 71

219. Problem. *Construct a triangle given the position of the nine-point center, and one angle, both in magnitude and position.*

Let ABC be the required triangle, with the given angle at A (Fig. 71). We have:

$$\angle B'NC' = 2\, B'A'C' = 2\, A;$$

hence the isosceles triangle $NB'C'$ has known angles. Thus if the known vertex N of this triangle is kept fixed, and the vertex B' describes the given fixed line AC, while the triangle remains similar to itself, the point C' will describe a straight line (§ 51) which will determine on the given line AB the point C'. Now draw the quadrilateral $AC'NB'$ so that angle $C'NB' = 2\,A$, and the solution is readily completed.

EXERCISES

1. Show that the complementary and the Euler triangles of a given triangle are congruent.
2. Show that the triangle $DB'C'$ is congruent to the Euler triangle (Fig. 68).
3. Show that the segment OP is bisected by the median AA' (Fig. 68).
4. If P is the symmetric of the vertex A with respect to the opposite side BC, show that HP is equal to four times the distance of the nine-point center from BC.
5. Show that the square of the tangent from a vertex of a triangle to the nine-point circle is equal to the altitude issued from that vertex multiplied by the distance of the opposite side from the circumcenter.
6. Prove that HA' passes through the diametric opposite of A on the circumcircle.
7. Through the vertices of a triangle ABC are drawn three parallel lines, of arbitrary direction, and the perpendiculars to these lines through the same vertices. Three rectangles are thus obtained of which the sides BC, CA, AB are diagonals, respectively. Prove that the three remaining diagonals of these three rectangles meet in a point on the nine-point circle of ABC.
8. Construct a triangle given, in position, a vertex, the orthocenter, and the nine-point center.
9. Construct a triangle given, in position, two vertices and the nine-point center.
10. Construct a triangle given, in position, the circumcenter and one vertex, and the distances of that vertex from the orthocenter and from the centroid.
11. Construct a triangle given, in position, the nine-point circle and the centroid, and the difference of two angles.
12. Construct a triangle given, in position, the nine-point center, a vertex, and the foot of the altitude to a side passing through that vertex.
13. Construct a triangle given, in position, the nine-point center, a vertex, and the projection of that vertex upon the opposite side.
14. Construct a triangle given, in position, the midpoint of the base, the midpoint of one of the arcs subtended by the base on the circumcircle, and the Euler point relative to the vertex opposite the base.
15. If a variable triangle has a fixed base and a constant circumradius, show that its nine-point circle is tangent to a fixed circle.
16. A variable triangle has a fixed vertex and a fixed nine-point circle; prove that the orthocenter describes a circle.
17. The orthocenter, the midpoint of the base, and the direction of the base of a variable triangle are fixed. Find the locus of the nine-point center.

18. The tangent to the nine-point circle at the midpoint of a side of the given triangle is antiparallel to this side with respect to the two other sides of the triangle.

G. THE ORTHOCENTRIC QUADRILATERAL

220. Definitions. If H is the orthocenter of a triangle ABC (Fig. 72), the four points A, B, C, H are such that each is the orthocenter of the triangle formed by the remaining three, as is readily seen from the figure. Four such points will be referred to as an *orthocentric group of points*, or an *orthocentric quadrilateral;* the four triangles formed by taking these points three at a time are an *orthocentric group of triangles.*

221. Theorem. *The four triangles of an orthocentric group have the same orthic triangle.*

This is immediately apparent from the figure.

222. COROLLARY I. *The four triangles of an orthocentric group have the same nine-point circle.*

Indeed, this is the circumcircle of their common orthic triangle (§ 207).

223. COROLLARY II. *The circumradii of the four triangles of an orthocentric group are equal.*

Indeed, the circumradii of these triangles are equal to the circumdiameter of their common nine-point circle (§ 209 a).

224. Theorem. *The circumcenters of an orthocentric group of triangles form an orthocentric quadrilateral.*

The circumcenters O, O_a, O_b, O_c of the four triangles ABC, BCH, CHA, HAB of an orthocentric group (Fig. 72) are the symmetrics of the orthocenters H, A, B, C of these triangles with respect to their common nine-point center N (§§ 222, 209 b); hence the proposition.

225. COROLLARY I. *An orthocentric group of triangles and the orthocentric group of their circumcenters have the same nine-point circle.*

The two orthocentric quadrilaterals $HABC, OO_aO_bO_c$ being symmetrical with respect to the nine-point center N of $HABC$, the nine-point circle of $OO_aO_bO_c$ is the symmetric of the nine-point circle (N) of $HABC$. Now the symmetric of a circle with respect to its center is the circle itself; hence the proposition.

226. COROLLARY II. *The four vertices of a given orthocentric group of triangles may be considered as the circumcenters of a second orthocentric group of triangles.*

Indeed, the two orthocentric groups $HABC$, $OO_aO_bO_c$ are symmetrical with respect to their common nine-point center N; hence the first group may be derived from the second in precisely the same manner in which the second has been derived from the first.

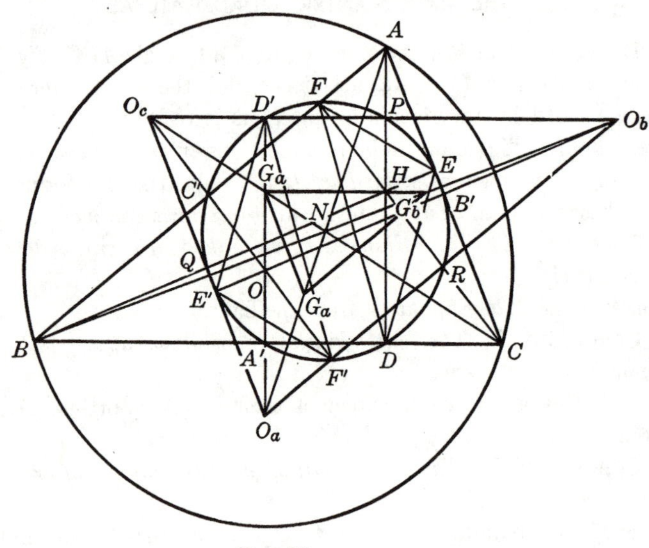

FIG. 72

227. Theorem. *The four centroids of an orthocentric group of triangles form an orthocentric group.*

The centroids G, G_a, G_b, G_c of the four triangles ABC, BCH, CHA_c HAB of an orthocentric group (Fig. 72) correspond to the orthocenters H, A, B, C of these triangles in the homothecy $(N, -\frac{1}{3})$ (§ 209); hence the proposition.

228. Theorem. *The nine-point circle of an orthocentric group of triangles is concentric with the nine-point circle of the orthocentric group formed by the centroids of the given group of triangles.*

Indeed, the nine-point circle (N_g) of the orthocentric group GG_aG_bG, (§ 227) corresponds to the nine-point circle (N) of the group $HABC$ in the homothecy $(N, -\frac{1}{3})$; hence N is the common center of the two circles (N) and (N_g).

EXERCISES

1. Show that the Euler lines of the four triangles of an orthocentric group are concurrent.
2. Show that the symmetric of the circumcenter of a triangle with respect to a side coincides with the symmetric of the vertex opposite the side considered with respect to the nine-point center of the triangle.
3. Show that the four vertices of an orthocentric group of triangles may be considered as the centroids of another orthocentric group of triangles, the two groups having the same nine-point center, this point being the center of similitude of the two groups, the ratio of similitude being -3.
4. Show that the circumcenters and the centroids of an orthocentric group of triangles form two orthocentric groups of points having the same nine-point center, this point being a homothetic center of the two groups, the homothetic ratio being 3.
5. Show that the circumcenters of the triangles HBC, HCA, HAB of an orthocentric group $HABC$ form a triangle congruent to ABC; the sides of the two triangles are parallel, and the point H is the circumcenter of the new triangle; the circumcenter of ABC is the orthocenter of the new triangle.
6. Show that the centroids of the triangles HBC, HCA, HAB of the orthocentric group $HABC$ form a triangle similar to ABC, and the sides of this new triangle are parallel to those of ABC; the orthocenter of the new triangle coincides with the centroid of ABC.
7. Construct a triangle given the feet of its altitudes.
8. Construct a triangle ABC given, in position, the nine-point center and the circumcenters of the triangles CAH, ABH, where H denotes the orthocenter of ABC.
9. Show that the algebraic sum of the distances of the points of an orthocentric group from any line passing through the nine-point center of the group is equal to zero.
10. The triangle ABC has a right angle at A, and AD is its altitude. The bisectors of the angles BAD, CAD meet BC in S, S', and the bisectors of the angles ABD, ACD meet AD in T, T'. If U, V, W are the incenters of the triangles ABC, ABD, ACD, show that (a) the points A, U, V, W form an orthocentric group; (b) the circumcenter of AVW lies on AD; (c) the points B, C, V, W are concyclic; (d) the points S, S', T, T' form an orthocentric group. State and prove other properties of the figure.

229. Theorem. *The four tritangent centers of a triangle form an orthocentric group.*

Indeed, the triangle $I_aI_bI_c$ formed by the external bisectors of the given triangle ABC has for its altitudes the internal bisectors of ABC; hence the incenter I of ABC is the orthocenter of the triangle $I_aI_bI_c$, which proves the proposition (§ 220).

230. Corollary. *A given triangle and its circumcircle are respectively the orthic triangle and the nine-point circle of the orthocentric group formed by the tritangent centers of the given triangle.*

231. *Remark I.* From the preceding propositions (§§ 229, 230) it follows that all the properties of an orthocentric group of points may be applied to the tritangent centers of a triangle. We thus obtain a series of propositions of which the following are examples.

(*a*) *The circumradius of a triangle formed by any three tritangent centers of a given triangle is equal to the circumdiameter of the given triangle* (§ 223).

(*b*) *The symmetric of a tritangent center of a given triangle with respect to the circumcenter of the triangle is the circumcenter of the triangle formed by the three remaining tritangent centers of the given triangle* (§ 224).

(*c*) *The circumcenters of the four triangles determined by the tritangent centers of a given triangle are the tritangent centers of the symmetric of the given triangle with respect to the circumcenter of this triangle.*

Indeed, the circumcenters of the orthocentric group $II_aI_bI_c$ (§ 229) are the symmetrics of these four points with respect to the nine-point center of the group (§ 224), and this nine-point center coincides with the circumcenter O of the triangle ABC (§ 230).

232. *Remark II.* The preceding propositions (§ 231) were obtained by considering the given triangle ABC as the orthic triangle of its tritangent centers, whereas in the study of the orthocenter (§ 175 ff.) and of the nine-point circle the triangle ABC was considered to be the basic triangle having its own orthic triangle and its own nine-point circle. This double role which a triangle may be made to play in connection with its orthocenter on the one hand, and its tritangent centers on the other hand, makes it possible to "translate" or "transform" properties established for the orthocenter into properties of the tritangent centers, and vice versa, without having to prove anew the properties thus obtained.

For instance, considering the triangle ABC as the basic triangle we have noticed that the circle having AH for diameter passes through the points E, F (§ 175) and has its center P on the nine-point circle of ABC (§ 207). Now the points A, H are tritangent centers of the orthic triangle DEF of ABC (§§ 192, 193). Hence if we consider ABC as the orthic triangle of its group of tritangent centers I, I_a, I_b, I_c, we may say that the circle having II_a for diameter will pass through the vertices B, C of ABC and have its center on the circumcircle of ABC. This proposition, however, was proved independently (§ 122).

The circle having I_bI_c for diameter corresponds to the circle having BC for diameter. Both these circles were considered before, independently (§§ 122, 175).

The student may find other examples of this kind.

233. Theorem. *The area of the triangle formed by the excenters of a given acute-angled triangle is equal to the product of the perimeter and the circumradius of the given triangle.*

234. Theorem. *In an acute-angled triangle the sum of the distances of the vertices from the corresponding sides of the orthic triangle is equal to the circumdiameter of the given triangle increased by the distance of the orthocenter from a side of the orthic triangle.*

These are "translations" of properties proved before (§§ 199, 146).

235. Theorem. *The segment joining the orthocenter of a triangle to the circumcenter of the triangle determined by the excenters of the given triangle is bisected by the incenter of the medial triangle.*

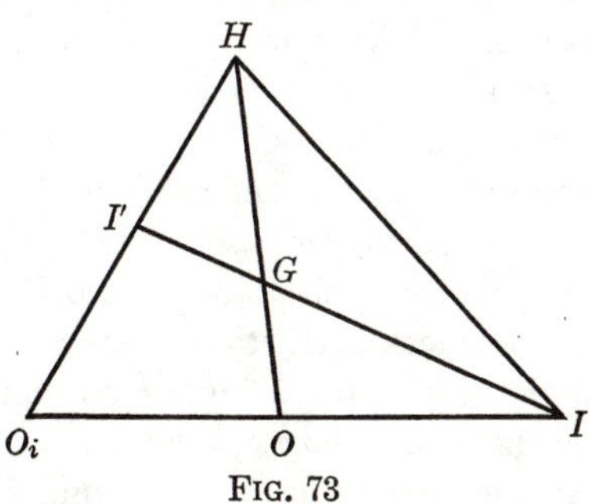

FIG. 73

The circumcenter O_i (Fig. 73) of the triangle $I_aI_bI_c$ formed by the excenters of the given triangle ABC is the symmetric of the incenter I of ABC with respect to the circumcenter O of ABC (§ 231 b). If H and G are the orthocenter and the centroid of ABC, the point G is also the centroid of the triangle HIO_i, for $HG = 2\,GO$ (§ 201); hence the line IG passes through the midpoint I' of HO_i; moreover, $IG = 2\,GI'$, i.e., I' corresponds to I in the homothecy $(G, -\frac{1}{2})$. Consequently I' is the point which corresponds in the medial triangle $A'B'C'$ of ABC to the incenter I of ABC (§ 98), which proves the proposition.

236. Theorem. *In an acute-angled triangle, the symmetrics of the sides of the orthic triangle with respect to the corresponding sides of the given triangle form a new triangle whose incenter coincides with the circumcenter of the given triangle.*

Let O_a be the symmetric of the circumcenter O of the given triangle ABC with respect to the side BC. The segments OO_a, AH (Fig. 72) are equal and parallel; hence OA, HO_a are equal and parallel.

The radius $OA = R$ is perpendicular to the side $d = EF$ of the orthic triangle DEF (§ 188); hence HO_a is also perpendicular to d, so that if HO_a meets d in the point L, the segment $HL = m$ is the inradius of the triangle DEF (§ 192).

Now if d' is the symmetric of d with respect to BC, the distance of O from d' is equal to the distance O_aL from d, for O and O_a are also symmetric with respect to BC; moreover, the distance $O_aL = R + m$ does not depend upon the choice of the side d of DEF; hence O is equidistant from the sides d', e', f' of the triangle $d'e'f'$ formed by d' and its two analogues e' and f'.

237. *Remark I.* We have proved incidentally that the inradius of the triangle $d'e'f'$ is equal to the sum of the circumradius of the given triangle and the inradius of the orthic triangle.

238. *Remark II.* The triangle ABC is the acute-angled triangle of the orthocentric group of triangles $ABCH$ of which DEF is the common orthic triangle. The preceding propositions (§§ 236, 237) may be modified so as to become applicable to the other triangles of the orthocentric group. The task is left to the reader.

239. *Remark III.* If in the preceding propositions (§§ 236, 237) DEF is taken to be the given triangle, the points B, C are two of its excenters; the line d is tangent to the corresponding excircles (B), (C); hence the symmetric d' of d with respect to the line of centers BC is also tangent to (B) and (C), i.e., d' is the fourth common tangent (besides the sides of DEF) of these two circles. The point O is the circumcenter of the triangle determined by the excenters of the given triangle DEF. Thus the above propositions (§§ 236, 237) may be "translated" (§ 232) as follows.

The fourth common tangents of the excircles, taken in pairs, of an acute-angled triangle form a second triangle whose incenter coincides with the circumcenter of the triangle formed by the excenters of the given triangle, and whose inradius is equal to the inradius of the given triangle increased by its circumdiameter.

EXERCISES

1. A variable triangle ABC has a fixed circumcircle and a fixed incircle. Show that the triangle $A'B'C'$ having for its vertices the traces A', B', C' of the external bisectors of ABC on the circumcircle of ABC has a fixed orthocenter, a fixed centroid, and a fixed nine-point circle.
2. Given, in position, the circumcenter O of a triangle ABC, and the circumcenters of the triangles II_bI_c, II_cI_a, construct the triangle ABC.
3. The inscribed circle of a triangle ABC touches the sides BC, CA, AB in D, E, F; I is the incenter of ABC; L, M, N are the orthocenters of the triangles IBC, ICA, IAB. Prove that (a) DL, EM, FN are equal to the exradii of ABC; (b) MN, NL, LM pass through D, E, F, respectively; (c) the triangles LMN, ABC are equivalent; (d) if the excircles relative to A, B, C touch BC, CA, AB in D_1, E_1, F_1, the lines MN, NL, LM are at right angles to AD_1, BE_1, CF_1, respectively; the lines MN, NL, LM meet AD_1, BE_1, CF_1, respectively, on the inscribed circle.

REVIEW EXERCISES

THE CIRCUMCIRCLE

1. Construct a triangle given m_a, t_a, $B - C$.
2. Construct a triangle given the traces on the circumcircle of the altitude, the median, and the bisector issued from the same vertex.
3. Construct a triangle given the difference of the base angles and the traces, on the base, of the altitude, the median, and the bisector issued from the opposite angle.
4. Show that the mediator of the bisector AU of the triangle ABC, the perpendicular to BC at U, and the circumdiameter of ABC passing through A are concurrent.
5. From the foot U of the bisector AU of the triangle ABC a perpendicular UQ is drawn to the circumradius AO of ABC meeting AC in P. Prove that AP is equal to AB.
6. On a given circle to find a point such that the lines joining it to two given points shall determine on the circle a chord of given length.
7. On a given circle to find a point which, with the points determined on the circle by the lines joining this point to two given points, shall form a triangle having a given angle. (Two cases: the given angle is, or is not, equal to the angle formed by the two sides passing through the two given points.)
8. In a given circle to inscribe a triangle having a given vertical angle, so that its base shall touch another given circle, and one side shall touch a third given circle.
9. A segment BC of constant length moves so that its ends remain on two fixed lines AB, AC. Show that the circumcircle of the triangle ABC is tangent to a fixed circle.
10. Construct a triangle given an angle, the altitude issued from the vertex of this angle, and the sum of the distances of the foot of this altitude from the other two sides of the triangle.

11. Show that the foot of the altitude to the base of a triangle and the projections of the ends of the base upon the circumdiameter passing through the opposite vertex of the triangle determine a circle having for center the midpoint of the base.

MEDIANS

12. The internal bisector of the angle B of the triangle ABC meets the sides $B'C'$, $B'A'$ of the medial triangle in the points A'', C''. Prove that AA'', CC'' are perpendicular to the bisector, and that $B'A'' = B'C''$. Similarly for the external bisector.

13. A', B', C' are the traces of the medians of the triangle ABC on the circumcircle. If $B'C' = a'$, $C'A' = b'$, $A'B' = c'$, show that, in the usual notation for triangle ABC:

$$am_a:a' = bm_b:b' = cm_c:c'.$$

14. Show that the parallels through the vertices A, B, C of the triangle ABC to the medians of this triangle issued from the vertices B, C, A, respectively, form a triangle whose area is three times the area of the given triangle.

15. Construct an isosceles triangle given the median and the altitude to a lateral side.

16. If R is the circumradius of the triangle ABC, centroid G, r', r'', r''' the circumradii of the triangles BCG, CAG, ABG, and r the circumradius of the triangle formed by the medians of ABC, show that $4\,Rr^2 = 3\,r'r''r'''$.

17. A variable circle passes through two fixed points A, B. If M is the second point of intersection of the circle with a fixed line passing through A, find the locus of the centroid of the triangle determined by the points A, M and the center of the circle; find the locus of the midpoint of the radius passing through M.

18. Through the centroid G of the triangle ABC the parallel is drawn to BC meeting AB, AC in A_b, A_c; the parallel to CA meeting BC, BA in B_c, B_a; the parallel to AB meeting CA, CB in C_a, C_b. Prove that the two triangles $A_bB_cC_a$, $A_cB_aC_b$ are equal.

19. The side BC of the triangle ABC is produced beyond C to the point A_1 so that $CA_1 = k \cdot BC$, and similarly $AB_1 = k \cdot CA$, $BC_1 = k \cdot AB$. Let A_2, B_2, C_2 be the midpoints of the segments AA_1, BB_1, CC_1. Show that (a) the three triangles ABC, $A_1B_1C_1$, $A_2B_2C_2$ have the same centroid; (b) the ratio of the areas of the triangles $A_2B_2C_2$ and ABC is equal to $\frac{1}{4}(1 + 3\,k + 3\,k^2)$. Consider the special cases when $k = 1$ and $\frac{1}{3}$. Find other properties of the figures mentioned in this exercise.

TRITANGENT CIRCLES

20. Construct the bisector of an angle whose vertex is inaccessible.

21. If the internal bisector of an angle of a triangle bisects the angle formed by the other two internal bisectors, show that the triangle is isosceles.

22. Construct a triangle given the area, an angle, and the length of the segment joining the incenter to the excenter relative to the given angle.

23. The circle determined by the foot D of the altitude AD and the points I, I_a of the triangle ABC meets AD again in L. Show that AL is equal to the circumdiameter of ABC. State and prove an analogous proposition involving two excenters.

24. Given the length of the side BC of the variable triangle ABC and the positions of I_b and I_c, show that (a) the vertex A describes a straight line; (b) the directions of AB and AC are fixed; (c) the circumradius of ABC is constant; (d) the circumcenter of ABC describes a circle.

Construct a triangle, given:

25. $a, r, b \pm c$. **26.** $a, h_a, b - c$. **27.** r, r_a, S.

28. Construct a triangle ABC given the angle A and the inradii of the two triangles ABI, ACI, where I is the incenter of ABC.

29. From the point of intersection of the base of a triangle with the internal bisector of the vertical angle the tangent is drawn to the incircle. Prove that the angle which this tangent makes with the base is equal to the difference of the base angles.

30. Construct a triangle given the circumradius, the sum of an exradius and the inradius, and the length of the line of centers of the two corresponding circles (R, $r_a + r$, $II_a = 2d$). *Hint.* The distance of any point K of the circumcircle (O) from the common chord of (O) and the circle (K, d) is equal to $\frac{1}{2}(r_a - r)$ (§ 144 a); hence r and r_a.

31. The sides of a given triangle intercept the segments n_a, n_b, n_c on the parallels to the sides of this triangle through the incenter. Show that:

$$(n_a:a) + (n_b:b) + (n_c:c) = 2, \quad 4S = (n_a h_a + n_b h_b + n_c h_c).$$

State and prove analogous relations regarding the excenters. *Hint.* $n_a:a = (h_a - r):h_a$.

32. Show that the distances of the incenter of a triangle from the medians of the triangle are equal to:

$$(b - c)r:2m_a, \quad (c - a)r:2m_b, \quad (a - b)r:2m_c.$$

State and prove analogous relations for the excenters.

Construct a triangle, given:

33. r, A, S. *Hint.* S and r determine p; hence the triangle AI_aZ_a.

34. $a - b, A, r$. *Hint.* From the triangle AIZ we have $AZ = \frac{1}{2}(b + c - a)$, which, together with $a - b$, determines the side c.

35. $a - c, A, r$.

36. a, h_b, r. *Hint.* h_b and a determine the angle C; hence the triangle IXC.

37. $b + c, A, r_b - r_c$. **38.** $b + c, B - C, r_b + r_c$. **39.** $2p, A, r_a - r_b$.

40. The line joining the foot U of the internal bisector AU of the angle A of the triangle ABC to the point of contact Y of the incircle with the side CA meets the perpendicular to CA at A in the point F. Prove that $AF = h_a$.

41. The line joining a vertex of a triangle to the point of contact of the opposite side with the incircle (excircle relative to this side) divides the triangle into two triangles whose incircles (excircles) touch the line considered in the same point.

42. (a) With the usual notations, show that:

$$AX^2 + AX_a^2 + AX_b^2 + AX_c^2 = 3(b^2 + c^2) - a^2.$$

(b) Show that the sum of the squares of the twelve distances of the vertices of a triangle from the points of contact of the respectively opposite sides with the four tritangent circles of the triangle is equal to five times the sum of the squares of the sides of the triangle.

43. Show that the projection of the vertex B of the triangle ABC upon the internal bisector of the angle A lies on the line joining the points of contact of the incircle with the sides BC, AC. State and prove an analogous proposition for the external bisectors.

44. Show that the midpoint of a side of a triangle, the foot of the altitude on this side, and the projections of the ends of this side upon the internal bisector of the opposite angle are four concyclic points. Does the proposition hold for the external bisector?

ALTITUDES

45. In order to draw a line parallel to a given line s through the inaccessible point of intersection X of two given lines p, q, draw any perpendicular to the line s, meeting p, q in A, B. If H is the point of intersection of the perpendiculars from A and B to q and p, respectively, the perpendicular from H to AB is the required line. Justify this construction.

46. The midpoint A' of the base and the feet E, F of the altitudes to the lateral sides of a variable triangle are fixed, and $A'E = A'F$. Find the locus of the vertices and of the orthocenter of the triangle.

47. Show that the symmetric of the orthocenter of a triangle with respect to a vertex, and the symmetric of that vertex with respect to the midpoint of the opposite side, are collinear with the circumcenter of the triangle.

48. A triangle varies so as to remain similar to itself. The orthocenter is fixed, and one vertex lies on a fixed line. Prove that the other two vertices lie on two fixed straight lines.

49. If through the midpoints of the sides of a triangle having its vertices on the altitudes of a given triangle, perpendiculars are dropped to the respective sides of the given triangle, show that the three perpendiculars are concurrent.

50. If D' is the second point of intersection of the altitude ADD' of the triangle ABC with the circumcircle, center O, and P is the trace on BC of the perpendicular from D' to AC, show that the lines AP, AO make equal angles with the bisector of the angle DAC.

51. The orthocenter, the midpoint of the base, and the direction of the base of a variable triangle are fixed. Show that its circumcircle passes through two fixed points. Find the locus of its centroid.

52. Construct a triangle given h_b, h_c, t_a.

53. Show that the triangle formed by the foot of the altitude to the base of a triangle and the midpoints of the altitudes to the lateral sides is similar to the given triangle; its circumcircle passes through the orthocenter of the given triangle and through the midpoint of its base.

54. With the usual notation, show that the angle formed by the lines $C'E$, $B'F$ is equal to $3A$, or its supplement.

55. Show that the symmetrics P, Q of a given point L with respect to the sides Ox, Oy of a given angle, and the points $P' = (LQ, Ox)$, $Q' = (LP, Oy)$ lie on a circle passing through O.

56. The sides of the anticomplementary triangle of the triangle ABC meet the circumcircle of ABC in the points P, Q, R. Show that the area of the triangle PQR is equal to four times the area of the orthic triangle of ABC.

57. If B', C' are the projections, upon the sides AB, AC of the triangle ABC, of a variable point of the circle having BC for diameter, show that the locus of the orthocenter of the variable triangle $AB'C'$ is a circle having for diameter the side EF of the orthic triangle relative to the vertex A.

58. Show that (a) if an acute-angled triangle is similar to its orthic triangle, the two triangles are equilateral; (b) if an obtuse-angled triangle is similar to its orthic triangle, its angles are $(180°:7)$, $(360°:7)$, $(720°:7)$.

59. Show that in a variable triangle having a fixed base and a fixed circumcircle the line joining the feet of the altitudes on the two variable sides is of constant length.

60. A is a fixed point on the larger of two concentric circles, and B, C are the points of intersection of this circle with a variable tangent BC to the smaller circle. Prove that the locus of the orthocenter of the variable triangle ABC is a circle.

61. Through the orthocenter of the triangle ABC parallels are drawn to the sides AB, AC, meeting BC in D, E. The perpendiculars to BC at D, E meet AB, AC in two points D', E' which are collinear with the diametric opposites of B, C on the circumcircle of ABC.

62. Let D, D' be the projections of the midpoint A' of BC upon the circumradii OB, OC of the triangle ABC, and E, E' and F, F' the analogous points relative to the points B' and C'. Prove that:

$$\sqrt{DD':a} + \sqrt{EE':b} + \sqrt{FF':c} = (R + r):R.$$

THE NINE-POINT CIRCLE

63. Through the midpoints of the sides of a triangle parallels are drawn to the external bisectors of the respectively opposite angles. Show that the triangle thus formed has the same nine-point circle as the given triangle.

64. Construct a triangle given, in position, the orthocenter, a point on the nine-point circle, the midpoint of the base, and the indefinite line containing the base.

65. The line joining the orthocenter of a triangle ABC to the midpoint of the side BC meets the circumcircle of ABC in the points A_1, A_2. Prove that the orthocenters of the three triangles ABC, A_1BC, A_2BC are the vertices of a right triangle.

66. The base BC and the opposite angle A of a variable triangle ABC are fixed. Show that (a) the line $B'Q$ has a fixed direction; (b) the nine-point circle is tangent to a fixed circle.

67. Show that the foot of the altitude of a triangle on a side, the midpoint of the segment of the circumdiameter between this side and the opposite vertex, and the nine-point center are collinear.

68. If A, B, C are the centers of three equal circles (A), (B), (C) having a point L in common, and D, E, F are the other points which the circles (B) and (C), (C) and (A), (A) and (B) have in common, show that the circle DEF is equal to the given circles, and that the center of this circle coincides with the orthocenter of the triangle ABC.

69. Show that the parallels to the internal bisectors of a triangle drawn through the respective Euler points are concurrent; the line joining their common point to the nine-point center of the given triangle is parallel to the line joining the circumcenter of the given triangle to its incenter. State and prove analogous propositions, using both internal and external bisectors.

70. Show that the nine-point center of the triangle IBC lies on the internal bisector of the angle A' of the complementary triangle $A'B'C'$ of the given triangle ABC. State and prove analogous propositions for the triangles I_aBC, I_bBC, I_cBC.

71. If O, H are the circumcenter and the orthocenter of the triangle ABC, show that the nine-point circles of the three triangles OHA, OHB, OHC have two points in common.

72. If A', B', C' are the midpoints of the sides BC, CA, AB of a triangle ABC, respectively, show that the nine-point centers of the triangles $AB'C'$, $BC'A'$, $CA'B'$ form a triangle homothetic with ABC in the ratio $1:2$. Find other properties of the figure.

MISCELLANEOUS EXERCISES

73. Show that the Euler lines of the three triangles cut off from a given triangle by the sides of its orthic triangle have a point in common, on the nine-point circle of the given triangle.

74. If a triangle (S) is inscribed in a triangle (T) and the two triangles are similar, show that the orthocenter of (S) coincides with the circumcenter of (T).

75. If the circle (BC) having for diameter the side BC of the triangle ABC meets CA, AB in E, F, show that the double chord determined by the two circles (BC) and (AEF) on any line through E (or F) has its midpoint on the nine-point circle of ABC.

76. Show that the symmetrics of the foot of the altitude to the base of a triangle with respect to the other two sides lie on the side of the orthic triangle relative to the base.

77. In the plane of a triangle to find two directions such that the three pairs of lines having these directions and passing through the vertices of the triangle shall meet the opposite sides in six concyclic points.

78. Show that the midpoint of an altitude of a triangle, the point of contact of the corresponding side with the excircle relative to that side, and the incenter of the triangle are collinear.

79. Show that the line joining the circumcenter O of a triangle to its incenter I passes through the orthocenter H' of the triangle formed by the points of contact of the sides of the triangle with the incircle, and that, moreover, $H'I:OI = r:R$. Is the proposition valid for the excenters?

80. If X' is the symmetric, with respect to the internal bisector of the angle A, of the point of contact of the side BC of the triangle ABC with the incircle, and A'

is the midpoint of BC, show that the line $A'X'$ and its two analogues $B'Y'$, $C'Z'$ have a point in common. Is the proposition valid for an excircle?

81. A line AD through the vertex A meets the circumcircle of the triangle ABC in D. If U, V are the orthocenters of the triangles ABD, ACD, respectively, prove that UV is equal and parallel to BC.

82. The internal bisectors of the angles B, C of the triangle ABC meet the line AX_a joining A to the point of contact of BC with the excircle relative to this side in the points L, M. Prove that $AL:AM = AB:AC$.

83. Given the length of the internal bisector of an angle of a triangle and the directions of all three bisectors, to construct the triangle. Consider analogous problems involving the external bisectors, or both internal and external bisectors.

84. Show that the product of the distances of the incenter of a triangle from the three vertices of the triangle is equal to $4\,Rr^2$. State and prove analogous formulas for the excenters.

85. AA', BB', CC' are the medians of a triangle ABC of which G and H are the centroid and the orthocenter. Upon $A'A$ a point U is taken so that $A'A \cdot A'U = A'B^2 = A'C^2$. Similarly for the points V and W with regard to BB' and CC'. Show that the triangle UVW has its sides proportional to the medians of ABC, and that GH is a diameter of its circumcircle.

86. A parallel to the median AA' of the triangle ABC meets BC, CA, AB in the points H, N, D. Prove that the symmetrics of H with respect to the midpoints of NC, BD are symmetrical with respect to the vertex A.

87. A variable transversal meets the sides AB, AC of the triangle ABC in the points P_1, Q_1. Let O_1 be the circumcenter of AP_1Q_1, and O_2 the circumcenter of the triangle determined by A and the isotomics of P_1, Q_1. Show that the line O_1O_2 passes through a fixed point and that the second point common to the two circles considered lies on a fixed circle.

88. If the lines AX, BY, CZ joining the vertices of a triangle ABC to the points of contact X, Y, Z of the sides BC, CA, AB with the incircle meet that circle again in the points X', Y', Z', show that:

$$AX \cdot XX' \cdot BC = BY \cdot YY' \cdot CA = CZ \cdot ZZ' \cdot AB = 4\,rS,$$

where r, S are the inradius and the area of ABC. State and prove analogous relations for the excircles.

89. The perpendiculars DP, DQ dropped from the foot D of the altitude AD of the triangle ABC upon the sides AB, AC meet the perpendiculars BP, CQ erected to BC at B, C in the points P, Q respectively. Prove that the line PQ passes through the orthocenter H of ABC.

90. Through the vertices of a given triangle are drawn the symmetrics of the corresponding sides of the anticomplementary triangle with respect to a given direction. Show that the three lines thus obtained meet in a point on the circumcircle of the triangle. State this proposition with reference to the medial triangle and the nine-point circle.

91. The altitudes AHD, BHE, CHF of the triangle ABC are produced beyond D, E, F to the points P, Q, R by the lengths AH, BH, CH, respectively. The parallels through P, Q, R to the sides BC, CA, AB form a triangle $A_1B_1C_1$. Show that (a) H is the centroid of the triangle $A_1B_1C_1$; (b) the homothetic center of the triangles ABC, $A_1B_1C_1$ is the circumcenter of each of these triangles.

92. The points H, O are the orthocenter and the circumcenter of the triangle ABC, and P, P' are two points symmetrical with respect to the mediator of BC. The perpendicular from P to BC meets BC in P_1 and OP' in P''; M is the midpoint of HP. Prove that (a) $2\,MP_1 = AP''$; (b) the symmetric of P_1 with respect to the midpoint of OM lies on AP'.

93. The side AB of a parallelogram $ABCD$ is produced to E so that $BE = AD$. The perpendicular to ABE at E meets the perpendicular from C to the diagonal BD in F. Show that AF bisects the angle A.

94. If AD, BE are altitudes of the triangle ABC, and BL the perpendicular from B to DF, show that if $LB^2 = LD \cdot LE$, the triangle is isosceles.

95. The point S is the homothetic center of two triangles ABC, $A'B'C'$, and A'', B'', C'' are the symmetrics of A', B', C' with respect to the mediators of BC, CA, AB. The lines AA'', BB'', CC'' meet the parallels through S to BC, CA, AB in A_1, B_1, C_1. Show that the circle $A_1B_1C_1$ passes through S and has its center on the Euler line of ABC.

96. If h, m, t are the altitude, the median, and the internal bisector issued from the same vertex of a triangle whose circumradius is R, show that:

$$4\,R^2h^2(t^2 - h^2) = t^4(m^2 - h^2).$$

97. The perpendiculars BB', CC' from the vertices B, C of the triangle ABC upon an arbitrary diameter d of the incircle (I) of ABC meet the lines XY, YZ joining the points of contact X, Y, Z of (I) with BC, CA, AB in the points B'', C''. Prove that (a) $BB':BB'' = CC':CC''$; (b) the lines d, $B''C''$ meet on BC; (c) the line $B''C''$ passes through the excenter of ABC relative to the side BC.

98. Construct a triangle given the base, the vertical angle, and the sum of the squares of the lines drawn from the vertical angle to bisect the segments of the base made by the foot of the altitude from the same vertex.

99. Construct a triangle given a side and the altitude to the base, so that the diameter of the incircle shall be equal to the side of the square inscribed in the triangle and having two of the vertices on the base of the triangle.

100. Construct a triangle given, in position, the feet of the altitude and of the median to the base, and the incenter (D, A', I).

101. Construct a triangle given the perimeter, the vertical angle, and the length of the line drawn from the vertical angle to meet the base and divide it in a given ratio.

102. Construct a triangle given the vertical angle, an adjacent side, and the ratio of the area of the required triangle to the area of the triangle formed by the bisectors of the given angle and the opposite side of the required triangle.

103. Construct a triangle ABC of which AD, BE, CF are the unknown altitudes having given the three products $BC \cdot CD$, $CA \cdot AE$, $AB \cdot BF$.

104. Construct a triangle given h_a, t_a, $b + c$.

105. Let A', B', C' and A'', B'', C'' be the traces of the internal and the external bisectors of the triangle ABC on its circumcircle. Construct ABC given the segments (a) $A''B$, $B''C$, $C''A$; (b) $A''B$, $B'C$, $C'A$.

106. In a given circle to inscribe a right triangle of given inradius, so that one side of the right angle shall pass through a given point.

107. Construct a triangle given three of the six points in which the three internal bisectors meet the incircle.

108. Construct a triangle ABC given the altitude AD and the ratios $AE:EC$, $AF:FB$, where E, F are the feet of the other two altitudes.

109. Construct a triangle given two Euler points and the centroid.

110. Construct a triangle given the base, the corresponding median, and the difference of the other two sides.

111. One vertex of a variable triangle inscribed in a fixed circle is fixed, and the opposite side passes through a fixed point. (a) Show that the orthocenter describes a circle. (b) Show that the nine-point circle is tangent to two fixed concentric circles.

112. The tangent to a circle (O) at a variable point M meets the fixed diameter AA' in the point T. Show that the locus of the incenter of the triangle OMT consists of four straight lines. Find the loci of the excenters.

113. The circumcircle and the directions of two sides of a variable triangle are fixed. Find the locus of its tritangent centers.

114. Show that the second of the common internal tangents to the incircles of the two triangles into which a given triangle is divided by a variable line through a vertex passes through the point of contact of the incircle of the given triangle relative to the vertex considered. State and prove an analogous property for the excircles relative to the vertex considered.

IV

THE QUADRILATERAL

A. THE GENERAL QUADRILATERAL

240. Theorem. *The midpoints of the sides of a quadrilateral are the vertices of a parallelogram.*

The segment PS joining the midpoints P, S of the two sides AB, AD of the given quadrilateral $ABCD$ (Fig. 74) and the segment QR joining

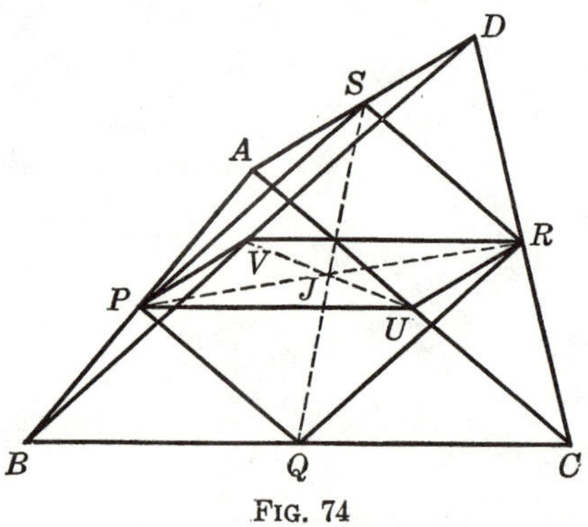

FIG. 74

the midpoints Q, R of the sides CB, CD are both parallel to, and equal to one-half of, the diagonal BD of the quadrilateral; hence the proposition.

241. COROLLARY. (*a*) *The perimeter of the parallelogram PQRS is equal to the sum of the diagonals of the given quadrilateral* (Fig. 74).

(*b*) *The area of the parallelogram is equal to one-half the area of the given quadrilateral.*

The proof is left to the student.

When will the parallelogram become a rectangle? A rhombus? A square?

242. Theorem. *The lines joining the midpoints of the two pairs of opposite sides of a quadrilateral and the line joining the midpoints of the diagonals are concurrent and are bisected by their common point.*

The lines PR, QS (Fig. 74) bisect each other, say, in J, for they are the diagonals of the parallelogram $PQRS$ (§ 240).

If U, V are the midpoints of the diagonals AC, BD, then $PURV$ is a parallelogram, for both PV and RU are parallel to, and equal to one-half of, AD; hence the diagonal UV of the parallelogram $PURV$ is bisected by the midpoint J of the diagonal PR.

243. Definition. The point J (§ 242) will be referred to as the *centroid* of the quadrilateral $ABCD$.

244. Theorem. *The four lines obtained by joining each vertex of a quadrilateral to the centroid of the triangle determined by the remaining three vertices are concurrent.*

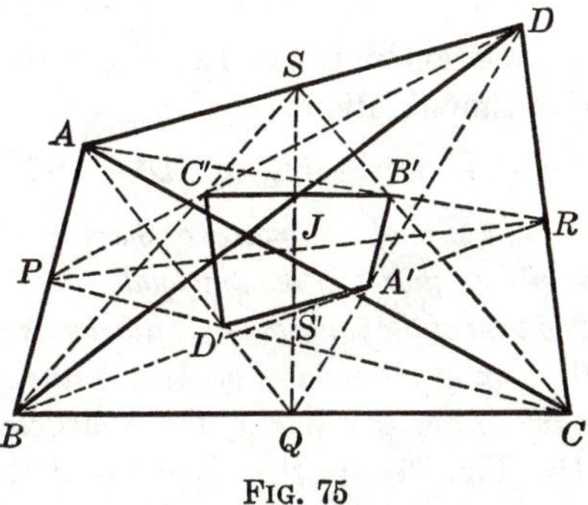

FIG. 75

Let $ABCD$ be the given quadrilateral (Fig. 75) and let A', B', C', D' be the centroids of the triangles BCD, CDA, DAB, ABC. We have (§ 88):

$$PC':PD = PD':PC = 1:3;$$

hence $C'D'$ is parallel to CD, and $C'D':CD = 1:3$.

Thus the sides of the quadrilateral $A'B'C'D'$ are parallel to the respective sides of $ABCD$ and proportional to these sides; hence the two quadrilaterals are homothetic, and the lines AA', BB', CC', DD' meet in the homothetic center of the two figures.

245. *Remark.* (a) The homothetic center of the two quadrilaterals (§ 244) coincides with the centroid J of $ABCD$.

The median QS of the triangle QAD (Fig. 75) bisects the segment $D'A'$ parallel to the base DA, in S'; hence S, S' are corresponding points in the two homothetic figures, and the line $SS'Q$ passes through the homothetic center of the two quadrilaterals. Similarly for the line PR. Hence the proposition.

(b) The senses of the two lines $A'D'$, AD are opposite, as is readily seen in the figure; hence the two quadrilaterals are inversely homothetic, so that corresponding points, like A and A', lie on opposite sides of the point J, and $JA:JA' = -3:1$.

246. Theorem. *The sum of the squares of the sides of a quadrilateral is equal to the sum of the squares of the diagonals increased by four times the square of the segment joining the midpoints of the diagonals.*

In the triangles ABD, CBD, VAC (Fig. 74) we have (§ 106):

$$2\,AV^2 = AB^2 + AD^2 - \tfrac{1}{2}\,BD^2,$$
$$2\,CV^2 = BC^2 + CD^2 - \tfrac{1}{2}\,BD^2,$$
$$2\,UV^2 = AV^2 + CV^2 - \tfrac{1}{2}\,AC^2.$$

Multiplying the last relation by two and adding to the sum of the first two we obtain, after simplification:

$$4\,UV^2 = AB^2 + AD^2 + BC^2 + CD^2 - AC^2 - BD^2.$$

247. COROLLARY. *The sum of the squares of the sides of a parallelogram is equal to the sum of the squares of the diagonals.*

248. Theorem. *The sum of the squares of the diagonals of a quadrilateral is equal to twice the sum of the squares of the two lines joining the midpoints of the two pairs of opposite sides of the quadrilateral.*

The lines PR, QS (Fig. 74) are the diagonals of the parallelogram $PQRS$ (§ 240), whose sides are equal to half the respective diagonals of $ABCD$; hence the proposition (§ 247).

249. Problem. *Construct a quadrilateral given its four sides and the line joining the midpoints of a pair of opposite sides.*

The sides of the quadrilateral $ABCD$ (Fig. 74) and the line PR being given, the parallelogram $PURV$ may be constructed, for we know its diagonal PR and its sides:

$$PU = RV = \tfrac{1}{2}\,BC, \quad PV = RU = \tfrac{1}{2}\,AD.$$

We thus locate the diagonal UV of the parallelogram $QUSV$, and $QUSV$ may now be constructed, in a similar manner.

The midpoints P, Q, R, S being located, the triangles SPA, PQB, ... may now be completed; hence the quadrilateral $ABCD$.

250. Problem. *Construct a quadrilateral given two opposite angles, the diagonals, and the angle between the diagonals.*

On the given diagonal AC, of the required quadrilateral $ABCD$, as chord construct the two arcs at the points of which AC subtends the given angles B, D. The second diagonal is known, both in length and direction, and the problem is reduced to placing this diagonal so that its ends shall lie on the two circles just constructed; hence the problem is solved (§ 8).

251. Theorem. *The internal bisectors of the angles of a quadrilateral form a cyclic quadrilateral.*

Let the internal bisectors of the angles A, D of the quadrilateral $ABCD$ meet in P, and those of the angles B, C in Q. We have:

$$\angle APD = 180° - \tfrac{1}{2}(A + D), \quad \angle BQC = 180° - \tfrac{1}{2}(B + C);$$

hence:

$$\angle APD + BQC = 360° - \tfrac{1}{2}(A + B + C + D) = 360° - 180° = 180°.$$

As an exercise, state and prove an analogous proposition for the external bisectors.

EXERCISES

1. Show that the internal bisectors of two consecutive angles of a quadrilateral form an angle equal to half the sum of the remaining two angles. State and prove an analogous proposition for the external bisectors.
2. $ABCD$ is a quadrilateral, P, Q, R, S the midpoints of its sides taken in order, U, V the midpoints of the diagonals, O any point; OP, OQ, OR, OS, OU, OV are divided in the same ratio in P', Q', R', S', U', V'. Prove that $P'R'$, $Q'S'$, $U'V'$ are concurrent.
3. Given the quadrilateral $ABCD$, show that the locus of the point M such that:
$$\text{area } MAB + \text{area } MCD = \text{area } MAC + \text{area } MBD$$
is the line joining the midpoints of the diagonals.
4. Construct a quadrilateral given the length of each of its sides and the length of the segment joining the midpoints of the diagonals.

B. THE CYCLIC QUADRILATERAL

252. Definition. A quadrilateral whose vertices lie on the same circle is said to be *inscriptible* or *cyclic*.

253. Theorem. *In a cyclic quadrilateral (a) the opposite angles are supplementary; (b) the angle between a side and a diagonal is equal to the angle between the opposite side and the other diagonal. Conversely.*

254. Theorem. *The line joining the midpoints of the two arcs which two opposite sides of a cyclic quadrilateral subtend on the circumcircle is perpendicular to the line joining the midpoints of the arcs subtended by the other two sides.*

Let E, F, G, H be the midpoints of the arcs subtended by the sides AB, BC, CD, DA of the cyclic quadrilateral $ABCD$, M the point of intersection of the diagonals AC, BD.

The midpoint E is taken on the opposite side of the chord AB from the vertices C, D of $ABCD$. Similarly for the points F, G, H.

The angle AMD is measured by half the sum of the arcs AD, BC. Now the line HF meets the diagonal BD in a point, S, inside the circle (O); hence the angle HSD is measured by half the sum of the arcs HD, BF, and is therefore equal to half of the angle AMD, i.e., the line FH is parallel to the internal bisector of the angle AMD.

Similarly the line EG is parallel to the internal bisector of the angle AMB. Hence the proposition.

255. Ptolemy's Theorem. *In a cyclic quadrilateral the product of the diagonals is equal to the sum of the products of the opposite sides. Conversely.*

Consider a quadrilateral $ABCD$ (Fig. 76 a and b) in which $AB = a$, $BC = b$, $CD = c$, $DA = d$, $AC = x$, $BD = y$, and let the line m be

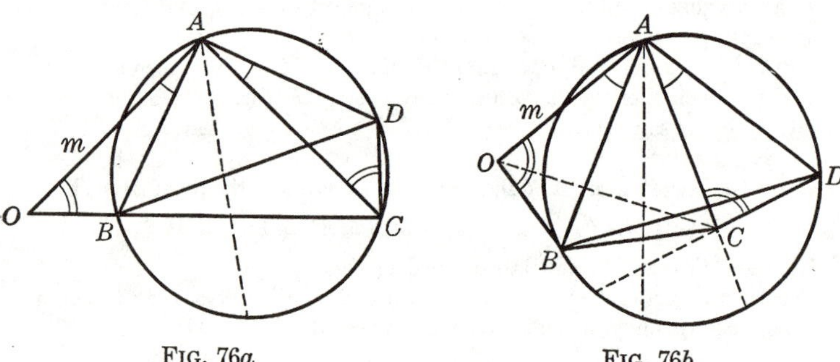

FIG. 76a FIG. 76b

the symmetric of the side AD with respect to the bisector of the angle BAC. The line m contains one and only one point O such that angle $AOB = ACD$. From the triangles ACD, AOB we have angle $ABO = ADC$. Thus, if $ABCD$ is cyclic, the point O lies on the side BC, and only in that case, for, conversely, if the point O lies on BC, it is readily seen that $ABCD$ is cyclic (Fig. 76 a).

The triangles AOB, ACD are similar, whether the point O lies on BC or not; hence:

(1) $$AO:AC = AB:AD = OB:CD.$$

Moreover, the triangles OAC, BAD are similar, for angle $OAC = BAD$, and the including sides are proportional, by (1); hence:

(2) $$OC:BD = AC:AD.$$

If $ABCD$ is cyclic, we have (Fig. 76 *a*):

$$OC = OB + BC.$$

Now $BC = b$, and from (1) and (2) we have:

$$OB = ac:d, \quad OC = xy:d;$$

hence, substituting and simplifying:

(3) $$xy = ac + bd,$$

which proves the direct proposition.

If $ABCD$ is not cyclic, we have (Fig. 76 *b*):

$$OC < OB + BC,$$

or, making the same substitutions as above:

$$xy < ac + bd;$$

i.e., in a (noncyclic) quadrilateral the product of the diagonals is smaller than the sum of the products of the opposite sides.

This proposition together with the proved direct theorem lead to the announced converse: *If the product of the diagonals of a quadrilateral is equal to the sum of the products of the opposite sides, the quadrilateral is cyclic.*

256. Problem. *Construct a cyclic quadrilateral given its sides a, b, c, d.*

The proof of the preceding theorem suggests the following construction.

Assuming that b, d are a pair of opposite sides, on a straight line lay off $CB = b$, and $BO = ac:d$ (Fig. 76 *a*). One locus for the vertex A is the circle (B, a). From (1) (§ 255) we have $AO = ax:d$; hence:

$$AO:AC = (ax:d):x = a:d,$$

which gives an Apollonian circle as a second locus for A. The triangle ACD is now readily completed, and $ABCD$ is the required quadrilateral. The point D is to be taken so that B, D shall lie on opposite sides of AC.

The two circles determining the point A intersect in two points, symmetric with respect to BC. Either of these points may be taken, but the two solutions will be symmetric with respect to BC, so that the problem has essentially only one solution.

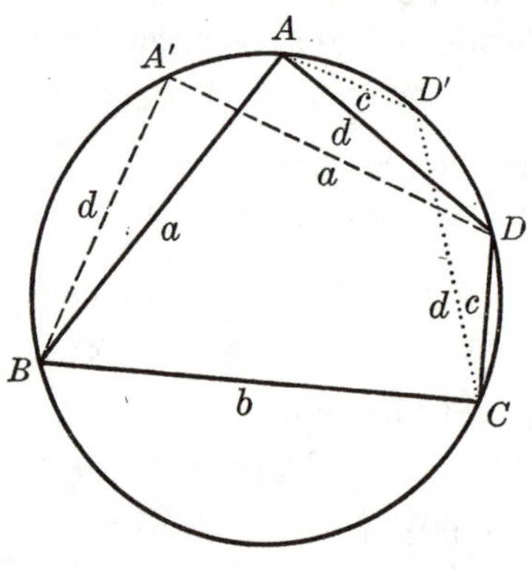

FIG. 77

257. *Remark.* (a) We assumed that the side d is the opposite to the side b. We obtain two other quadrilaterals, if the sides a, c, are, in turn, taken to be the opposite sides of b. However, if the first quadrilateral $ABCD$ is constructed, the quadrilaterals $A'BCD$ and $ABCD'$ corresponding to the other two cases are readily obtained (Fig. 77), and these quadrilaterals solve the problem, for the solution is in each case unique (§ 256).

(b) The arcs BAD', CDA' being equal, by construction (Fig. 77), we have:

$$BD' = CA' = z.$$

Thus the three quadrilaterals $ABCD$, $ABCD'$, $A'BCD$ have only three different diagonals x, y, z.

Applying Ptolemy's theorem to the quadrilaterals $ABCD'$, $A'BCD$, we have:

$$xz = ad + bc, \quad yz = ab + cd;$$

hence:

$$x:y = (ad + bc):(ab + cd).$$

The student may give a verbal expression for this formula.

This formula combined with (3) (§ 255) readily yields a formula for x and one for y.

(c) The triangle BCD (Fig. 77) is the common part of the quadrilaterals $ABCD$, $A'BCD$, and the sides of the two triangles ABD, $A'BD$ are respectively equal; hence the two quadrilaterals are equivalent.

Similarly for the quadrilaterals $DABC$, $D'ABC$. Thus the three quadrilaterals $ABCD$, $A'BCD$, $ABCD'$ are equivalent.

(d) Let R, S be the common circumradius and the common area of the three quadrilaterals $ABCD$, $A'BCD$, $ABCD'$, and we have (§ 255):

$$xy = ac + bd;$$

hence:

$$xyz = acz + bdz = 4R(\text{area } A'CD + \text{area } A'CB)(\S\ 86),$$

or:

$$xyz = 4RS.$$

258. Theorem. *The perpendiculars from the midpoints of the sides of a cyclic quadrilateral to the respectively opposite sides are concurrent.*

Let O, J be the circumcenter and the centroid (§ 243) of the cyclic quadrilateral $ABCD$, and P, Q, R, S the midpoints of the sides AB, BC, CD, DA (Fig. 78).

Let the perpendicular SS' from S upon BC meet OJ in M. The lines SS', OQ are parallel, and J bisects the line QS; hence J bisects OM.

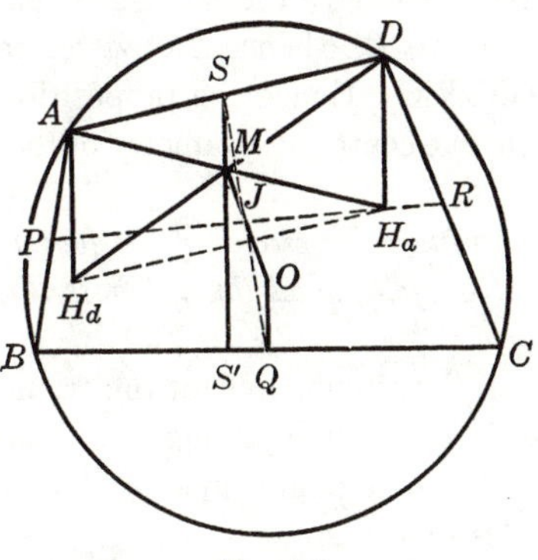

FIG. 78

Thus SS' passes through the symmetric M of the circumcenter O with respect to the centroid J of the quadrilateral, i.e., through a point which does not depend upon the choice of the perpendicular we started with. Hence the proposition.

259. *Remark.* The perpendicular from the midpoint of each diagonal upon the other diagonal also passes through M, for the line joining their midpoints is also bisected by the centroid J (§ 242).

260. Definition. The symmetric of the circumcenter of a cyclic quadrilateral with respect to the centroid is called the *anticenter* of the cyclic quadrilateral.

261. Theorem. *The four lines obtained by joining each vertex of a cyclic quadrilateral to the orthocenter of the triangle formed by the remaining three vertices bisect each other.*

Let H_d, H_a be the orthocenters of the triangles ABC, DBC (Fig. 78). We have (§ 195):

$$AH_d = 2\,OQ = DH_a,$$

and the lines AH_d, DH_a are both perpendicular to BC; hence ADH_aH_d is a parallelogram, and the diagonals AH_a, DH_d bisect each other.

Similarly DH_d is bisected by the analogous lines BH_b, CH_c, and, in turn, bisects each of these lines. Hence the proposition.

The common point X of these four lines is thus a center of symmetry of the two quadrilaterals $ABCD$, $H_aH_bH_cH_d$.

262. *Remark.* The point X (§ 261) coincides with the anticenter M of the quadrilateral.

Indeed, in the triangle DAH_d the line SX is parallel to AH_d and therefore perpendicular to BC; hence SX passes through M (§ 258). Similarly for PX, QX, RX. Hence the proposition.

The point M is thus the center of symmetry of the two quadrilaterals $ABCD$, $H_aH_bH_cH_d$.

263. Theorem. *The nine-point circles of the four triangles determined by the four vertices of a cyclic quadrilateral pass through the anticenter of the quadrilateral.*

The point D lies on the circumcircle of the triangle ABC (Fig. 78); hence the midpoint M of the line joining D to the orthocenter H_d of ABC (§ 261) lies on the nine-point circle of ABC (§ 210). Similarly for the other triangles of the group.

264. Theorem. *The sum of the squares of the distances of the anticenter of a cyclic quadrilateral from the four vertices is equal to the square of the circumdiameter of the quadrilateral.*

In the triangle MAD (Fig. 78) we have (§ 106):

$$MA^2 + MD^2 = 2\,MS^2 + \tfrac{1}{2}\,AD^2. \qquad (1)$$

Now $MS = OQ$, as opposite sides of a parallelogram; hence (1) becomes, considering the right triangle OQB and denoting by R the circumradius of $ABCD$:

$$MA^2 + MD^2 = 2R^2 + \tfrac{1}{2}AD^2 - \tfrac{1}{2}BC^2.$$

The sides AB, BC, CD of $ABCD$ give rise to three analogous relations, and adding the four relations we readily obtain the stated result.

265. Theorem. *The incenters of the four triangles determined by the vertices of a cyclic quadrilateral form a rectangle.*

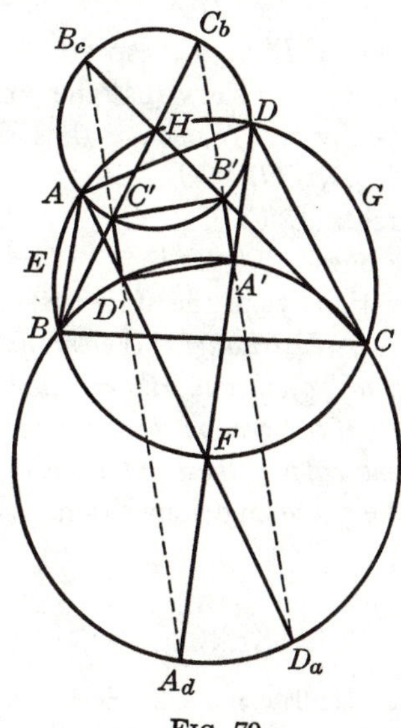

FIG. 79

Let E, F, G, H be the midpoints of the arcs subtended by the sides AB, BC, CD, DA of the cyclic quadrilateral $ABCD$ on the circumcircle (Fig. 79).

The incenters D', A' of the triangles ABC, DBC lie on the bisectors AF, DF of the angles BAC, BDC and on the circle (F, FB) (§§ 116, 122); thus the triangle $FA'D'$ is isosceles, and the base $A'D'$ is perpendicular to the bisector FH of the angle F. Similarly the line FH is perpendicular to the line $B'C'$ joining the incenters B', C' of the triangles CAD, BAD.

Similarly, the line EG is perpendicular to the lines $A'B'$, $C'D'$. Now the lines FH, EG are perpendicular (§ 254); hence the proposition.

266. Theorem. *If of the four triangles determined by the vertices of a cyclic quadrilateral three triangles are considered having a common vertex, the three excenters relative to this vertex in the three triangles are three vertices of a rectangle, the fourth vertex of which is the incenter of the fourth triangle.*

Let D_a, D_b, D_c; A_b, A_c, A_d; B_a, B_c, B_d; C_a, C_b, C_d be the excenters of the triangles ABC, BCD, CDA, DAB, respectively, the excenter A_b being relative to the vertex B in the triangle BCD, etc. (Fig. 79).

The two pairs of points D', D_a; A', A_d are diametrically opposite on the circle (F, FB) (§ 122); hence $D'A'D_aA_d$ is a rectangle, and therefore A_d, D_a lie on the lines $C'D', B'A'$, respectively (§ 265). Considering the circle (H, HA) we may show that the points B_c, C_b lie on the lines $D'C', A'B'$, respectively. We have thus the two sets of four collinear points: $A'B'C_bD_a$, $C'D'A_dB_c$.

Considering the circles (E, EA), (G, GC) we obtain two additional points on each of the lines $A'D', B'C'$.

Analogous considerations yield the stated result.

267. Theorem. *The sixteen tritangent centers of the four triangles determined by the vertices of a cyclic quadrilateral lie by fours on eight lines. These eight lines consist of two groups of four parallel lines, and the lines of one group are perpendicular to those of the other group.*

This summarizes the preceding propositions (§§ 265, 266).

EXERCISES

1. Show that in a cyclic quadrilateral the distances of the point of intersection of the diagonals from two opposite sides are proportional to these sides.
2. In the cyclic quadrilateral $ABCD$ the perpendicular to AB at A meets CD in A', and the perpendicular to CD at C meets AB in C'. Show that the line $A'C'$ is parallel to the diagonal BD.
3. In a given circle to inscribe a quadrilateral given the diagonals and the angle between them.
4. Show that the perpendicular from the point of intersection of two opposite sides, produced, of a cyclic quadrilateral upon the line joining the midpoints of the two sides considered passes through the anticenter of the quadrilateral.
5. Show that the anticenter of a cyclic quadrilateral is the orthocenter of the triangle having for vertices the midpoints of the diagonals and the point of intersection of those two lines.
6. Show that the anticenter of a cyclic quadrilateral is collinear with the two symmetrics of the circumcenter of the quadrilateral with respect to a pair of opposite sides.

7. If H_a, H_b, H_c, H_d are the orthocenters of the four triangles determined by the vertices of the cyclic quadrilateral $ABCD$, show that the vertices of $ABCD$ are the orthocenters of the four triangles determined by the points H_a, H_b, H_c, H_d.

8. Show that the product of the distances of two opposite sides of a cyclic quadrilateral from a point on the circumcircle is equal to the product of the distances of the other two sides from the same point.

9. Show that the four lines obtained by joining each vertex of a cyclic quadrilateral to the nine-point center of the triangle formed by the remaining three vertices, are concurrent.

10. Show that the four nine-point centers of the four triangles determined by the vertices of a cyclic quadrilateral form a cyclic quadrilateral.

11. Three vertices of a variable cyclic quadrilateral are fixed. Find the locus of (a) its centroid; (b) its anticenter.

12. If a, b, c, d are the sides and S the area of a cyclic quadrilateral, show that:

$$S^2 = (p - a)(p - b)(p - c)(p - d),$$

where $p = \frac{1}{2}(a + b + c + d)$ [Formula of Brahmagupta].

C. OTHER QUADRILATERALS

268. Definition. A quadrilateral is said to be *circumscriptible* if its sides are tangent to the same circle.

269. Theorem. *In a circumscriptible quadrilateral the sums of the two pairs of opposite sides are equal.*

If p, q, r, s are the lengths of the tangents from the vertices of the given quadrilateral to its inscribed circle, it is readily seen that the sum of two opposite sides is equal to $p + q + r + s$.

270. Converse Theorem. *If the sum of two opposite sides of a quadrilateral is equal to the sum of the remaining two sides, the quadrilateral is circumscriptible.*

Suppose that in the given quadrilateral $ABCD$ (Fig. 80) we have:

$$AB + CD = AD + BC. \tag{1}$$

If two adjacent sides of $ABCD$ are equal, the remaining two sides are also equal, by (1), and the proof becomes trivial.

Assuming that AB is greater than BC, we have, by (1),

$$AB - BC = AD - CD. \tag{2}$$

Mark the point P between A, B and the point Q between A, D so that we shall have:

$$BP = BC, \quad DQ = DC,$$

and therefore, by (2):

$$AP = AQ.$$

We have thus three isosceles triangles APQ, BCP, DCQ; hence the internal bisectors of the angles A, B, D are the mediators of the sides of the triangle PQC, and therefore meet in the circumcenter O of PQC.

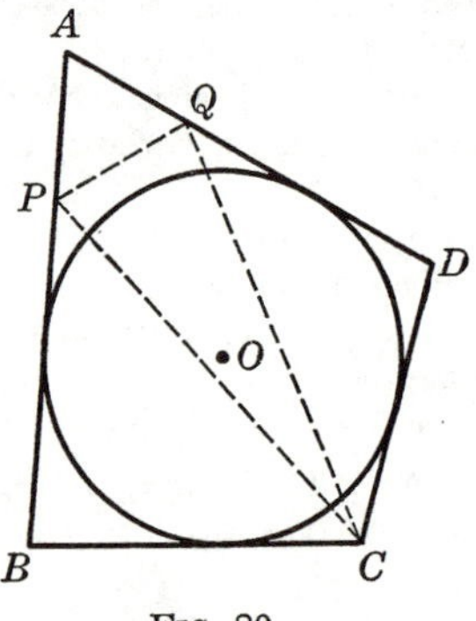

FIG. 80

The point O common to the internal bisectors of the angles A, B, D of the quadrilateral $ABCD$ necessarily lies inside of $ABCD$ and is, furthermore, equidistant from the sides of $ABCD$; hence the proposition.

271. Definition. A quadrilateral is said to be *orthodiagonal* if its diagonals are perpendicular to each other.

272. Theorem. *In an orthodiagonal quadrilateral the sum of the squares of two opposite sides is equal to the sum of the squares of the remaining two sides, and conversely.*

If the diagonals AC, BD of the quadrilateral $ABCD$ are rectangular, we have, denoting by O their point of intersection:

$$AB^2 + CD^2 = AO^2 + BO^2 + CO^2 + DO^2,$$
$$AD^2 + BC^2 = AO^2 + DO^2 + BO^2 + CO^2;$$

hence:

$$AB^2 + CD^2 = AD^2 + BC^2. \tag{1}$$

Conversely, if (1) holds, we have:

$$AB^2 - AD^2 = BC^2 - CD^2;$$

hence the line joining the points A and C is perpendicular to BD (§ 11, locus 12).

273. Theorem. *In an orthodiagonal quadrilateral the two lines joining the midpoints of the pairs of opposite sides are equal.*

Indeed, the parallelogram determined by the midpoints of the sides of the quadrilateral (§ 240) is in the present case a rectangle, and the two lines considered are the diagonals of this rectangle.

274. Corollary. *In an orthodiagonal quadrilateral the midpoints of the sides lie on a circle having for center the centroid of the quadrilateral* (§§ 242, 243).

275. Theorem. *If an orthodiagonal quadrilateral is cyclic, the anticenter coincides with the point of intersection of its diagonals.*

If O, J are the circumcenter and the centroid of the cyclic quadrilateral $ABCD$ (Fig. 81), and the diagonals AC, BD are orthogonal, the orthocenter H_d of the triangle ABC lies on the perpendicular BD to the base AC; now the line DH_d passes through the anticenter M of the quadrilateral (§§ 261, 262); hence M lies on the diagonal BH_dD.

Similarly M lies on the diagonal AC; hence the proposition.

276. Theorem of Brahmagupta (c. 628). *In a quadrilateral which is both orthodiagonal and cyclic the perpendicular from the point of intersection of the diagonals to a side bisects the side opposite.*

Indeed, the perpendicular RP' from the midpoint R of the side CD (Fig. 81) to the opposite side AB passes through the point M (§§ 258, 275).

277. Corollary. *In a cyclic orthodiagonal quadrilateral the projections of the point of intersection of the diagonals upon the four sides lie on the circle passing through the midpoints of the sides.*

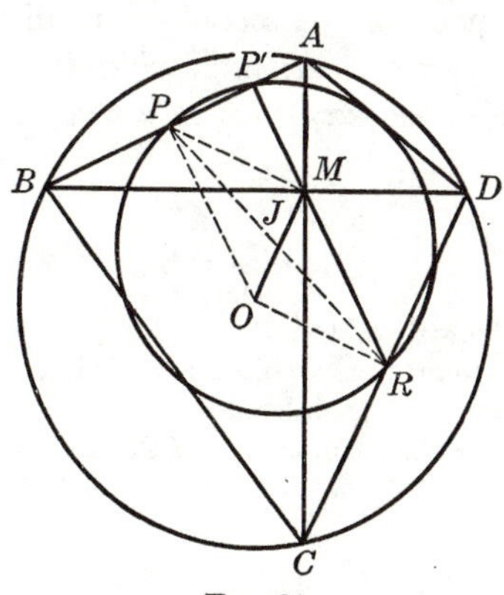

Fig. 81

Indeed, if P (Fig. 81) is the midpoint of the side AB, the diameter PR of the circle considered (§ 273) subtends a right angle at the point P'. Similarly for the other projections of M.

278. Theorem. *In a cyclic orthodiagonal quadrilateral the distance of a side from the circumcenter of the quadrilateral is equal to half the opposite side.*

The line OM (Fig. 81) bisects, in the centroid J, the line PR joining the midpoints P, R of the opposite sides AB, CD (§ 242); hence $PMRO$ is a parallelogram, and $OP = MR$. Now OP is perpendicular to the chord AB, and the median MR of the right triangle CMD is equal to half the hypotenuse CD; hence the proposition.

279. Theorem. *If a quadrilateral is both cyclic and orthodiagonal, the sum of the squares of two opposite sides is equal to the square of the circumdiameter of the quadrilateral.*

Indeed, in the right triangle AOP (Fig. 81) we have:

$$OA^2 = AP^2 + OP^2 = \tfrac{1}{4}AB^2 + \tfrac{1}{4}CD^2 \text{ (§ 278).}$$

280. COROLLARY I. *In the parallelogram PMRO* (Fig. 81) *we have* (§ 245):

$$PR^2 + OM^2 = OP^2 + PM^2 + MR^2 + RO^2 = 2(OP^2 + OR^2);$$

hence, by the preceding proposition (§ 279):

$$PR^2 = \tfrac{1}{2}(AB^2 + CD^2) - OM^2 = 2OA^2 - OM^2.$$

This formula gives the diameter of the circle considered before (§ 277) and shows that this circle depends only upon the radius of the given circle and the point of intersection of the diagonals.

281. COROLLARY II. *In a quadrilateral which is both cyclic and orthodiagonal the sum of the squares of the sides is equal to eight times the square of the circumradius.*

EXERCISES

1. Show that the line joining the midpoints of the diagonals of a cyclic orthodiagonal quadrilateral is equal to the distance of the point of intersection of the diagonals from the circumcenter of the quadrilateral.
2. If the diagonals of a cyclic quadrilateral $ABCD$ are orthogonal, and E is the diametric opposite of D on its circumcircle, show that $AE = CB$.

SUPPLEMENTARY EXERCISES

1. Construct a quadrilateral given the four sides and the sum of two opposite angles.
2. Construct a quadrilateral given, in position, the projections of the point of intersection of the diagonals upon the four sides.
3. About a given quadrilateral circumscribe a rhombus similar to a given rhombus.

4. Construct a rectangle similar to a given rectangle so that its sides shall be tangent respectively to four given circles.

5. A variable triangle ABC has a fixed base BC and a constant opposite angle. On the internal bisector of that angle A a segment $AL = \frac{1}{2}(AB + AC)$ is laid off. Show that the locus of L is a circle.

6. Show that the square of the area of a bicentric quadrilateral (that is, one that is both cyclic and circumscriptible) is equal to the product of its four sides.

V

THE SIMSON LINE

282. Theorem. (*a*) *The feet of the three perpendiculars to the sides of a triangle from a point of its circumcircle are collinear.*

(*b*) *Conversely. If the feet of the perpendiculars from a point to the sides of a triangle are collinear, the point lies on the circumcircle of the triangle.*

Let P be a point inside the angle ABC of the triangle ABC and let L, M, N be the projections of P upon the sides BC, CA, AB (Fig. 82). The circles (PA), (PC) described on PA, PC as diameters pass through the pairs of points M and N, M and L, respectively, and we have:

$$(1)\quad \angle APN = AMN, \quad \angle CPL = CML, \quad \angle B + NPL = 180°.$$

Now suppose that P lies on the circle ABC. The angle APC is then supplementary to B and we have:

$$(2)\qquad \angle APC = NPL;$$

hence, subtracting these equal angles in turn from the angle NPC, we have:

$$(3)\qquad \angle APN = CPL,$$

and therefore, from (1):

$$(4)\qquad \angle AMN = CML.$$

Thus the three points L, M, N are collinear.

Conversely, if L, M, N are collinear (4) holds, and from (4) and (1) we have (3), and therefore also (2), i.e., angle NPL is supplementary to B, and the point P is concyclic with A, B, C.

Note. The proof of the direct theorem is valid, whether the given point P lies inside the angle B, as it was assumed, or inside either of the other two angles of ABC. The proof of the converse theorem is also valid in these cases. But in the case of the converse theorem, should the point P be supposed to lie in the vertical angle of an angle of the triangle ABC, it is evident that in such a case the points L, M, N, could not be collinear.

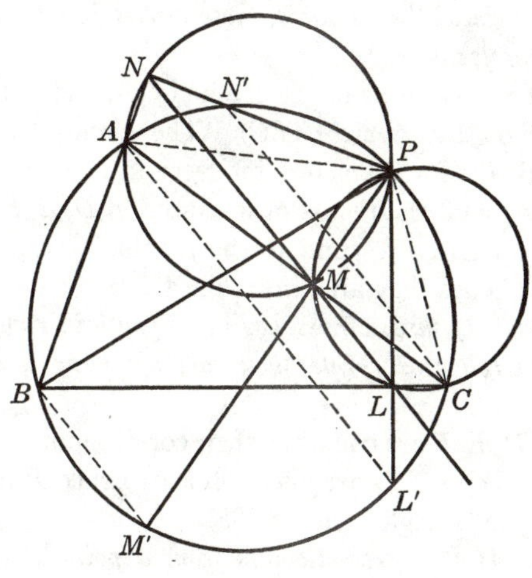

FIG. 82

It may be observed that the line LMN cuts the circumcircle, for it passes through the point M which lies inside the circle.

283. Definition. The line LMN (§ 282) is called the *Simson line*, or, more briefly, the *simson* of the point P with respect to the triangle ABC, or for the triangle ABC; sometimes LMN is referred to as the *pedal line* of P.

The line LMN will sometimes be indicated by the symbol $P(ABC)$.

The point P is called the *pole* of the line LMN for the triangle ABC.

284. Theorem. *If three chords drawn through a point of a circle are taken for diameters of three circles, these circles intersect, in pairs, in three new points, which are collinear.*

Indeed, If PA, PB, PC (Fig. 82) are three chords of the same circle, the three circles having these chords for diameters pass through the feet of the perpendiculars from P to the lines BC, CA, AB, and these three points are collinear (§ 282 a).

285. Conversely. *If the three circles having for diameters the three segments PA, PB, PC intersect, in pairs, in three collinear points, the point P is concyclic with A, B, C.*

Indeed, these circles pass in pairs through the feet of the perpendiculars from P to the lines BC, CA, AB, and if these three points are collinear the four points P, A, B, C are concyclic (§ 282 b).

286. Theorem. *If three circles pass through the same point of the circumcircle of the triangle of their centers, these circles intersect, in pairs, in three collinear points.*

Let A, B, C be the centers of the given circles, and P their common point, situated on the circle (ABC). The circles (A'), (B'), (C') having PA, PB, PC for diameters intersect in three collinear points D', E', F' (§ 284), and the points of intersection D, E, F of the given circles taken in pairs correspond to the points D', E', F' in the homothecy $(P, 2)$; hence they are also collinear.

287. Conversely. *If three circles having a point in common intersect in pairs in three collinear points, their common point is concyclic with their centers.*

If the points D, E, F are collinear, their corresponding points D', E', F' in the homothecy $(P, 2)$ are also collinear; hence P belongs to the circumcircle of the triangle ABC (§ 285).

288. Theorem. *If the perpendiculars from a point P of the circumcircle (O) of a triangle ABC to the sides BC, CA, AB meet (O) again in the points L', M', N', the three lines AL', BM', CN' are parallel to the simson of P for ABC.*

Indeed, we have (Fig. 82):

$$\angle L'AC = L'PC = LPC = LMC;$$

hence AL' is parallel to NLM.

Similarly for BM' and CN'.

289. Problem. *Find the point whose Simson line for a given triangle has a given direction.*

Through any vertex, say B, of the given triangle ABC draw a line having the given direction, meeting the circumcircle (O) again in K (Fig. 83). The perpendicular KM from K to the side AC, opposite the vertex B, meets (O) in the required point P, and the parallel through M to KB is the simson of P (§ 288).

290. Theorem. *The Simson line bisects the line joining its pole to the orthocenter of the triangle.*

Let the altitude BE (Fig. 83) meet (O) again in T, and let the parallel to BK through the orthocenter H of ABC meet the line PMK in U.

From the parallelogram $BHUK$ and the isosceles trapezoid $PTBK$ we have:

$$HU = BK, \quad PT = BK;$$

hence $HUPT$ is an isosceles trapezoid. Now E is the midpoint of HT (§ 178); hence the mediator EA of the base HT of $HUPT$ bisects the

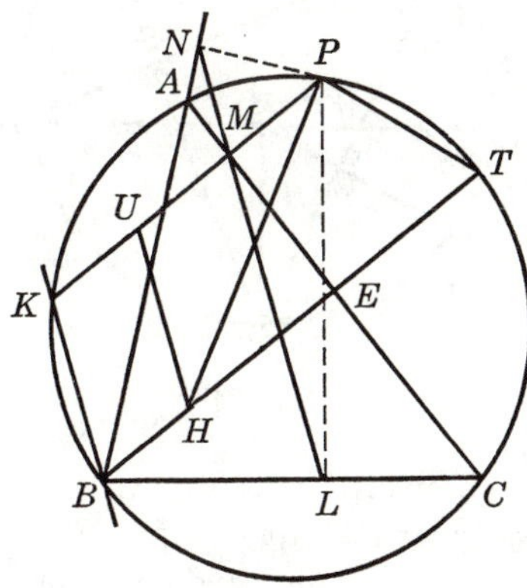

FIG. 83

other base UP, in M. Thus the simson LMN passes through the midpoint M of the side PU of the triangle PUH and is parallel to UH; hence LMN bisects the third side HP.

291. *Remark.* The midpoint of the segment HP lies on the nine-point circle (N) of ABC (§ 210).

292. De la Hire's Problem. *Construct a triangle given an angle, the sum of the including sides, and the altitude to the third side* $(A, b + c = 2\,s, h_a)$.

Let ABC (Fig. 84) be the required triangle, W, P the traces of the internal bisector AW of the angle A on BC and on the circumcircle of ABC, and R, S the projections of P upon AB, AC. We have:

$$(1) \quad AR + AS = (AB + BR) + (AC - SC) = (AB + AC) + (BR - SC).$$

Now $PB = PC$ (§ 122) and $PR = PS$; hence, from the two pairs of congruent right triangles PAR and PAS, PRB and PSC, we have $AR = AS$ and $BR = CS$, so that, from (1):

$$AR = AS = \tfrac{1}{2}(AB + AC) = s.$$

Thus in the quadrilateral $ASPR$ we know the angle $SAR = A$, the sides AR, AS, and the angles ARP, ASP are right angles; hence the quadrilateral may be constructed, and AP is the mediator of RS.

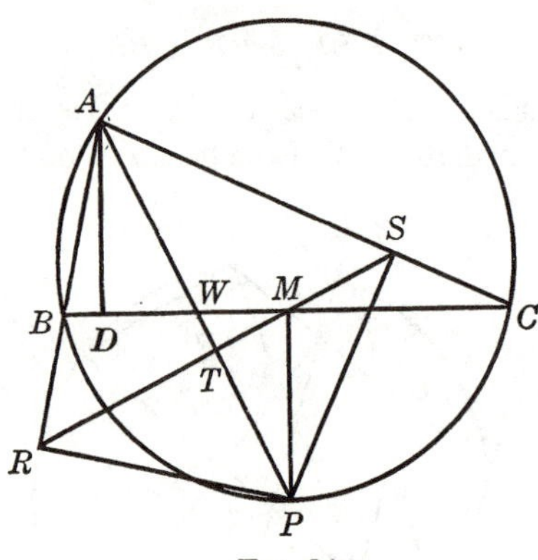

FIG. 84

The line RS is the simson of P for ABC; hence RS passes through the foot M of the perpendicular PM from P upon BC. Let T be the trace of RS on AWP. From the similar right triangles PMT, ADW, we have:

$$(2) \qquad PT:AD = PM:AW = PM:(AP - PW),$$

and in the right triangles PMW, APS we have:

$$(3) \qquad PM^2 = PT \cdot PW, \quad PS^2 = PT \cdot PA;$$

hence, from (2) and (3):

$$(4) \qquad PM^2 + AD \cdot PM - PS^2 = 0.$$

The segment $AD = h_a$ is given, and PS is known from the quadrilateral $ASPR$ already constructed; hence PM may be constructed (§ 67).

Draw the circle (P, PM) meeting RS in the point M. The perpendicular to PM at M will meet the sides AR, AS of the quadrilateral $ASPR$ in the vertices B, C of the required triangle ABC.

The problem has no solution if the altitude h_a is greater than the segment $AT = s \sin \frac{1}{2} A$. When h_a is equal to AT, the triangle ABC is isosceles. When AD is smaller than AT, the construction yields two solutions, symmetrical with respect to AP.

EXERCISES

1. Show that the diametric opposite of a vertex of a triangle on its circumcircle has for Simson line, with respect to this triangle, the side of the triangle opposite the vertex considered.
2. Show that the simson of the point where an altitude cuts the circumcircle again passes through the foot of the altitude and is antiparallel to the corresponding side of the triangle with respect to the other two sides.
3. Are there any points which lie on their own Simson lines with respect to a given triangle?
4. If the Simson line $P(ABC)$ meets BC in L and the altitude AD in K, show that the line PK is parallel to LH, where H is the orthocenter of ABC.
5. If the Simson line $P(ABC)$ is parallel to the circumradius OA, show that the line PA is parallel to BC.
6. If the perpendiculars dropped from a point P of the circumcircle (O) of the triangle ABC upon the sides, meet (O) in the points A', B', C', show that the two triangles ABC, $A'B'C'$ are congruent and symmetrical with respect to an axis.
7. If L, M, N are the feet of the perpendiculars from a point P of the circumcircle of a triangle ABC upon its sides BC, CA, AB, prove that (a) the triangles PLN, PAC are similar; (b) $PL \cdot MN$, $PM \cdot NL$, $PN \cdot LM$ are proportional to BC, CA, AB; (c) $PA \cdot PL = PB \cdot PM = PC \cdot PN$.
8. In a given circle to inscribe a triangle so that the Simson line for this triangle of a given point of the circle shall coincide with a given line. One vertex may be chosen arbitrarily.
9. Construct a triangle given A, $b - c$, h_a.
10. Let P, Q, R be the traces of the internal bisectors of the triangle ABC on the circumcircle (O). Show that the Simson lines of P, Q, R for ABC are the external bisectors of the medial triangle (T') of ABC. Consider the simsons, for ABC, of the traces P', Q', R' of the external bisectors of ABC on (O). Consider the traces of any three concurrent bisectors of ABC, or of any three bisectors forming a triangle.
11. Show that the symmetrics, with respect to the sides of a triangle, of a point on its circumcircle lie on a line passing through the orthocenter of the triangle.
12. If the simson of a point P passes through the diametric opposite of P on the circumcircle of the triangle considered, show that it also passes through the centroid of that triangle.

293. Theorem. *The angle formed by the Simson lines of two points for the same triangle is measured by half the arc between the two points.*

Let P, P' be the given points (Fig. 85) and K, K' the two points where the perpendiculars from P, P' to the side AC meet the circumcircle of ABC. The simsons of P, P' are parallel to the lines BK, BK'; therefore the angle between the simsons is equal to the angle KBK'.

Now the inscribed angle KBK' is measured by half the arc KK', and arc KK' = arc PP'; hence the proposition.

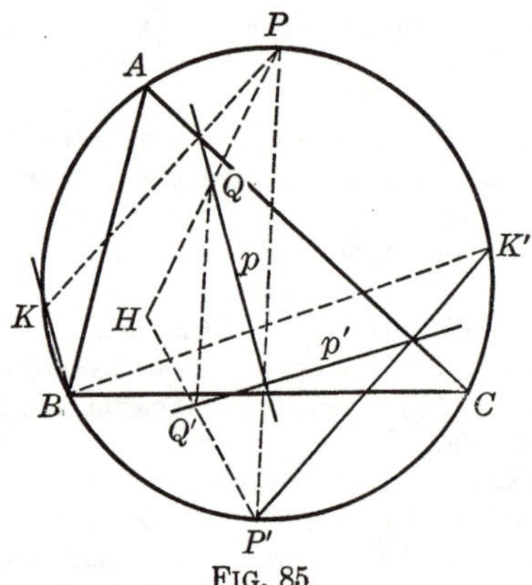

FIG. 85

294. COROLLARY. *The perpendiculars (or parallels) through the points P, P' to the Simson lines $P'(ABC)$, $P(ABC)$, respectively, intersect on the circumcircle of the triangle ABC.*

295. Theorem. *The Simson lines of two diametrically opposite points are perpendicular to each other and intersect on the nine-point circle of the triangle.*

If the points P, P' (Fig. 85) are diametrically opposite, the angle between their simsons p, p' is measured by one fourth of the circumference (§ 293).

The midpoints Q, Q' of the segments HP, HP' correspond to the points P, P' in the homothecy $(H, \frac{1}{2})$; hence the line QQ' corresponds to PP' in this homothecy, and therefore QQ' is a diameter of the nine-point circle (N) of ABC (§ 210). Now the two rectangular simsons p, p' pass through the points Q, Q', respectively (§ 290), hence they intersect on the circle (N).

296. *Remark. Two rectangular Simson lines meet a side of the triangle in two isotomic points.*

For these two points are the projections upon the side considered of the ends of a diameter of the circumcircle of the triangle.

297. COROLLARY. *The nine-point circle of a triangle is the locus of the points of intersection of the pairs of Simson lines corresponding to pairs of diametrically opposite points on the circumcircle of the triangle.*

298. Theorem. *The angle formed by the two Simson lines of the same point with respect to two triangles inscribed in the same circle is the same for all positions of the point on the circle.*

Let ABC, DEF be the two triangles inscribed in the same circle (O) (Fig. 86) and let the perpendiculars to AC and DF from any point P

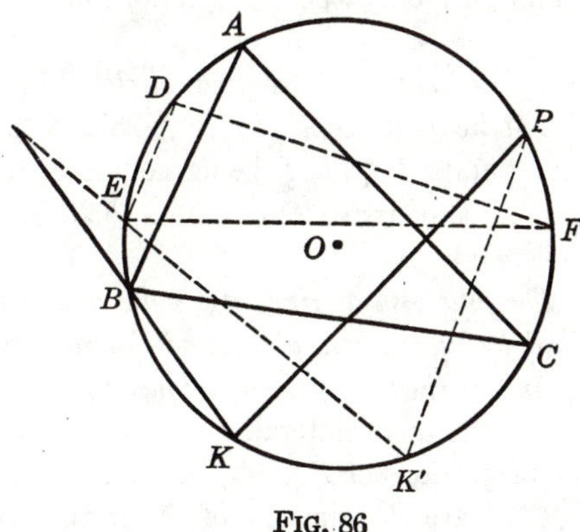

FIG. 86

of (O) meet (O) again in the points K and K'. The simsons p, p' of P for ABC, DEF are parallel to the lines BK, EK', respectively (§ 288). Now the angle between BK and EK' is measured by:

(1) $\frac{1}{2}(\text{arc } KK' - \text{arc } BE) = \frac{1}{2}(\text{arc } AD + \text{arc } CF - \text{arc } BE).$

Since this expression does not depend upon P, the proposition is proved.

Note. In the figure considered the lines BK, EK' meet outside the circle (O), and the lines AC, DF meet inside of (O). It may, however, be shown that the proposition remains valid if either of these intersections, or both, change to the opposite.

299. Theorem. *The four circumcircles of the four triangles determined by four given straight lines, taken three at a time, have a point in common.*

Let p, q, r, s be the four given lines. The circumcircles (pqr), (pqs) of the two triangles pqr, pqs have the point pq in common; hence they meet in a second point, M. Now if P, Q, R, S are the projections of M upon the lines p, q, r, s, the points in each of the two triads P, Q, R and P, Q, S are collinear (§ 282 a); hence the four points P, Q, R, S lie on the same line.

The projections of the point M upon the sides of the triangle qrs are thus collinear; hence M lies on the circumcircle of this triangle (§ 282 b), and also on the circumcircle of prs, for similar reasons.

300. Definition. The point M (§ 299) is called the *Miquel point* of the four lines p, q, r, s.

301. *Remark.* Given four lines (no two parallel, no three concurrent), there is a point, and only one, whose projections upon these lines are collinear.

302. COROLLARY. *The centers of the four circles* (pqr), (qrs), (rsp), (spq) *and the point* M *lie on the same circle* (§§ 287, 299).

Indeed, the four points P, Q, R, S being collinear, the point M is concyclic with any three of these four centers (§ 287); hence the five points are concyclic.

303. Theorem. *The four Simson lines of a point of a circle for the four triangles determined by the vertices of a quadrilateral inscribed in that circle, admit the point considered for their Miquel point.*

Let $ABCD$ be a cyclic quadrilateral, P a point on its circumcircle, and A', B', C' the projections of P upon the lines DA, DB, DC. The lines $A'B'$, $B'C'$, $C'A'$ are the simsons of P for the triangles DAB, DBC, DCA, and the point P is concyclic with the vertices of the triangle $A'B'C'$ formed by these Simson lines, for the points A', B', C' lie on the circle having PD for diameter.

Similarly for any other three of the four lines considered. Hence the proposition (§§ 299, 300).

304. Theorem. *The four Simson lines of four points of a circle, each taken for the triangle formed by the remaining three points, are concurrent.*

Let $ABCD$ be a cyclic quadrilateral. The Simson line $D(ABC)$ passes through the midpoint of the segment joining D to the orthocenter of ABC (§ 290), i.e., through the anticenter of $ABCD$ (§ 262). Similarly for the simsons $A(BCD)$, $B(CDA)$, $C(DAB)$.

305. Theorem. *The orthocenters of the four triangles formed by four lines taken three at a time are collinear.*

The segments joining the Miquel point M of the four lines p, q, r, s (§ 300) to the orthocenters of the four triangles formed by these lines are bisected (§ 290) by the common Simson line $PQRS$ of M for these four triangles (§ 299); hence these orthocenters lie on the line which corresponds to the line $PQRS$ in the homothecy $(M, 2)$.

EXERCISES

1. If two triangles are inscribed in the same circle and are symmetrical with respect to the center of that circle, show that the two simsons of any point of the circle for these triangles are rectangular.
2. The vertices B, C and the circumcircle (O) of the variable triangle ABC are fixed, and P, P' are two fixed points of (O). Show that the point of intersection of the Simson lines $P(ABC)$, $P'(ABC)$ describes a circle.
3. Show that the Simson lines of the three points where the altitudes of a triangle cut the circumcircle again form a triangle homothetic to the orthic triangle, and its circumcenter coincides with the orthocenter of the orthic triangle.
4. The chords PA, PB, PC of a given circle are the diameters of three circles (PA), (PB), (PC). The circle (PA) meets the lines PB, PC in the points A', A''; the circle (PB) meets PA, PC in B', B''; the circle (PC) meets PA, PB in C', C''. Show that the three lines $A'A''$, $B'B''$, $C'C''$ are concurrent.

SUPPLEMENTARY EXERCISES

1. A variable triangle has a fixed circumcircle and a fixed centroid. Prove that the simson, for this triangle, of a point on its circumcircle passes through a fixed point.
2. A variable triangle ABC is inscribed in a fixed circle. The vertex A and the direction of the internal bisector of the angle A are fixed. Show that the Simson line of a given point P for ABC has a fixed direction.
3. A variable circle passing through two fixed points A, D cuts two fixed lines through the point A in B and C. Find the locus of the orthocenter of the triangle ABC.
4. A variable triangle has a fixed circumcircle and a fixed vertex, and the variable opposite side passes through a fixed point. Prove that the Simson line for this triangle of a given point on its circumcircle passes through a fixed point.
5. A variable circle having its center on the base BC of an isosceles triangle ABC and passing through A, cuts the sides AB, AC in Q, R. Find the locus of the midpoint of QR.
6. In a given circle an infinite number of triangles may be inscribed such that the Simson line, with respect to these triangles, of a given point of the circle will coincide with a given line. Find the locus of the centroid of these triangles.
7. The perpendiculars dropped upon the sides BC, CA, AB of the triangle ABC from a point P on its circumcircle meet these sides in L, M, N and the circle in A', B', C'. The simson LMN meets $B'C'$, $C'A'$, $A'B'$ in L', M', N'. Prove that the lines AL', BM', CN' are concurrent.
8. Find the locus of the projection of the foot of the altitude to the base of a triangle upon the line joining the projections of a variable point of the base upon the other two sides of the triangle.
9. Prove that if the Simson line of a point for a triangle passes through the diametric opposite of the given point, it passes also through the centroid of the triangle. Conversely.

10. Construct an equilateral triangle so that its sides shall pass, respectively, through three given collinear points, and its circumcircle shall pass through a fourth given point.

11. The circumradius OP of the triangle ABC meets the sides of the triangle in the points A', B', C'. Show that the projections A'', B'', C'' of the points A', B', C' upon the lines AP, BP, CP lie on the Simson line of P for ABC.

12. Let L, M, N be the projections of the point P of the circumcircle of the triangle ABC upon the sides BC, CA, AB, and let the Simson line LMN meet the altitudes AD, BE, CF in the points L', M', N'. Show that (a) the segments LM, $L'M'$ are equal to the projection of the side AB upon the Simson line; (b) the projections of the segments AL', BM', CN' upon LMN are equal.

13. Through two diametrically opposite points of a circle parallels are drawn to the simsons of these points with respect to a given triangle inscribed in the circle. Show that the midpoints of the segments which these parallels determine on the three sides of the triangle are collinear.

14. Prove that if three triangles are inscribed in the same circle, the three simsons of any point of the circle form a triangle of a fixed species. Consider some special cases.

VI

TRANSVERSALS

A. INTRODUCTORY

306. Directed Segments. When two or more segments are considered on the same line, we can more readily account for their mutual relations if we take into consideration not only the length of each segment but its sense as well.

A point may traverse a straight line in two opposite senses. We choose, arbitrarily, one of these senses, say the one from left to right, as the positive sense, and the other as the negative sense. If on this line a segment AB, say 5 inches long, has for its extremities the points A, B, and the sense from A to B is the positive sense of the line, we shall say that the segment is $+5$ inches long, but if the sense from A to B is the negative sense of the line, we shall say that the segment AB is equal to -5 inches.

Consequently, in the first case considered the segment BA would be equal to -5 inches, and in the second case considered the segment BA would be equal to $+5$ inches. We thus distinguish between the two extremities A, B of the segment. When we read the segment as AB, A is its beginning and B its end, but when we read the segment as BA, then B is the beginning and A is the end.

We shall refer to a segment whose magnitude and sense are both considered, as a *directed segment.*

It is clear that in the case of directed segments we have

$$AB + BA = 0. \tag{1}$$

307. Three Collinear Points. Given two points A, B on a line, a third point C may be marked on the line, either between the points

A, B, or beyond the point A, or beyond the point B. If the position of the point C relative to the points A and B is to be accounted for in terms of the lengths alone of the segments AB, AC, BC, we arrive at three different equalities corresponding to the three different positions of C enumerated. The use of directed segments enables us to cover all three cases with one formula, namely:

$$(2) \qquad AB = AC + CB, \quad \text{or} \quad AB + BC + CA = 0.$$

EXERCISES

1. Show that $AB + BC + CD = AD$.
2. Show that if M is the midpoint of the segment AB, we have, both in magnitude and in sign: $AB = 2\,AM = 2\,MB$.
3. Show that if O, A, B are collinear, we have, both in magnitude and in sign:
$$OA^2 + OB^2 = AB^2 + 2\,OA \cdot OB.$$
4. Show that if A, B, P are any three collinear points, and M is the midpoint of the segment AB, we have, both in magnitude and in sign, $PM = \frac{1}{2}(PA + PB)$.
5. Show that if $OA + OB + OC = 0$, and P is any point of the line AB, then $PA + PB + PC = 3\,PO$.
6. Show that if we have on the same line $OA + OB + OC = 0$, $O'A' + O'B' + O'C' = 0$, then $AA' + BB' + CC' = 3\,OO'$.
7. Show that $AB \cdot CD + AC \cdot DB + AD \cdot BC = 0$. *Hint.* $CD = CA + AD$, $DB = DA + AB$, $BC = BA + AC$. Substitute and simplify.

B. STEWART'S THEOREM

308. Theorem. *If A, B, C are three collinear points, and P any other point, we have, both in magnitude and in sign:*

$$(1) \qquad PA^2 \cdot BC + PB^2 \cdot CA + PC^2 \cdot AB + BC \cdot CA \cdot AB = 0.$$

(a) If P lies outside of the line ABC, let E be the foot of the perpendicular from P on ABC. We have:

$$\begin{aligned} PA^2 &= PE^2 + EA^2 = PE^2 + (EC + CA)^2 \\ &= PE^2 + EC^2 + 2\,EC \cdot CA + CA^2, \\ PB^2 &= PE^2 + EB^2 = PE^2 + (EC + CB)^2 \\ &= PE^2 + EC^2 + 2\,EC \cdot CB + CB^2. \end{aligned}$$

Now $PE^2 + EC^2 = PC^2$ and $CB = -BC$; hence, substituting, we obtain:

$$\begin{aligned} PA^2 &= PC^2 + CA^2 + 2\,EC \cdot CA, \\ PB^2 &= PC^2 + BC^2 - 2\,EC \cdot BC, \end{aligned}$$

or, multiplying these two relations by BC, CA, respectively, and adding:

$$\begin{aligned} PA^2 \cdot BC + PB^2 \cdot CA &= PC^2(BC + CA) + CA^2 \cdot BC + BC^2 \cdot CA \\ &= PC^2(BC + CA) + BC \cdot CA(CA + BC) \\ &= (PC^2 + BC \cdot CA)(BC + CA) \\ &= (PC^2 + BC \cdot CA)(-AB), \end{aligned}$$

which is the required formula (1).

(b) If P lies on the line ABC, we have, applying the formula (1) to any point Q of the perpendicular to ABC at the point P:

$$(2) \qquad QA^2 \cdot BC + QB^2 \cdot CA + QC^2 \cdot AB + BC \cdot CA \cdot AB = 0.$$

Now $QA^2 = QP^2 + PA^2$, $QB^2 = QP^2 + PB^2$, $QC^2 = QP^2 + PC^2$. Substituting into (2) and remembering that $AB + BC + CA = 0$ we obtain the formula (1).

EXERCISES

1. Use Stewart's formula to find the length of the medians of a triangle, of the internal bisectors, etc.

2. Show that the sum of the squares of the distances of the vertex of the right angle of a right triangle from the two points of trisection of the hypotenuse is equal to $\frac{5}{9}$ the square of the hypotenuse.

3. If H, G, O, I are the orthocenter, the centroid, the circumcenter, and the incenter of a triangle, show that:

$$HI^2 + 2\,OI^2 = 3(IG^2 + 2\,OG^2), \quad 3(IG^2 + \tfrac{1}{2}\,HG^2) - IH^2 = 2\,R(R - 2\,r),$$

where R and r are the circumradius and the inradius.

4. If a line through the vertex A of an equilateral triangle ABC meets BC in F and the circumcircle in M, show that:

$$\frac{1}{MF} = \frac{1}{MB} + \frac{1}{MC}.$$

C. MENELAUS' THEOREM

309. *Observation.* A transversal may meet two sides of a triangle and the third side produced (Fig. 87 *a*), or all three sides produced (Fig. 87 *b*).

The trace of the transversal on a side and the two vertices lying on that side determine two segments. The six segments thus determined by a transversal on the sides of a triangle may be divided into two sets each containing three nonconsecutive segments, i.e., no two segments of the same set having a common end.

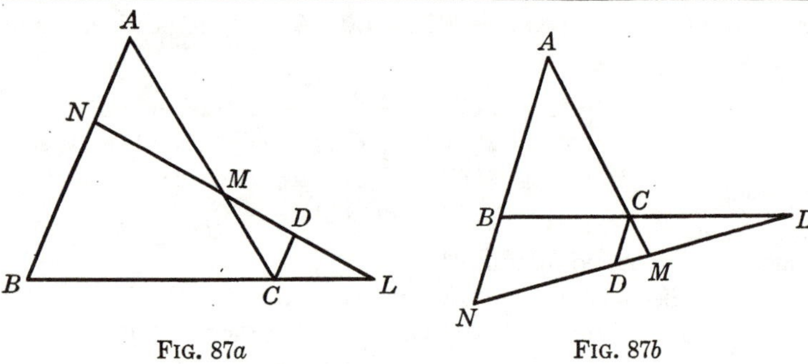

FIG. 87a FIG. 87b

310. Menelaus' Theorem. *The six segments determined by a transversal on the sides of a triangle are such that the product of three nonconsecutive segments is equal to the product of the remaining three.*

Let L, M, N be the traces of the transversal LMN on the sides BC, CA, AB of the triangle ABC (Fig. 87 a and b), and let the parallel through C to the opposite side AB meet LMN in D. From the two pairs of similar triangles NBL and DCL, MAN and MDC we have:

$$BL \cdot DC = LC \cdot NB, \quad CM \cdot AN = MA \cdot DC;$$

hence, multiplying and simplifying:

$$BL \cdot CM \cdot AN = LC \cdot NB \cdot MA,$$

which proves the proposition.

311. *Remark I.* It is often more convenient to write this formula in the following form:

(m) $$\frac{AN}{NB} \cdot \frac{BL}{LC} \cdot \frac{CM}{MA} = 1.$$

The formula (m) is readily written down by following the perimeter of the triangle in a definite sense of rotation. The first segment is taken from a vertex to a point of section by the transversal, and the second segment from this point of section to the second vertex lying on the side considered, and so on, until the initial vertex is reached.

312. *Remark II.* If the segments in the formula (m) are considered as directed segments and are read in the order suggested, one of the three ratios will always be negative, and the other two either both positive, or both negative (§ 309). Thus we have, both in magnitude and in sign:

(m′) $$\frac{AN}{NB} \cdot \frac{BL}{LC} \cdot \frac{CM}{MA} = -1.$$

313. Converse Theorem. *If three points are taken, one on each side of a triangle, so that these points divide the sides into six segments such that the products of the segments in each of the two sets of nonconsecutive segments are equal in magnitude and opposite in sign, the three points are collinear.*

Let L, M, N be three points on the three sides of the triangle ABC such that the formula (m′) holds. If the three points are not collinear, let K be the point on the side BC collinear with M and N. According to the direct theorem we have:

$$\frac{AN}{NB}\cdot\frac{BK}{KC}\cdot\frac{CM}{MA}=-1.$$

Comparing this formula with the formula (m′), we have, both in magnitude and in sign:

$$BL:LC = BK:KC,$$

i.e., the points K and L divide the side BC in the same ratio, both in magnitude and in sign; hence they coincide, which proves the proposition.

314. Theorem. (*a*) *The external bisectors of the angles of a triangle meet the opposite sides in three collinear points.*

(*b*) *Two internal bisectors and the external bisector of the third angle meet the opposite sides in three collinear points.*

(a) If U', V', W' are the traces of the external bisectors of the angles A, B, C on the opposite sides, we have:

$$BU':U'C = c:b, \quad CV':V'A = a:c, \quad AW':W'B = b:a.$$

The product of the three ratios is thus equal to one; hence the proposition (§ 313).

The proof of (b) is analogous.

315. *Remark.* The six bisectors of the angles of a triangle determine on the opposite sides six points which lie by threes on four straight tion.

316. Corollary. *The sides of the orthic triangle meet the sides of the given triangle in three collinear points* (§ 193).

317. Definition. The line joining the three points (§ 316) is called the *orthic axis* of the triangle.

318. Theorem. *The lines tangent to the circumcircle of a triangle at the vertices meet the opposite sides in three collinear points.*

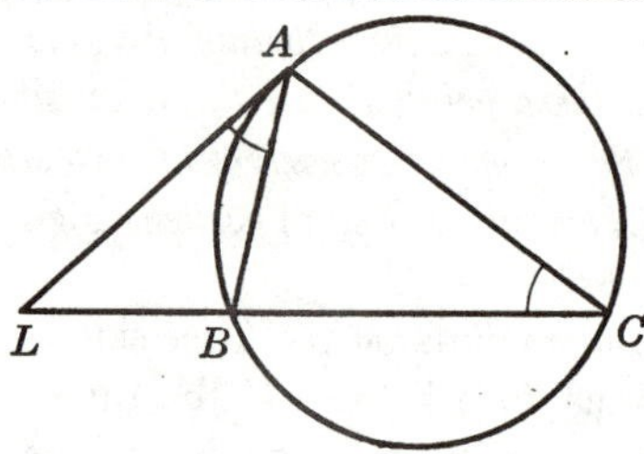

FIG. 88

Let L, M, N (Fig. 88) be the traces on BC, CA, AB of the tangents to the circumcircle of ABC at the points A, B, C. The angles LAB and C being measured by half of the same arc AB, the two triangles ALB, ALC are similar; hence:

$$AL:LC = AB:AC, \quad \text{or} \quad AL^2:LC^2 = AB^2:AC^2.$$

Now $AL^2 = LB \cdot LC$; hence:

$$BL:LC = c^2:b^2.$$

Similarly:

$$CM:MA = a^2:c^2, \quad AN:NB = b^2:a^2;$$

hence the points L, M, N are collinear (§ 313).

319. *Remark.* We have proved incidentally that the tangent to the circumcircle at the vertex of a triangle divides the opposite side externally in the ratio of the squares of the two adjacent sides.

320. Definition. The triangle formed by the tangents to the circumcircle of a given triangle at the vertices is called the *tangential triangle* of the given triangle.

321. Theorem. *The isotomic points of three collinear points are collinear.*

If L, M, N are three collinear points on the sides BC, CA, AB of the triangle ABC, we have (§ 310):

$$\frac{BL}{LC} \cdot \frac{CM}{MA} \cdot \frac{AN}{NB} = -1.$$

Now if L', M', N' are the isotomic points of L, M, N, we have:
$CL':L'B = BL:LC, \quad BN':N'A = AN:NB, \quad AM':M'C = CM:MA$;
hence the proposition (§ 313).

322. Definition. Two transversals like LMN and $L'M'N'$ are sometimes referred to as *reciprocal transversals.*

323. Problem. *Draw two lines, of given directions, so that they shall be reciprocal transversals for a given triangle.*

Let ABC (Fig. 89) be the given triangle and let the line through A, having the given direction d, meet the lines through B and C, having the given direction d', in the points P, Q. If B', C' are the midpoints

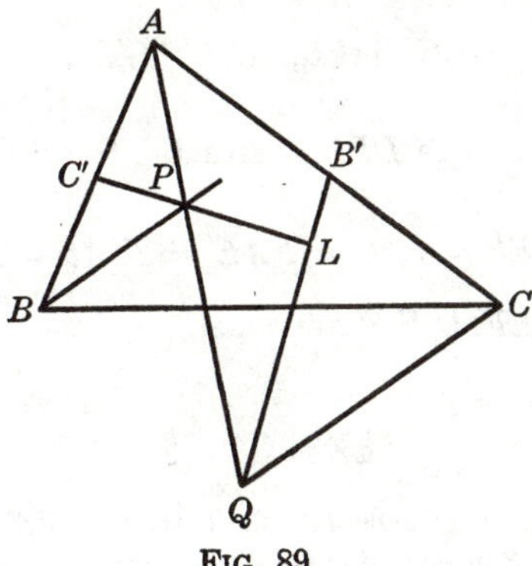

FIG. 89

of AC, AB, the point $L = (PC', QB')$ is common to the two required transversals. Indeed, the point L being collinear with Q and B', the parallels through L to QA, QC will meet the side AC in two isotomic points; similarly for the side AB.

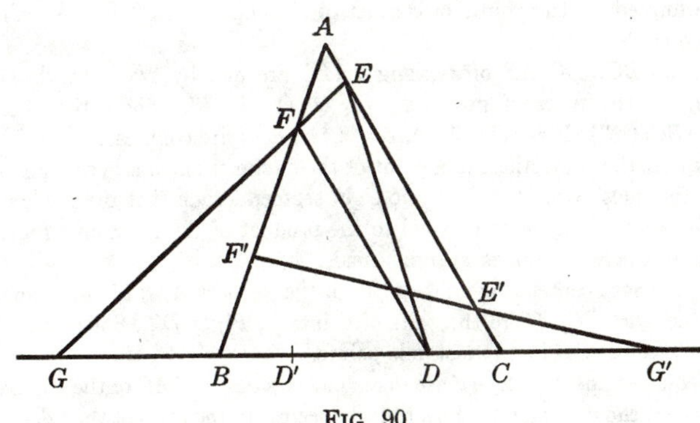

FIG. 90

324. Theorem. *If the pairs of points D, D'; E, E'; F, F' are isotomic on the sides BC, CA, AB of the triangle ABC, the two triangles DEF, $D'E'F'$ are equivalent.*

Let the lines EF, $E'F'$ (Fig. 90) meet BC in G, G'. The distances of the points C, D from the base EF are proportional to GC, GD; hence:

$$\text{area } CEF : \text{area } DEF = GC : GD,$$

and, similarly:

$$\text{area } BE'F' : \text{area } D'E'F' = G'B : G'D'.$$

From the reciprocal transversals EF, $E'F'$ we have:

$$GC = G'B, \quad GD = G'D';$$

hence:

$$\text{area } CEF : \text{area } DEF = \text{area } BE'F' : \text{area } D'E'F'.$$

Now:

$$\text{area } CEF = FEC = FAE' = E'AF = E'F'B;$$

hence the areas DEF, $D'E'F'$ are equal.

EXERCISES

1. The median AA' of the triangle ABC meets the side $B'C'$ of the medial triangle $A'B'C'$ in P, and CP meets AB in Q. Show that $AB = 3\,AQ$.
2. Show that, if a line through the centroid G of the triangle ABC meets AB in M and AC in N, we have, both in magnitude and in sign, $AN \cdot MB + AM \cdot NC = AM \cdot AN$.
3. Prove that two orthogonal Simson lines are reciprocal transversals.
4. Prove that the triangle formed by the points of contact of the sides of a given triangle with the excircles corresponding to these sides is equivalent to the triangle formed by the points of contact of the sides of the triangle with the inscribed circle.
5. The sides BC, CA, AB of a triangle ABC are met by two transversals PQR, $P'Q'R'$ in the pairs of points P, P'; Q, Q'; R, R'. Show that the points $X = (BC, QR')$, $Y = (CA, RP')$, $Z = (AB, PQ')$ are collinear.
6. Show that the projections of a point of the circumcircle of a cyclic quadrilateral upon the sides divide the sides into eight segments such that the product of four nonconsecutive segments is equal to the product of the remaining four. *Hint.* Consider the Simson lines as transversals.
7. A circle whose center is equidistant from the vertices A, B of the triangle ABC cuts the sides BC, AC in the pairs of points P, P'; Q, Q'. Show that the lines PQ, $P'Q'$ meet AB in two isotomic points.
8. Two equal segments AE, AF are taken on the sides AB, AC of the triangle ABC. Show that the median issued from A divides EF in the ratio of the sides AC, AB.

D. CEVA'S THEOREM

325. *Observation.* If a point S is taken inside the triangle ABC, the points of intersection L, M, N of the sides BC, CA, AB with the lines AS, BS, CS divide those sides internally (Fig. 91 *a*); if the point S is taken outside of ABC, one of the points L, M, N divides the corre-

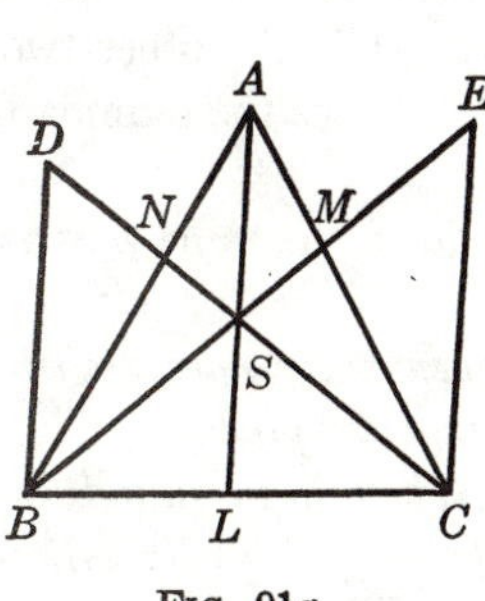

FIG. 91a

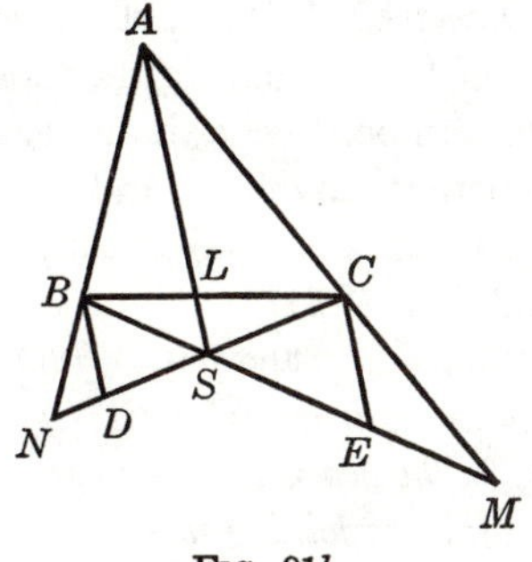

FIG. 91b

sponding side internally, and the remaining two points divide the sides externally (Fig. 91 b).

The six segments into which the points L, M, N divide the sides may be separated into two groups each containing three nonconsecutive segments (§ 309).

326. Ceva's Theorem. *The lines joining the vertices of a triangle to a given point determine on the sides of the triangle six segments such that the product of three nonconsecutive segments is equal to the product of the remaining three segments.*

Let the parallels BD, CE to the line ASL (Fig. 91 a and b) through B, C meet CN, BM in D, E. From pairs of similar triangles we have:

$$\begin{aligned} AN:NB &= AS:BD, \\ BL:BC &= LS:CE, \\ BC:LC &= BD:LS, \\ CM:MA &= CE:AS; \end{aligned}$$

hence, multiplying and simplifying:

$$\text{(c)} \qquad \frac{AN}{NB}\cdot\frac{BL}{LC}\cdot\frac{CM}{MA} = 1,$$

which proves the proposition.

OTHERWISE. If we cut the two triangles ALC, ALB (Fig. 91) by the two transversals BSM, CSN, respectively, we have, by Menelaus' theorem, both in magnitude and in sign:

$$\begin{aligned} LB\cdot CM\cdot AS &= -BC\cdot MA\cdot SL, \\ AN\cdot BC\cdot LS &= -NB\cdot CL\cdot SA. \end{aligned}$$

Multiplying and simplifying we obtain the formula (c).

327. *Remark.* If directed segments are considered, one of the three ratios of the formula (c) is always positive, while the other two are either both positive, or both negative (§ 325); hence the formula (c) is valid both in magnitude and in sign.

The formula (c) may readily be written down the same way as the formula (m) (§ 311).

328. Converse Theorem. *If three points taken on the sides of a triangle determine on these sides six segments such that the products of the segments in the two nonconsecutive sets are equal, both in magnitude and in sign, the lines joining these points to the respectively opposite vertices are concurrent.*

The proof is similar to the proof of the converse of Menelaus' theorem (§ 313).

329. Definitions. If a line is drawn through a vertex of a triangle, the segment included between the vertex and the opposite side is called a *cevian.* The segments AL, BM, CN (Fig. 91) are cevians.

The triangle LMN (Fig. 91) may be called the *cevian triangle* of the point S for the triangle ABC.

330. Theorem. *The lines joining the vertices of a triangle to the points of contact of the opposite sides with the inscribed circle are concurrent.*

If X, Y, Z are the points of contact of the sides BC, CA, AB of the triangle ABC with the incircle, we have (§ 157):

$$AZ = AY, \quad BX = BZ, \quad CY = CX;$$

hence:

$$AZ \cdot BX \cdot CY = BZ \cdot CX \cdot AY,$$

which proves the proposition (§ 328).

331. Definition. The point common to the lines AX, BY, CZ (§ 330) is often referred to as the *Gergonne point* of the triangle.

332. Theorem. *The lines joining the points of contact of an excircle with the sides of a triangle to the vertices opposite the respective sides, are concurrent.*

333. Theorem. *The lines joining the vertices of a triangle to the points of contact of the opposite sides with the excircles relative to those sides are concurrent.*

The proofs of these two propositions are analogous to the proof of the preceding proposition (§ 330).

334. Definition. The point common to the three lines of the last proposition (333) is often referred to as the *Nagel point* of the triangle.

EXERCISES

1. Use Ceva's theorem to prove that in a triangle (a) the medians are concurrent; (b) the internal bisectors are concurrent; (c) the altitudes are concurrent. *Hint.* $AF:AE = b:c$.
2. A parallel to the side BC of the triangle ABC meets AB, AC in B', C'. Prove that the lines BC', $B'C$ meet on the median from A.
3. Given a triangle ABC and two points P, P'. The parallels through P' to PA, PB, PC meet BC, CA, AB in the points D', E', F', and the parallels through P to $P'A$, $P'B$, $P'C$ meet BC, CA, AB in D, E, F. Show that if the lines AD', BE', CF' are concurrent, the same holds for the lines AD, BE, CF.
4. With a point M of the side BC of the triangle ABC as center circles are drawn passing through B and C, respectively, meeting AB, AC again in N, P. For what position of M will the lines AM, BP, CN be concurrent? *Hint.* Consider the traces on BC of the mediators of AB, AC.

335. Theorem. *If the three lines joining three points marked on the sides of a triangle to the respectively opposite vertices are concurrent, the same is true of the isotomics of the given points.*

The proof is similar to the proof relative to reciprocal transversals (§ 321).

336. Definition. The points common to the two sets of concurrent lines considered (§ 335) may be referred to as *isotomic conjugates* for the triangle, or, more briefly, as *isotomic* for the triangle.

337. Theorem. *The incenter of a triangle is the Nagel point of the medial triangle.*

Let L be the point where the line IX, produced, meets the line AX_a (Fig. 92). From the two pairs of similar triangles IUX and UX_aI_a, AIL and AI_aX_a we have:

$$XI:X_aI_a = IU:UI_a, \quad IL:X_aI_a = IA:AI_a.$$

Now the right-hand sides of these two equalities are equal. Indeed, the two pairs of points A, U and I, I_a are harmonic (120); hence $XI = IL$, and $XA' = A'X_a$ (§ 160). Consequently the lines $A'I$ and AX_a are parallel.

The lines $A'I$, AX_a may be considered corresponding elements in the two homothetic figures formed by the triangle ABC and its medial triangle $A'B'C'$ (§ 98), for these two lines are parallel and pass through the homothetic points A, A'. Hence the line $A'I$ and its analogous lines $B'I$, $C'I$ meet in the Nagel point of $A'B'C'$, i.e., the Nagel point of $A'B'C'$ coincides with I.

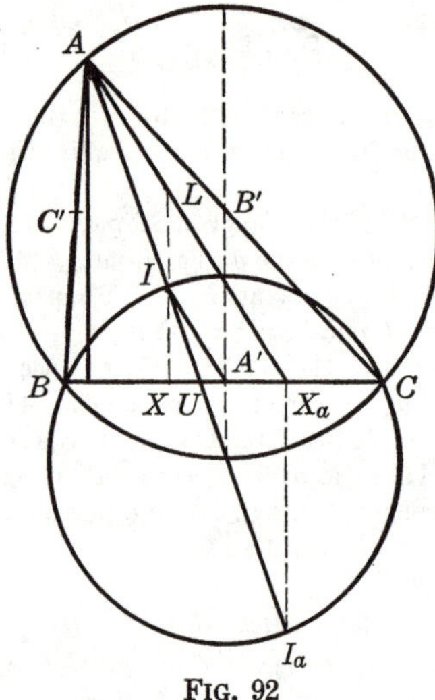

FIG. 92

338. COROLLARY I. *The Nagel point M of ABC and I, its incenter, are collinear with the centroid G of ABC, and* $2\ IG = GM$.

339. COROLLARY II. *The Nagel point of a triangle is the incenter of the anticomplementary triangle.*

340. Theorem. *If LMN is the cevian triangle of the point S for the triangle ABC, we have:*

$$\frac{SL}{AL} + \frac{SM}{BM} + \frac{SN}{CN} = 1.$$

The segments SL, AL are proportional to the distances of the points S, A from the side BC; hence:

$$SL:AL = \text{area } SBC:\text{area } ABC,$$

and similarly:

$$SM:BM = \text{area } SCA:\text{area } ABC, \quad SN:CN = \text{area } SAB:\text{area } ABC.$$

Adding these three equalities and observing that:

$$\text{area } SBC + SCA + SAB = \text{area } ABC,$$

we obtain the announced relation.

The formula is valid both in magnitude and in sign, whatever the position of the point S with respect to ABC may be.

341. COROLLARY. *We have:*

$$\frac{AS}{AL} + \frac{BS}{BM} + \frac{CS}{CN} = 2.$$

Indeed, we have:

$$\frac{AS}{AL} = \frac{AL - SL}{AL} = 1 - \frac{SL}{AL},$$

and similarly for the other two ratios. Adding the three we readily obtain the stated formula.

342. Theorem. *If LMN is the cevian triangle of the point S for the triangle ABC, we have:*

$$\frac{AS}{SL} = \frac{AM}{MC} + \frac{AN}{NB}.$$

If the parallels through C, B to ASL meet BM, CN in E, D, we have from similar triangles:

$$\begin{aligned} AS:EC &= AM:MC, \\ AS:DB &= AN:NB, \\ SL:EC &= BL:BC, \\ SL:DB &= LC:BC. \end{aligned}$$

Dividing the sum of the first two of these equalities by the sum of the last two we readily arrive at the stated formula.

343. Desargues' Theorem. *Given two triangles ABC, $A'B'C'$, if the three lines AA', BB', CC' meet in a point, S, the three points $P \equiv (BC, B'C')$, $Q \equiv (CA, C'A')$, $R \equiv (AB, A'B')$ lie on a line, s. Conversely.*

Cutting the three triangles SBC, SCA, SAB (Fig. 93) by the three transversals $PB'C', QC'A', RA'B'$, respectively, we have by Menelaus' theorem, both in magnitude and in sign:

$$\frac{SB'}{B'B}\cdot\frac{BP}{PC}\cdot\frac{CC'}{C'S} = -1, \quad \frac{SC'}{C'C}\cdot\frac{CQ}{QA}\cdot\frac{AA'}{A'S} = -1, \quad \frac{SA'}{A'A}\cdot\frac{AR}{RB}\cdot\frac{BB'}{B'S} = -1.$$

Multiplying these three equalities, we obtain, after simplification:

$$\frac{BP}{PC}\cdot\frac{CQ}{QA}\cdot\frac{AR}{RB} = -1,$$

which proves the proposition (§ 313).

Conversely, suppose that the three points P, Q, R are collinear. If the lines BB', CC' meet in S, it is required to prove that the line AA' also passes through S.

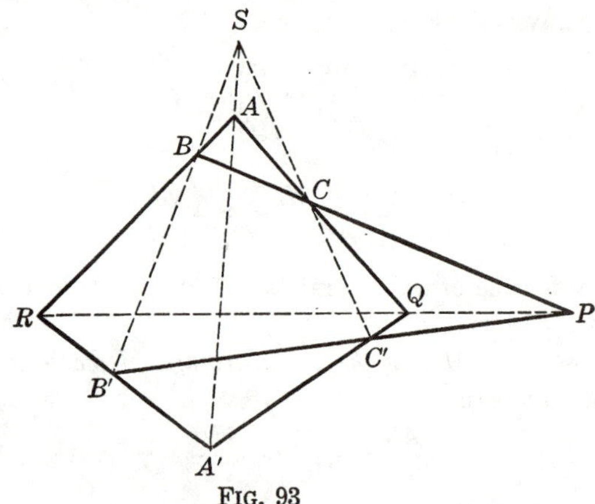

FIG. 93

By assumption, the lines $QR, CB, C'B'$ joining corresponding vertices in the two triangles QCC', RBB' are concurrent; hence by the direct theorem the three points $S \equiv (CC', BB')$, $A \equiv (QC, RB)$, $A' \equiv (C'Q, B'R)$ are collinear.

344. Definitions. The two triangles ABC, $A'B'C'$ are said to be *homological*, or *perspective.*

The point S is the *center*, and the line PQR the *axis*, *of homology, or of perspectivity* of the two triangles.

EXERCISES

1. If the incircle of the triangle ABC touches the sides BC, CA, AB in the points X, Y, Z, and M is the Gergonne point of ABC, show that $AM:MX = a(p-a):(p-b)(p-c)$; prove that $(AM:MX)\cdot(BM:MY)\cdot(CM:MZ) = 4\,R:r$. State and prove similar results for the excircles.
2. Prove that the sum of the reciprocals of the three cevians passing through the circumcenter of a triangle is equal to twice the reciprocal of the circumradius.
3. The lines AP, BQ, CR through the vertices of a triangle ABC parallel, respectively, to the lines OA', OB', OC' joining any point O to the points A', B', C' marked, in any manner whatever, on the sides of BC, CA, AB meet these sides in the points P, Q, R. Show that:
$$\frac{OA'}{AP} + \frac{OB'}{BQ} + \frac{OC'}{CR} = 1.$$
4. If the altitudes AD, BE, CF of the triangle ABC meet the circumcircle of ABC again in P, Q, R, show that we have $(AP:AD) + (BQ:BE) + (CR:CF) = 4$.
5. The parallels to the sides of a triangle ABC through the same point, M, meet the respective medians in the points P, Q, R. Prove that we have, both in magnitude and in sign, $(GP:GA) + (GQ:GB) + (GR:GC) = 0$.

6. The circumdiameters AP, BQ, CR of a triangle ABC meet the sides BC, CA, AB in the points K, L, M. Show that $(KP:AK) + (LQ:AQ) + (MR:AM) = 1$.
7. Show that the line joining the incenter of the triangle ABC to the midpoint of the segment joining A to the Nagel point of ABC is bisected by the median issued from A.
8. Use Desargues' theorem to prove that (a) the sides of the orthic triangle meet the sides of the given triangle in three collinear points; (b) the lines joining the points of contact of the sides of a triangle with the inscribed circle meet the corresponding sides of the triangle in three collinear points; (c) the lines joining the vertices of a triangle to the corresponding vertices of the tangential triangle are concurrent. State and prove other similar propositions.
9. By means of Desargues' theorem, construct, with ruler alone, the line joining a given point to the inaccessible point of intersection of two given lines.

SUPPLEMENTARY EXERCISES

1. Use Menelaus' theorem to prove the Simson line theorem (§ 282).
2. Show that (a) the twelve lines projecting from the vertices of a triangle and joining the points of contact of the opposite sides with the tritangent circles intersect by threes in eight points which may be grouped into four pairs of isotomic points for the triangle; (b) four of these points are the tritangent centers of the anticomplementary triangle of the given triangle.
3. If two triangles are symmetrical with respect to a point, show that the reciprocal transversals of the sides of one triangle with respect to the other are concurrent.
4. If the cevian triangle of a point for a given triangle has its sides antiparallel to the sides of the given triangle, show that the point coincides with the orthocenter of the given triangle.
5. Three parallel lines are cut by three parallel transversals in the points A, B, C; A', B', C'; A'', B'', C''. Show that the lines $B''C$, $C'A''$, AB' are concurrent.
6. Show that the lines joining the vertices of a given equilateral triangle to the images of a given point in the respectively opposite sides are concurrent.
7. The triangle $(Q) = DEF$ is inscribed in the triangle $(P) = ABC$, and the triangle $(R) = KLM$ is inscribed in (Q). Show that if any two of these triangles are perspective to the third, they are perspective to each other.
8. Through the vertices of a triangle ABC lines are drawn intersecting in O and meeting the opposite sides in D, E, F. Prove that the lines joining A, B, C to the midpoints of EF, FD, DE are concurrent.
9. If (Q) is the cevian triangle of a point M for the triangle (P), show that the triangle formed by the parallels through the vertices of (P) to the corresponding sides of (Q) is perspective to (P).
10. Prove that the lines joining the midpoints of three concurrent cevians to the midpoints of the corresponding sides of the given triangle are concurrent.
11. The parallels through the vertices of a given triangle (Q) to the respective sides of a second given triangle (P') form a triangle (P), and the parallels through the vertices of (P') to the respective sides of (Q) form a triangle (Q'). Show that if the triangles in one of the two pairs (P), (Q); (P'), (Q') are perspective, the same is true of the triangles of the second pair.

VII

HARMONIC DIVISION

345. Definition. Let the points C, D divide the segment AB internally and externally in the same ratio (§ 54). If we consider the segments on the line $ABCD$ as directed segments (§ 306) and agree to write the ratios in which the segment AB is divided by the points C, D in the following form:

$$AC:CB, \quad AD:DB,$$

the first ratio will be positive and the second negative. Hence we have, both in magnitude and in sign:

(a) $$(AC:CB) + (AD:DB) = 0,$$

(b) $$\frac{AC}{CB} = -\frac{AD}{DB}, \quad \text{or} \quad \frac{AC}{CB}:\frac{AD}{DB} = -1.$$

The formula (b) is sometimes represented by the symbol $(ABCD) = -1$, and the four points involved are said to form a *harmonic range.*

346. Theorem. *If the segments of the harmonic range* $(ABCD) = -1$ *are measured from the midpoint* O *of the segment* AB, *we have, both in magnitude and in sign:*

(c) $$OC \cdot OD = OA^2.$$

From the formula (b) we have:

(d) $$\frac{AC + CB}{AC - CB} = \frac{AD - DB}{AD + DB}.$$

Now making use of the formulas in §§ 306, 307, we have, both in magnitude and in sign:

$$\text{(e)} \quad \begin{cases} AC + CB = AB = AD + DB = 2\,AO, \\ AC - CB = (AO + OC) - (CO + OB) = 2\,OC, \\ AD - DB = (AO + OD) - (DO + OB) = 2\,OD. \end{cases}$$

Substituting these values into (d) we obtain (c).

347. Converse Theorem. *If O is the midpoint of the segment AB, and if C, D are two points of the line AB such that $OC \cdot OD = OA^2$, both in magnitude and in sign, then $(ABCD)$ is a harmonic range.*

Indeed, from the given formula, by making use of the relations (e), we readily obtain (d), from which follows (b).

348. Theorem. *If, in the harmonic range $(ABCD)$, all segments are measured from one point of the range, say B, we have, both in magnitude and in sign:*

$$\text{(f)} \qquad \frac{2}{BA} = \frac{1}{BC} + \frac{1}{BD}.$$

The formula (b) (§ 345) may be put in the form

$$\frac{AB + BC}{CB} = -\frac{AB + BD}{DB};$$

hence:

$$-1 + (BA:BC) = 1 - (BA:BD),$$

and (f) follows immediately.

349. Converse Theorem. If A, B, C, D are four collinear points and the directed segments BA, BC, BD satisfy the relation (f), the points form a harmonic range.

This converse proposition is proved by taking the above steps in reverse order.

350. Problem. *Find a pair of points which separates harmonically each of two given pairs of collinear points.*

The two given pairs of points A, B; C, D (Fig. 94) and any point X outside the given line determine two circles XAB, XCD meeting in a second point Y. The circle having for center the trace R of XY on AB and for radius the length of the tangent RZ from R to the circle XAB meets the line AB in the two required points P, Q.

Indeed, we have:

$$RP^2 = RZ^2 = RX \cdot RY = RA \cdot RB = RC \cdot RD;$$

hence the points P, Q satisfy the two conditions of the problem (§ 346).

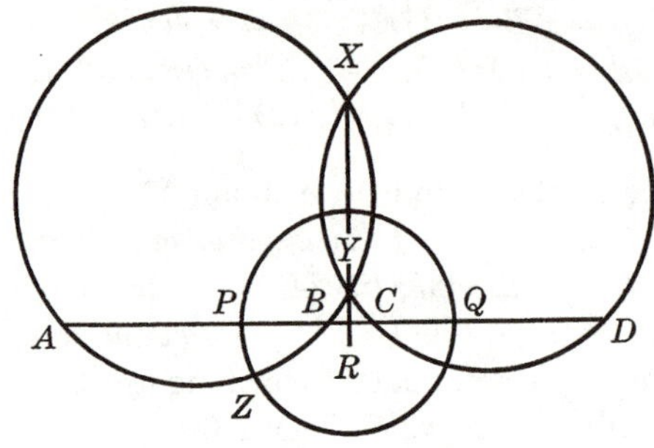

FIG. 94

DISCUSSION. In order that the problem may have a solution it is necessary that the point R shall lie outside the circles XAB, XCD. This will happen when the segments AB, CD lie entirely outside one another, or when one lies entirely inside the other, but not when the two segments overlap. The line XY will be parallel to AB if the two segments AB, CD have the same midpoint.

When the problem has a solution, the solution is unique.

EXERCISES

1. Show that if the four points O, A, C, D are collinear, and $OA^2 = OC \cdot OD$, both in magnitude and in sign, the symmetric of A with respect to O coincides with the harmonic conjugate of A with respect to C, D.
2. Given the parallelogram $MDOM'$, the vertex O is joined to the midpoint C of MM'. If the internal and external bisectors of the angle COD meet MD in A and B, show that $MD^2 = MA \cdot MB$.
3. If the points C, D divide harmonically the segment AB, whose midpoint is O, prove that:
$$OC^2 + OD^2 = CD^2 + 2\,OA^2.$$
4. If $(ABCD) = -1$, and A', B' are the harmonic conjugates of D with respect to the pairs of points A, C and B, C, prove that $(A'B'CD) = -1$. *Hint.* Use formula (f) (§ 348).
5. If $(ABCD) = -1$, and M is the midpoint of AB, show that:
$$\frac{1}{CA \cdot CB} + \frac{1}{DA \cdot DB} = \frac{1}{MA \cdot MB}.$$
6. If $(ACBD) = -1$, show that:
$$\frac{1}{BC} = \frac{1}{AB} + \frac{1}{AD} + \frac{1}{CD}.$$
7. If $(ABCD) = -1$, show that $DA \cdot DB = DC \cdot DO$, where O is the midpoint of the segment AB.

8. If A, B, C, D are four points on a straight line taken in order, show that the locus of a point at which the segments AB, CD subtend equal angles is a circle having for the ends of a diameter the pair of points separating harmonically each of the two pairs of points A, D; B, C.

351. Theorem. *Given* $(ABCD) = -1$ *and a point* O *outside the line* AB, *if a parallel through* B *to* OA *meets* OC, OD *in* P, Q, *we then have* $PB = BQ$.

From the two pairs of similar triangles OAC and PCB, OAD and BQD (Fig. 95) we have:

$$\text{(p)} \qquad AO:PB = AC:CB, \quad AO:BQ = AD:BD.$$

Now since $(ABCD) = -1$, the right-hand sides of these two equalities are equal; hence $PB = BQ$.

352. Converse Theorem. *Given four collinear points* A, B, C, D *and a point* O *outside that line, if the parallel through* B *to* OA *meets* OC, OD *in* P, Q, *and* $PB = BQ$, *then* $(ABCD) = -1$.

Indeed, from similar triangles we have the relations (p), and since $PB = BQ$, the left-hand sides of the equalities (p) are equal; therefore the right-hand sides are equal; hence $(ABCD) = -1$.

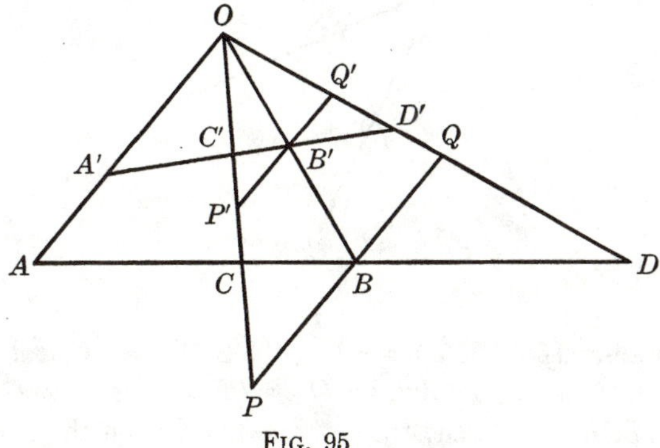

FIG. 95

353. Theorem. *Given* $(ABCD) = -1$ *and a point* O *outside the line* AB, *any transversal cuts the four lines* OA, OB, OC, OD *in four harmonic points.*

Let $A'B'C'D'$ be the transversal (Fig. 95) and let the parallels to OA through B and B' meet OC, OD in P, Q and P', Q'. From similar triangles we have:

$$PB:P'B' = OB:OB' = BQ:B'Q'.$$

Now since $(ABCD) = -1$, we have (§ 351) $PB = BQ$; hence $P'B' = B'Q'$, and therefore $(A'B'C'D') = -1$ (§ 352).

354. Definitions. Four concurrent lines OA, OB, OC, OD which are cut by one transversal, and therefore by every transversal (§ 353), in four harmonic points, are said to form a *harmonic pencil;* each of the four lines is called a *ray* of the pencil, and O is called its *center*, or *vertex;* OA and OB, OC and OD are said to be the *pairs of conjugate rays*, and OC, OD are said to be *harmonically conjugate* with respect to OA, OB, or *harmonically separated* by OA, OB. The pencil is often represented by $O(ABCD) = -1$, or by $O(AB, CD)$.

355. Theorem. *If two conjugate rays of a harmonic pencil are rectangular, they are the bisectors of the angles formed by the other two rays of the pencil.*

Given $O(ABCD) = -1$ (Fig. 96) let a parallel to OA meet OB, OC, OD in F, G, H. We have (§ 351) $GF = FH$, and if OA is perpendicular to OB, the triangle OGH is isosceles; hence the proposition.

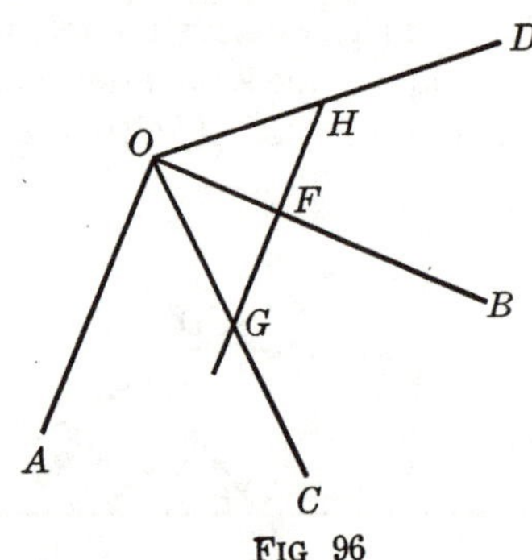

FIG 96

356. Theorem. *If* $(ABCD) = -1$, $(A'B'C'D') = -1$, *and the lines* AA', BB', CC' *meet in a point, say* O, *then* DD' *passes through* O.

Since $(ABCD) = -1$, the trace of the line OD on the line $A'B'$ is the harmonic conjugate of C' with respect to A', B', and therefore coincides with D'; hence the proposition.

357. COROLLARY. *If* $(ABCD) = -1$, $(AB'C'D') = -1$, *and the lines* AB, AB' *are distinct, the lines* BB', CC', DD' *are concurrent.*

EXERCISES

1. Construct the harmonic conjugate of the ray OC with respect to the rays OA, OB.
2. In the triangle ABC prove that $A(OHII_b) = -1$.
3. Show that on a given line there is, in general, one point such that the lines joining that point to two given points form an angle which is bisected by the given line.

4. Given $(ABCD) = -1$. The perpendiculars to CD at the points C, D meet a transversal through A in P, Q. Show that AB bisects the angle PBQ.
5. Prove that (a) the parallels through a given point to the four lines of a harmonic pencil also form a harmonic pencil; (b) the perpendiculars from a given point to the four rays of a harmonic pencil also form a harmonic pencil; (c) the symmetrics, with respect to a given axis, of the four rays of a harmonic pencil also form a harmonic pencil. *Hint.* Prove by superposition.
6. Show that the lines joining a point of a circle to the ends of a chord divide harmonically the diameter perpendicular to that chord.
7. In the parallelogram $ABCD$, AE is drawn parallel to BD; show that $A(ECBD) = -1$.
8. If A', B', C' are the midpoints of the sides of the triangle ABC, prove that $A'A$ is the harmonic conjugate of $A'C$ with respect to $A'B'$, $A'C'$.
9. AD, AA' are the altitude and the median of the triangle ABC; the parallels through A' to AB, AC meet AD in P, Q; show that $(ADPQ) = -1$.
10. With the usual notations for the triangle ABC, if DF meets BE in K, show that $(BHKE) = -1$; also if EF meets BC in M, show that $(BCDM) = -1$.
11. Through a given point to draw a transversal so that its three points of intersection with three given lines (concurrent or not) shall form with the given point a harmonic range.
12. The tangent to a circle at the point C meets the diameter AB, produced, in T; prove that the other tangent from T to the circle is divided harmonically by CA, CB, CT, and its point of contact.
13. The circle having for diameter the median AA' of the triangle ABC meets the circumcircle in L: show that $A(LDBC) = -1$, where AD is the altitude.

SUPPLEMENTARY EXERCISES

1. Given three collinear points A, B, C, the points A', B', C' are constructed so that $(AA'BC) = -1$, $(BB'CA) = -1$, $(CC'AB) = -1$. Prove that we also have $(A'AB'C') = -1$, $(B'BC'A') = -1$, $(C'CA'B') = -1$.
2. Given two fixed lines OX, OY, a variable secant through the fixed point A of OX meets OY in B, and the circle (B, BO) meets AB in C, D. Show that the locus of the point M such that $(AMCD) = -1$ is a circle.
3. Two variable points P, Q are taken on the base BC of the triangle ABC so that $(BCPQ) = -1$. Show that the locus of the circumcenter of the triangle APQ is a straight line.
4. The sides AB, AC intercept the segments DE, FG on the parallels to the side BC through the tritangent centers I and I_a. Show that:

$$\frac{2}{BC} = \frac{1}{DE} + \frac{1}{FG}.$$

5. A secant through the vertex A of the parallelogram $ABCD$ meets the diagonal BD and the sides BC, CD in the points E, F, G. Show that:

$$\frac{1}{AE} = \frac{1}{AF} + \frac{1}{AG}.$$

VIII

CIRCLES

A. INVERSE POINTS

358. Definition. Two points collinear with the center of a given circle and the product of whose distances from the center is equal to the square of the radius are said to be *inverse points* with respect to, or for, the circle (Fig. 97).

Two inverse points lie on the same side of the center of the circle.

Of two inverse points one lies inside and the other outside the circle.

If a point is taken on the circle, its inverse coincides with the point considered.

359. Theorem. *Two inverse points divide the corresponding diameter harmonically. Conversely.*

Let P, Q (Fig. 97) be the two inverse points, A, B the ends of the diameter passing through P, Q, and O the center of the circle. We have, by hypothesis, $OP \cdot OQ = OB^2$; hence $(ABPQ) = -1$ (§ 346).

Conversely, if $(ABPQ) = -1$, we have (§ 347) $OP \cdot OQ = OB^2$; hence the points P, Q are inverse for the circle.

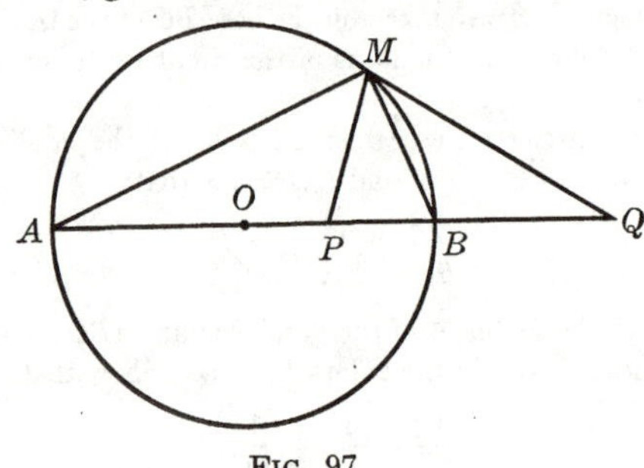

FIG. 97

360. Theorem. *The ratio of the distances of a variable point of a circle from two given inverse points is constant.*

Let the inverse points P, Q lie on the diameter AB (Fig. 97). The range $(ABPQ)$ is harmonic (§ 359), hence also the pencil $M(ABPQ)$, where M is any point on the circle. Now angle $AMB = 90°$; hence MA, MB bisect the angle PMQ (§ 355) and therefore:

$$MP:MQ = PB:BQ.$$

361. Definition. The triangle formed by the feet of the perpendiculars from a given point upon the sides of a given triangle is called the *pedal triangle* of the point with respect to, or for, the given triangle.

362. Theorem. (*a*) *The pedal triangles, for a given triangle, of two points inverse for the circumcircle of the given triangle are similar.*

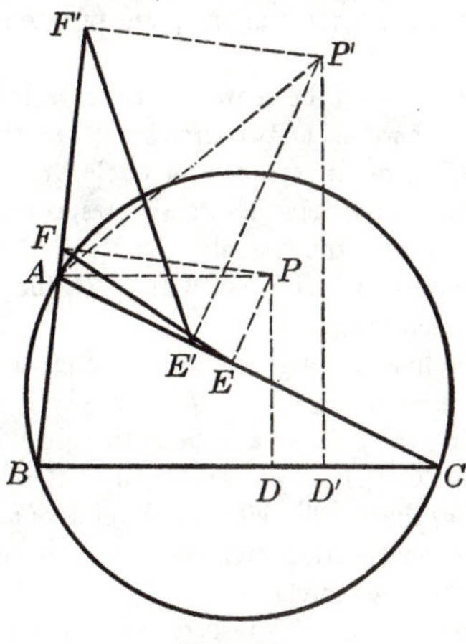

Fig. 98

(*b*) *Conversely. If the pedal triangles of two points for a given triangle are similar, the two points are inverse with respect to the circumcircle of the given triangle.*

Let $DEF, D'E'F'$, be the pedal triangles of the points P, P' with respect to the triangle ABC. In the two circles having AP, AP' for diameters the two chords $EF, E'F'$ (Fig. 98) subtend the same angle A; hence the two isosceles triangles having $EF, E'F'$ for bases and the centers of the respective circles for opposite vertices are similar. Thus:

$$EF:E'F' = AP:AP'.$$

Similarly:

$$FD:F'D' = BP:BP', \quad DE:D'E' = CP:CP'.$$

Now if P, P' are inverse points for the circle ABC, the right-hand sides of these proportions are equal (§ 360); hence the same holds for the left-hand sides, which proves the direct proposition.

Conversely, if $DEF, D'E'F'$ are similar, the left-hand sides of these propositions are equal; hence the same holds for the right-hand sides; consequently, by locus 11 (§ 11) the points P, P' divide harmonically a diameter of the circle ABC, which proves the converse proposition.

EXERCISES

1. Prove that two pairs of inverse points with respect to the same circle are cyclic, or collinear.
2. From a point P outside a given circle, center O, the tangents are drawn to the circle. Show that P is the inverse of the point of intersection of OP with the chord of contact.
3. If the circle (B) passes through the center A of the circle (A), and a diameter of (A) meets the common chord of the two circles in F and the circle (B) again in G, show that the points F, G are inverse for the circle (A).
4. Show that (a) if P, Q are two inverse points with respect to a circle, center O, and M is any point on the circle, the triangles OMP, OMQ are similar; (b) if R, S is another pair of inverse points collinear with P, Q, the angles PMR, QMS are either equal or supplementary.
5. Show that (a) the two lines joining any point of a circle to the ends of a given chord meet the diameter perpendicular to that chord in two inverse points; (b) the two lines joining any point of a circle to two given inverse points meet the circle again in the ends of a chord perpendicular to the diameter containing the given inverse points; (c) the four lines joining the ends of a chord of a circle to two inverse points located on the diameter perpendicular to the chord considered intersect in pairs on the given circle.
6. If P, Q are two inverse points with respect to a circle and CD is a chord perpendicular to the diameter containing P, Q, the angles which the segments CP, DQ subtend at any point M of the circle are either equal or supplementary.

B. ORTHOGONAL CIRCLES

363. Definition. Two circles are said to be *orthogonal* if the square of their line of centers is equal to the sum of the squares of their radii.

364. Theorem. (*a*) *In two orthogonal circles the two radii passing through a common point of the two circles are rectangular.*

(*b*) *Conversely. If the two radii passing through a point common to two circles are rectangular, the two circles are orthogonal.*

365. Theorem. (*a*) *If two circles are orthogonal the circle having for diameter their line of centers passes through their common points.*

(*b*) *Conversely. If the circle having for diameter the line of centers of two given circles passes through the points common to these circles, the two given circles are orthogonal.*

366. Theorem. *If two circles are orthogonal, the radius of one circle passing through a point common to the two circles is tangent to the second circle.*

Let C be a point common to the two orthogonal circles (A), (B). The tangent to (B) at C is perpendicular to the radius BC; hence this tangent coincides with the radius AC of (A) (§ 364 a).

367. Converse Theorem. *Given two intersecting circles, if the radius of one circle passing through a point common to the two circles is tangent to the second circle, the two circles are orthogonal.*

The two radii passing through the common point considered are rectangular; hence the proposition (§ 364 b).

368. Theorem. *If two circles are orthogonal, any two points of one of them collinear with the center of the second circle are inverse for that second circle.*

Let (A), (B) (Fig. 99) be two orthogonal circles, let E, F be two points of (B) collinear with the center A of (A), and let C, D be the ends of the diameter AEF of (A).

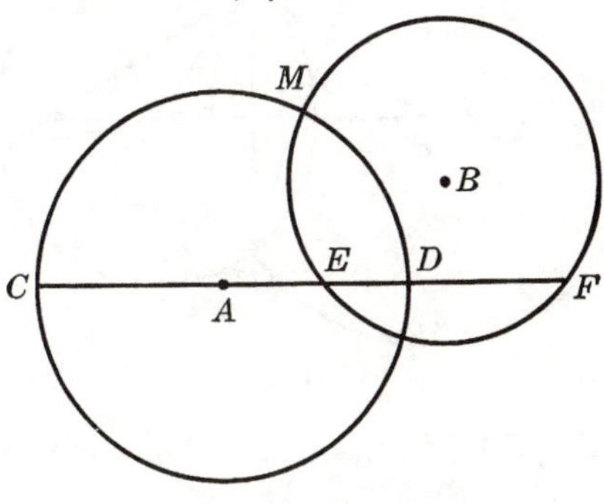

FIG. 99

If M is a point common to (A) and (B) we have $AM^2 = AE \cdot AF$. But $AM = AC = AD$; hence (§ 347) $(CDEF) = -1$, which proves the proposition (§ 359).

369. Converse Theorem. *If two points of one circle are inverse for a second circle, the two circles are orthogonal.*

By hypothesis, the two points E, F of (B) (§ 368) are collinear with A (Fig. 99) and are inverse with respect to (A); therefore:

$$AC^2 = AE \cdot AF; \quad \text{hence} \quad AM^2 = AE \cdot AF.$$

Thus AM is tangent to (B); hence the circles are orthogonal (§ 367).

370. Corollary. *If $(ABCD) = -1$, the circle on AB as diameter is orthogonal to any circle through C and D.*

371. Problem. *Through two given points to pass a circle orthogonal to a given circle.*

The two given points A, B and the inverse A' of one of them, say A, with respect to the given circle (O) determine the required circle (S).

The problem has, in general, one solution, for the inverse B' of B will also lie on (O) (§ 368).

If the points A, B are collinear with the center O of (O), the problem either has an infinite number of solutions, or is impossible, according as the points A, B are, or are not, inverse for (O).

372. Theorem. *The two lines joining the points of intersection of two orthogonal circles to a point on one of the circles meet the other circle in two diametrically opposite points.*

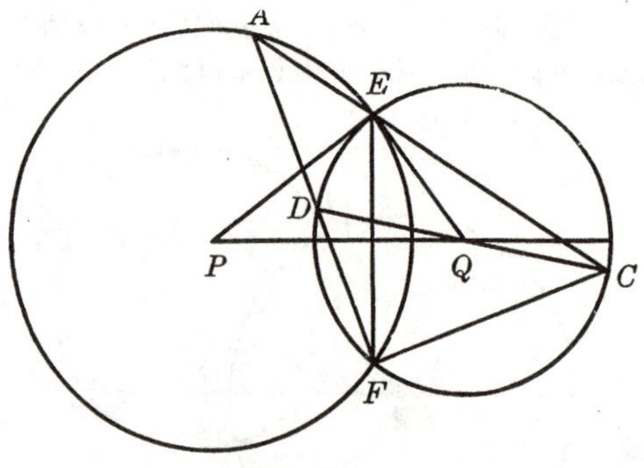

Fig. 100

Let the lines EA, FA joining the points of intersection E, F of the two orthogonal circles (P), (Q) to the point A of (P) meet (Q) again in C, D (Fig. 100). We have:

$$\angle EAF = \tfrac{1}{2}\, EPF = EPQ, \quad \angle ECF = \tfrac{1}{2}\, EQF = EQP;$$

hence:

$$\angle EAF + ECF = EPQ + EQP = 90^\circ,$$

and therefore the triangle CAF is right-angled at F. Thus CD subtends a right angle at F, which proves the proposition.

EXERCISES

1. With a given point as center to draw a circle orthogonal to a given circle.
2. Show that if the quadrilateral having for vertices the centers and the points of intersection of two circles is cyclic, the circles are orthogonal.
3. Show that if the sum of the angles in the segments of two intersecting circles on opposite sides of the common chord is equal to 90° or 270°, the two circles are orthogonal.
4. Through a given point to draw a circle orthogonal to two given circles.
5. Show that the product of the radii of two orthogonal circles is equal to one half the product of the common chord and the line of centers of the two circles.
6. Draw a circle orthogonal to a given circle and including within it a third of the circumference of the given circle.
7. (a) With the vertices of a given triangle ABC as centers, three circles are described so that each is orthogonal to the other two. Find their radii in terms of the sides of the triangle. (b) Show that the squares of the radii of these circles are respectively equal to $AH \cdot AD$, $BH \cdot BE$, $CH \cdot CF$.
8. Show that (a) in a triangle ABC the circles on AH and BC as diameters are orthogonal; (b) the circle IBC is orthogonal to the circle on I_bI_c as diameter.
9. Show that the locus of the centers of the circles passing through a given point and orthogonal to a given circle is a straight line.
10. Show that if AB is a diameter and M any point of a circle, center O, the two circles AMO, BMO are orthogonal.
11. Through a given point outside a given circle to draw a secant so that the product of the distances of its points of intersection with the circle to the diameter passing through the given point shall have a given value.
12. Show that (a) given two perpendicular diameters of two orthogonal circles, the lines joining an end of one of these diameters to the ends of the other pass through the points common to the two circles; (b) the ends of two perpendicular diameters of two orthogonal circles form an orthocentric group. Conversely.

C. POLES AND POLARS

373. Theorem. *The harmonic conjugate of a fixed point with respect to a variable pair of points which lie on a given circle and are collinear with the fixed point, describes a straight line.*

On the given circle (O) take two points E, F collinear with the given fixed point P (Fig. 101 a and b). Let M be the harmonic conjugate of P for E, F, and Q the foot of the perpendicular from M upon the diameter AB of (O) passing through P.

The circle PMQ has PM for diameter, and the points E, F divide PM harmonically; hence the circles (O) and PQM are orthogonal (§ 369). Consequently the points P, Q in which the circle PMQ is

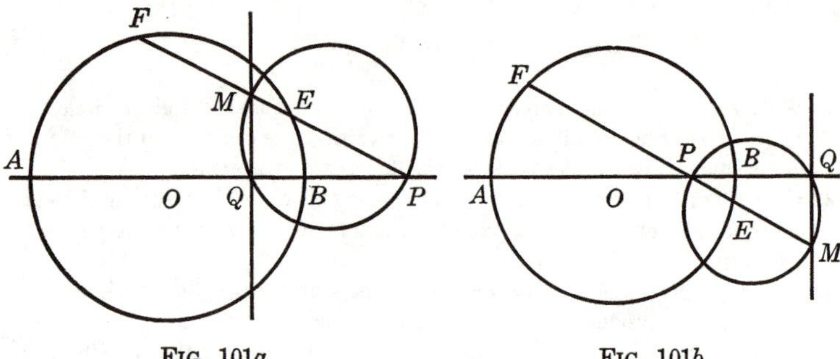

FIG. 101a FIG. 101b

met by the diameter AB of (O) are inverse with respect to (O), i.e., the point Q is fixed, regardless of the choice of the pair of points E, F of (O), which proves the proposition.

374. Definition. The line MQ (§ 373) is said to be the *polar line* of the point P with respect to, or for, the circle, and the point P is said to be the *pole* of the line MQ.

375. CONSEQUENCES. (*a*) *The polar of a given point for a circle is the perpendicular to the diameter passing through the given point, at the inverse of this point for the circle.*

(*b*) *The pole of a given line with respect to a circle is the inverse of the foot of the perpendicular from the center of the circle upon the given line.*

376. *Remark I.* The polar of a point on the circle is the tangent at that point to the circle, and the pole of a tangent is its point of contact.

For the inverse of a point on the circle coincides with the point considered.

377. *Remark II.* Every point in the plane has a polar for a given circle, except the center of the circle.

Every line in the plane has a pole for a given circle, except the lines passing through the center of the circle.

378. *Remark III.* If the pole lies inside the circle, the polar does not cut the circle.

If the pole lies outside the circle, its polar coincides with the line joining the points of contact of the tangents from the pole to the circle.

Indeed, the trace P' (Fig. 102) of the chord of contact TT' on the diameter OP passing through the pole P is the inverse of P with respect to the circle, inasmuch as in the right triangle OTP we have $OT^2 = OP \cdot OP'$.

379. Theorem. *If the polar of the point P passes through the point Q, the polar of Q passes through the point P.*

If the line PQ meets the circle in two points C, D, the proposition is almost obvious. Indeed, if the polar of P passes through Q, we

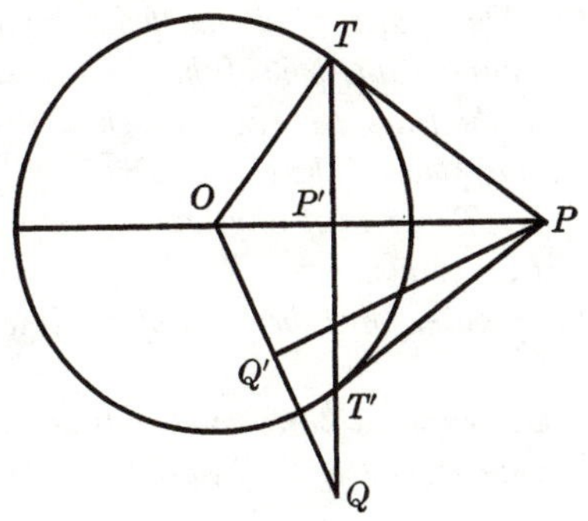

FIG. 102

have $(PQCD) = -1$ (§ 373); hence P is the harmonic conjugate of Q for C, D, and the polar of Q will pass through P.

The following proof is valid, whether or not the line PQ meets the circle.

The two points P, Q and the their inverses P', Q' for the given circle (O, R) are cyclic, for $OP \cdot OP' = R^2 = OQ \cdot OQ'$; hence the angles $PP'Q$, $PQ'Q$ of the cyclic quadrilateral $PQQ'P'$ (Fig. 102) are equal. But QP' is the polar of P, since it passes through Q and through the inverse P' of P; hence angle $PP'Q = 90°$, and therefore angle $PQ'Q = 90°$. Thus the line PQ' passes through the inverse Q' of Q and is perpendicular to OQ, i.e., PQ' is the polar of Q(§ 375), and since PQ' passes through P, the proposition is proved.

380. Definition. Two points such that the polar of one passes through the other are called *conjugate points* with respect to the circle.

A given point has an infinite number of conjugate points, namely, all the points of its polar.

If two conjugate points are collinear with the center of the circle, they are inverse points for the circle.

381. Theorem. *If the pole of the line p lies on the line q, the pole of the line q lies on the line p.*

Let P, Q be the poles of p, q. By hypothesis P lies on q, i.e., the polar q of Q passes through P; hence (§ 379) the polar p of P passes through Q.

382. Definition. Two lines such that the pole of one lies on the other are said to be *conjugate lines* for the circle.

A given line has an infinite number of conjugate lines, namely, all the lines through the pole of the given line.

383. Theorem. (*a*) *The polars of all the points of a given line pass through a fixed point, namely, the pole of the given line.*

(*b*) *The poles of all the lines passing through a given point lie on a straight line, namely, the polar of the given point.*

384. Corollary. (*a*) *The pole of any line is the point of intersection of the polars of two of its points.*

(*b*) *The polar of any point joins the poles of two lines passing through the point.*

385. Theorem. *If two conjugate lines meet outside the circle, they are harmonic conjugates with respect to the two tangents from their point of intersection to the circle.*

Let the conjugate lines PQ, PR meet the chord of contact TT' of the tangents from P to the circle in the points Q, R (Fig. 103). The pole of PR lies on PQ (§ 382) and on the polar TT' of P (§ 379), since PR passes through P; hence Q is the pole of PR. Thus (§ 373) $(TT'QR) = -1$, and therefore the pencil $P(TT'QR)$ is harmonic.

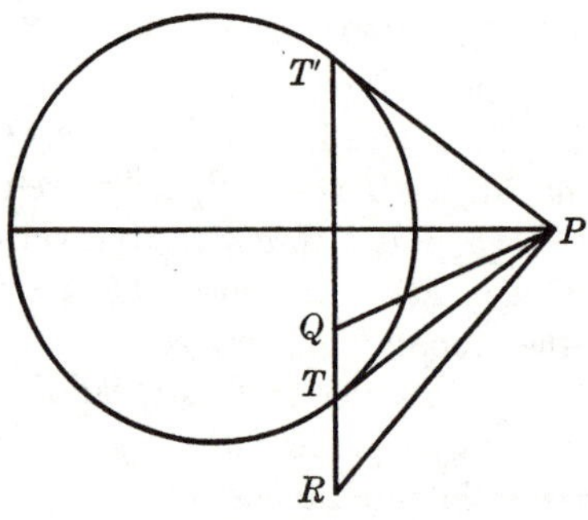

Fig. 103

386. Theorem. *If two circles are orthogonal, any two diametrically opposite points of one circle are conjugate with respect to the other circle.*

The line PC (Fig. 104) joining P to the center C of (C) meets (C') again in the inverse S of P with respect to (C) (§ 368); hence the polar of P with respect to (C) is perpendicular to PC at S, and therefore meets (C') again in the diametric opposite R of P on (C').

387. Converse Theorem. *If two diametrically opposite points of one circle are conjugate with respect to another circle, the two circles are orthogonal.*

By assumption, the polar of P for (C) passes through R (Fig. 104); it is also perpendicular to PC (§ 375); hence this polar coincides with

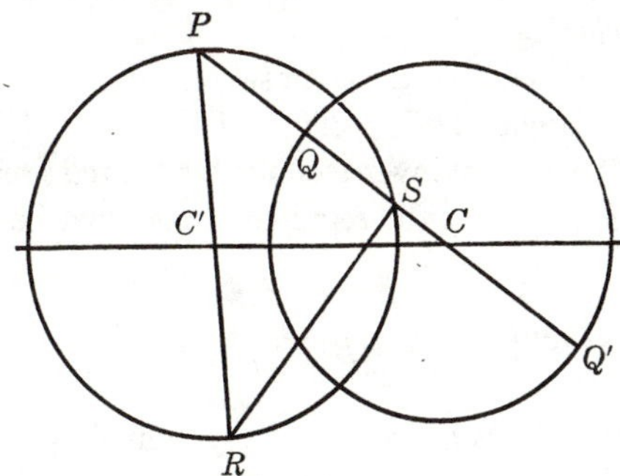

FIG. 104

RS, i.e., the point S is the inverse of P with respect to (C) (§ 375); hence the two circles are orthogonal (§ 369).

388. Definition. A triangle is said to be *self-conjugate*, or *conjugate*, or *self-polar*, or *polar* with respect to a given circle, when each side of the triangle is the polar of the opposite vertex.

It is readily seen that given the circle (O), an infinite number of such triangles may be constructed. Let P be an arbitrary point in the plane, and Q an arbitrary point on the polar p of P for (O). The polar q of Q will pass through P (§ 379) and meet p in the third vertex R of the required triangle PQR. The lines RQ, RP are the polars of P, Q, by construction, and the pole of PQ coincides with R (§ 384 a).

389. *Remark.* Of the three vertices of a polar triangle one lies inside the circle, and the other two vertices lie outside the circle.

If the vertex P of the triangle PQR, polar to the circle (O), lies inside of (O), the other two vertices Q, R lie on the polar of P, and therefore outside of (O) (§ 378).

If P is taken outside of (O), the polar p of P cuts (O) in two points, say, E and F (§ 378). The vertices Q, R are conjugate with respect to (O), hence they are harmonically separated by E, F, and therefore one of them lies inside and the other outside of (O).

390. Theorem. *If a triangle is polar with respect to a circle, the center of the circle coincides with the orthocenter of the triangle.*

If the triangle PQR is polar for the circle (O), the perpendiculars from P, Q, R upon their polars QR, RP, PQ pass through the center of (O) (§ 375); hence the proposition.

391. Problem. *Given a triangle, find a circle with respect to which the triangle is self-polar.*

The center of the required circle must coincide with the orthocenter H of the given triangle ABC (§ 390). The points A, B, C and the feet D, E, F will be pairs of inverse points for the required circle; hence the square of the radius of the required circle must be equal to each of the products:

$$(1) \qquad HA \cdot HD, \quad HB \cdot HE, \quad HC \cdot HF,$$

and these three products are equal in any triangle (§ 177). But the pairs of inverse points A, D; B, E; C, F must lie on the same side of the center H of the required circle (§ 358), which condition is satisfied only if ABC is obtuse-angled.

Thus the problem has no solution if ABC is acute-angled. If ABC is obtuse-angled, the problem has a unique solution, the center of the required circle (H) being the orthocenter H of ABC, the square of the radius of (H) being equal to one of the products of (1).

392. Definition. The circle (H) (§ 391) is called the *polar circle*, or the *conjugate circle*, of the triangle ABC.

EXERCISES

1. Show that the angle between two lines is equal to the angle which the segment determined by their poles with respect to a given circle subtends at the center of the circle.

2. Show that the locus of the point whose polars with respect to two given circles are orthogonal is the circle described on the line of centers of the given circles as diameter.

3. (a) TP, TQ are the tangents at the extremities of a chord PQ of a circle. The tangent at any point R of the circle meets PQ in S; prove that TR is the polar of S. (b) Given two fixed points R, S, an arbitrary circle (O) is drawn tangent to RS at R, and through S an arbitrary secant is drawn meeting (O) in P, Q. If the tangents to (O) at P, Q meet RS in U, V, show that the sum of the reciprocals of RU and RV is constant, independently of the choice of (O) and of the secant RPQ.

4. If the vertices of a triangle (P) have for their polars, with respect to a circle (O), the sides of the triangle (Q), show that the vertices of the triangle (Q) have for their polars the sides of the triangle (P).
5. Through a given point to draw a line passing through the inaccessible center of a given circle.
6. A variable circle passes through a fixed point C, and two fixed points H, K are conjugate with respect to the circle; prove that the center of the circle lies on a straight line perpendicular to the line joining C to the midpoint of HK.
7. On a given line to find two points which shall be equidistant from a given point on the line, and conjugate with respect to a given circle.
8. Show that (a) the two poles of the common chord of two orthogonal circles with respect to these circles coincide with the centers of the given circles; (b) if AB, CD are two harmonic segments, the harmonic conjugate of the midpoint of AB with respect to the pair C, D coincides with the harmonic conjugate of the midpoint of CD with respect to the pair A, B.
9. Show that the polar circle of a triangle cuts the sides of the triangle harmonically.
10. Prove that the pole of the line joining two conjugate points is the orthocenter of the triangle determined by the two given points and the center of the circle.
11. If X is any point on the side BC of the triangle ABC, show that the circle on AX as diameter is orthogonal to the polar circle of the triangle.
12. A line meets two given circles in two pairs of points. Find two points on this line such that each shall be the point of intersection of the two polars of the other with respect to the two given circles.
13. An infinite number of triangles may be drawn polar with respect to a given circle and having a given point for a common vertex. Show that (a) the centroids of these triangles lie on a straight line; (b) their orthocenter is fixed; (c) their circumcenters lie on a fixed line.
14. Given two chords AB, CD of a circle and their midpoints P, Q, prove that if AB bisects the angle CPD, then CD bisects the angle AQB.
15. A point M of a given circle (O) is projected upon two rectangular diameters u, v in the points A, B. Show that the line joining the projections upon u, v of the pole of the line AB for (O) is tangent to the circle.
16. If P, Q are any two points and PM, QN are the perpendiculars from each upon the polar of the other with respect to the same circle, center O, we have $OP:PM = OQ:QN$.

SUPPLEMENTARY EXERCISES

1. Show that (a) the two perpendiculars to the mediator of a side of a triangle at the points where this mediator cuts the other two sides of the triangle pass through vertices of the tangential triangle of the given triangle; (b) the line AI meets the sides XY, XZ in two points P, Q inverse with respect to the incircle $(I) = XYZ$, and the perpendiculars to AI at P, Q pass through the vertices B, C of the given triangle ABC.

2. Show that the perpendiculars dropped from the orthocenter of a triangle upon the lines joining the vertices to a given point meet the respectively opposite sides of the triangle in three collinear points.
3. Show that the perpendiculars from the vertices of a triangle to the lines joining the midpoints of the respectively opposite sides to the orthocenter of the triangle meet these sides in three points of a straight line perpendicular to the Euler line of the triangle.
4. Show that the perpendiculars to the internal bisectors of a triangle at the incenter meet the respective sides in three points lying on a line perpendicular to the line joining the incenter to the circumcenter of the triangle.
5. If A', B', C' are the traces of a transversal on the sides BC, CA, AB of a triangle ABC whose orthocenter is H, prove that the perpendiculars from A, B, C upon the lines HA', HB', HC', respectively, are concurrent in a point on the perpendicular from H to $A'B'C'$.
6. Show that the polars of a harmonic range form a harmonic pencil, and conversely.
7. Through the point of intersection of the tangents DB, DC to the circumcircle (O) of the triangle ABC a parallel is drawn to the line touching (O) at A. If this parallel meets AB, AC in E, F, show that D bisects EF.
8. Show that (a) if a circle (P) touches the sides AB, AC of the triangle ABC in E, F, the line EF, the perpendicular from the center P of (P) to BC, and the median of ABC issued from A are concurrent; (b) if X, Y, Z and X_a, Y_a, Z_a are the points of contact of the sides BC, CA, AB of the triangle ABC with the incircle (I) and the excircle (I_a), the points of intersection of YZ, Y_aZ_a with radii IX, I_aX_a, respectively, are collinear with A and the midpoint A' of BC.

D. CENTERS OF SIMILITUDE

393. Converse Problem. Given a circle, a homothetic center, and a homothetic ratio we constructed a circle homothetic to the given circle (§ 47). We found that the ratio of the radii of the two circles is equal to the given homothetic ratio, and that the homothetic center divides the line of centers in the given ratio.

Let us now consider the converse problem. *Two circles* (A), (B) (Fig. 105) *being given, to find a point* M *and a ratio* k *such that* (B) *shall correspond to* (A) *in the homothecy* (M, k).

From the properties just recalled it follows that k must be equal to the ratio $a:b$ of the radii a, b of the given circles (A), (B), and that the point M must coincide with one of the points S, S' which divide the line of centers AB of (A), (B) externally and internally in the ratio $a:b$.

On the other hand, it is clear that in either of the two homothecies $(S, a:b)$ and $(S', -a:b)$ the circle which corresponds to (A) coincides with (B). Hence: *Two circles are homothetic in two and only two ways.*

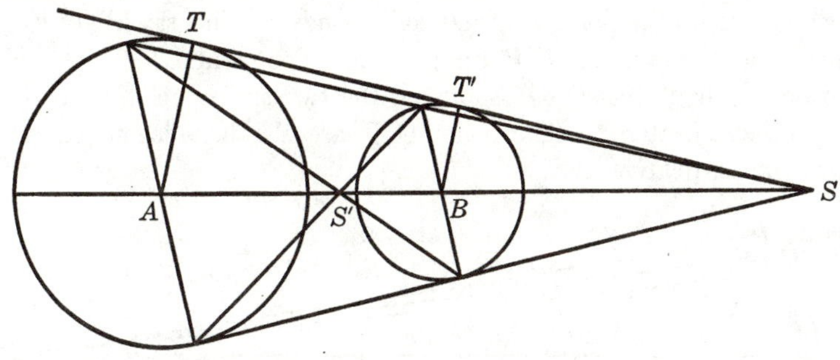

Fig. 105

394. Definitions. The point S (§ 393) is called the *external* or *direct center of similitude*, and S' the *internal* or *indirect center of similitude* of the two circles.

The external center of similitude is collinear with the ends of any two parallel radii directed in the same sense, or, what is the same thing, lying on the same side of the line of centers of the two circles. The internal center of similitude is collinear with the ends of any two parallel radii directed in opposite senses, or, what amounts to the same thing, lying on opposite sides of the line of centers.

The centers of the two circles and the two centers of similitude of the two circles are two pairs of harmonic points.

395. *Remark.* (a) If two circles are tangent to each other, their point of contact is a center of similitude of the two circles.

(b) Two equal circles have only one center of similitude, namely, the midpoint of their line of centers.

(c) If two circles are concentric, their common center is their only center of similitude.

396. Theorem. (*a*) *If two circles have external common tangents, these tangents pass through the external center of similitude of the two circles* (Fig. 105).

(*b*) *If two circles have internal common tangents, these tangents pass through the internal center of similitude.*

For in either case a common tangent to the two circles joins the ends of two parallel radii.

397. Definitions. (a) Let a line through one of the centers of similitude, say the external center S, of two circles (O), (O') (Fig. 106) meet (O) in P, Q, and (O') in P', Q'. Since S is a homothetic center of the two circles, to the point P of (O) corresponds a point, say P', of (O') such that the radii $OP, O'P'$ are parallel. The points P, P' are said to be a pair of *homologous points* on the two circles, relative to the homothetic center S. The points Q, Q' are also homologous on the two circles, relative to S.

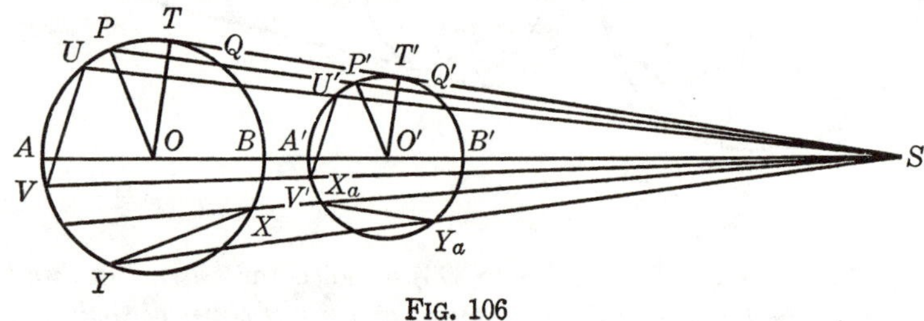

FIG. 106

Let U, V be any two points of (O) and U', V' their homologous points on (O'), relative to the same center of similitude. The chords $UV, U'V'$ are said to be *homologous chords* of the two circles.

Two homologous chords are parallel, for they are corresponding lines in two homothetic figures, unless they lie on a line passing through the center of similitude, as, for instance, the chords PQ, $P'Q'$ (Fig. 106).

(b) Two points on two circles collinear with a center of similitude of the circles and such that the radii passing through these points are not parallel are said to be *antihomologous points* with respect to the center of similitude considered. Thus P and Q', P' and Q (Fig. 106) are pairs of antihomologous points relative to S.

If X, Y are two points of (O), and X_a, Y_a their antihomologous points on (O') relative to the same center of similitude, the two chords XY, X_aY_a are said to be *antihomologous chords* of the two circles, with respect to the center of similitude considered.

If two circles intersect, two antihomologous points coincide in a point common to the two circles.

398. Theorem. *The product of the two segments determined by two pairs of homologous points in which a secant through a center of similitude of two circles cuts these circles, is constant.*

We have (Fig. 106):

$$\frac{SP}{SA} = \frac{SP'}{SA'} = \frac{SP - SP'}{SA - SA'} = \frac{PP'}{AA'}$$

and, similarly:

$$SQ:SB = QQ':BB';$$

hence, multiplying:

$$\frac{SP \cdot SQ}{SA \cdot SB} = \frac{PP' \cdot QQ'}{AA' \cdot BB'}.$$

Now the left-hand side of this equality is equal to unity; hence:

$$PP' \cdot QQ' = AA' \cdot BB'. \tag{1}$$

Since the right-hand side of (1) does not depend upon the choice of the secant SPP', the proposition is proved.

399. *Remark.* If the circles have a common tangent passing through S and touching the circles in T, T', we may consider the tangent STT' as a limiting position of the secant SP, for which the two segments PP', QQ' coincide in TT'; hence:

$$TT'^2 = PP' \cdot QQ'.$$

This formula may also be derived directly, in the same manner as formula (1).

400. Theorem. *The product of the distances of a center of similitude of two circles from two antihomologous points relative to the center of similitude considered, is constant.*

We have (Fig. 106):

$$SP:SP' = SQ:SQ', \quad \text{or} \quad SP \cdot SQ' = SP' \cdot SQ.$$

On the other hand we have:

$$SP \cdot SQ = SA \cdot SB, \quad SP' \cdot SQ' = SA' \cdot SB';$$

hence:

$$SA \cdot SB \cdot SA' \cdot SB' = SP \cdot SQ \cdot SP' \cdot SQ' = SP^2 \cdot SQ'^2.$$

Since the left-hand side of this equality does not depend upon the particular pair of antihomologous points P, Q' considered, the proposition is proved.

401. Corollary. *Two pairs of antihomologous points relative to the same center of similitude are cyclic, or collinear.*

402. Definition. The circle having for diameter the segment determined by the centers of similitude of two circles is called the *circle of similitude* of the two given circles.

403. Theorem. *The ratio of the distances of any point on the circle of similitude of two given circles from the centers of these circles is equal to the ratio of the radii of the given circles.*

The centers of similitude divide the line of centers of the two given circles in the ratio of the radii (§ 393); hence the proposition (locus 11, § 11).

404. COROLLARY. *The circle of similitude of two intersecting circles passes through the points common to the given circles.*

405. Theorem. *The six centers of similitude of three circles taken in pairs lie by threes on four straight lines.*

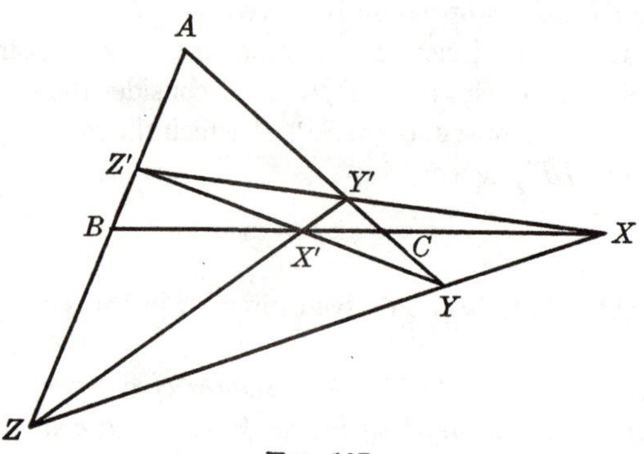

FIG. 107

Let X, Y, Z be the external centers of similitude of the pairs of given circles (B, b), (C, c); (C, c), (A, a); (A, a), (B, b) and let X', Y', Z' be the corresponding internal centers of similitude (Fig. 107). We have (§ 393):

$$BX:CX = b:c, \quad CY:AY = c:a, \quad AZ:BZ = a:b;$$

hence:

$$BX \cdot CY \cdot AZ = CX \cdot AY \cdot BZ,$$

and therefore X, Y, Z are collinear (§ 313).

In the same way it may be shown that the triads of points X, Y', Z'; Y, Z', X'; Z, X', Y' are collinear.

406. Definition. The four lines on which the six centers of similitude are situated are called the *axes of similitude*, or the *homothetic axes*, of the three given circles.

407. COROLLARY. *A line joining two of the centers of similitude of three circles taken in pairs passes through a third center of similitude.*

408. Definition. A circle is said to touch two given circles *in like manner*, if both contacts are external, or both are internal. The contacts are said to be *alike*.

A circle is said to touch two given circles *in unlike manner*, if one of the contacts is internal and the other external. The contacts are said to be *unlike*.

409. Theorem. *If a circle touches two given circles, the points of contact are antihomologous points on the two given circles.*

The points of contact U, V of a circle (M) with the two given circles (A), (B) are centers of similitude of the pairs of circles (M), (A) and (M), (B) (§ 395 a); hence the line UV passes through a center of similitude of the circles (A), (B) (§ 407).

The line UV passes through the external, or the internal, center of similitude of (A), (B) according as the contacts U, V are alike, or unlike (§ 408).

EXERCISES

1. Show that any circle through the centers of two given circles is orthogonal to the circle of similitude of the two given circles.
2. Show that the tangents to two circles at a point of intersection of the circles are equidistant from either center of similitude of the circles.
3. Show that (a) the centroid and the orthocenter of a triangle are the centers of similitude of the circumcircle and the nine-point circle of the triangle; (b) the incenter of a triangle is a center of similitude of the circumcircle and the circle determined by the excenters of the triangle.
4. Through a given point to draw a secant so that the two chords determined on it by two given circles shall have the same ratio as the radii of the circles.
5. The three pairs of circles (E), (F); (F), (D); (D), (E) touch each other in the points A, B, C. The lines BA, BC meet the circle (E) again in P, Q. Show that PQ passes through the center E of (E) and is parallel to the line DF.
6. Show that the twelve centers of similtude of the four tritangent circles of a triangle taken in pairs are (a) the six traces on the sides of the triangle of the bisectors of the respectively opposite angles and (b) the vertices of the given triangle each counted twice.
7. Show that the circle having for diameter an internal bisector of an angle of a triangle is the circle of similitude of the incircle and the excircle relative to the vertex considered. Find the circle of similitude of two excircles of a triangle.
8. The side BC of the triangle ABC touches the incircle (I) in X and the excircle (I_a) relative to BC in X_a. Show that the line AX_a passes through the diametric opposite X' of X on (I). State a similar proposition about the diametric opposite of X_a on (I_a).
9. (a) With the notations of the preceding exercise, show that if the line $A'I$ meets the altitude AD of ABC in P, then AP is equal to the inradius of ABC. State

and prove a similar proposition for the excircle. (b) If the perpendicular from A' to AI meets AD in Q, show that the line QX is perpendicular to $A'I$.

10. With the same notations as in the preceding exercise, show that the point X_a is the external center of similitude of the incircle and the circle having the altitude AD for diameter.

11. With the same notations as in the preceding exercise, show that the second tangent from A' to (I), the tangent to (I) at X', and the line YZ joining the points of contact of (I) with BA, CA, are concurrent. State and prove a similar proposition for (I_a).

12. With the same notations as in the preceding exercise, if the parallels to AX_a through B, C meet the bisectors CI, BI in L, M, show that the line LM is parallel to BC.

SUPPLEMENTARY EXERCISES

1. If PQ is a variable diameter of a circle, center A, and B, C are two fixed points collinear with A, show that the locus of the point $M = (CP, BQ)$ is a circle.
2. Construct a line whose distances from the vertices of a given triangle shall be proportional to three given segments p, q, r.
3. AD, CF are the external common tangents of two circles ABC, DEF; BE is one of the internal tangents. The diameters through B and E cut AC and DF respectively in G and H. Show that GH bisects BE. Show that the proposition is valid if the terms "internal" and "external" are interchanged.
4. The circle (O') passes through the center O of the circle (O). The common tangents of the two circles touch (O') in A, B. Show that AB is tangent to (O).
5. Show that the segment joining the ends of two parallel radii of two circles subtends a constant angle at either point of intersection of the two circles.

E. THE POWER OF A POINT WITH RESPECT TO A CIRCLE

410. Theorem. *The product of the distances of a given point, taken either inside or outside a given circle, from any two points which are collinear with the given point and lie on the circle, is a constant.*

Let E, F be any two points on the given circle (O) collinear with the given point L (Fig. 108 *a* and *b*), and A, B the ends of the diameter of (O) passing through L. The two triangles LAF, LBE are equiangular and therefore similar; hence:

$$LA:LF = LE:LB \quad \text{or} \quad LE \cdot LF = LA \cdot LB.$$

Now the right-hand side of the last equality does not depend upon the choice of the points E, F; the proposition is therefore proved.

411. *Remark.* When L lies outside of (O), and T is the point of contact of a tangent from L to (O) (Fig. 108 *b*), it may readily be shown that $LE \cdot LF = LT^2$.

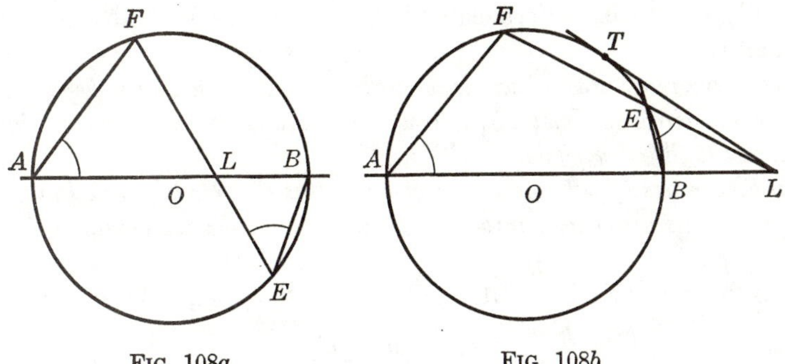

FIG. 108*a* FIG. 108*b*

412. Definition. The constant (§ 410) is called the *power of the point* with respect to the circle, or, more briefly, for the circle.

413. CONSEQUENCE. *If $OA = OB = r$, and $LO = d$, we have in Fig. 108 a:*

$$LA \cdot LB = (LO + OA)(OB - OL) = r^2 - d^2,$$

and in Fig. 108 b:

$$LA \cdot LB = (LO + OA)(LO - OB) = d^2 - r^2.$$

If we agree to consider negative magnitudes, these two results may be stated as follows: *The power of a point with respect to a circle is equal, both in magnitude and in sign, to the square of the distance of the point from the center of the circle diminished by the square of the radius of the circle.*

414. *Remark I.* (a) The power of a point for a circle is positive, zero, or negative according as the point lies outside, on, or inside the circle.

(b) When the point lies outside the circle, the square of the tangent from the point to the circle is equal to the power of the point for the circle.

(c) When the point lies inside the circle, the negative of the square of half the chord perpendicular to the diameter passing through the given point is equal to the power of the point for the circle.

(d) The power of the center of the circle for the circle is equal to the negative of the square of the radius.

415. *Remark II.* If the point L and the center O of the circle (O) remain fixed, while the radius r of (O) decreases so as to become zero, the circle (O) becomes a *point-circle*, but the notion of the power of a point for a circle, as formulated in § 413, remains applicable to this point-circle.

416. Theorem. (*a*) *If two circles are orthogonal, the square of the radius of either circle is equal to the power of the corresponding center with respect to the other circle.*

(*b*) *Conversely. If the square of the radius of a circle is equal to the power of its center with respect to a second circle, the two circles are orthogonal.*

If the two given circles (A, a), (B, b) are orthogonal, we have (§ 363):

$$AB^2 = a^2 + b^2;$$

hence:

$$a^2 = AB^2 - b^2, \quad b^2 = AB^2 - a^2,$$

which proves the direct proposition.

Conversely, if we have:

$$a^2 = AB^2 - b^2, \quad \text{then} \quad AB^2 = a^2 + b^2,$$

which proves the circles to be orthogonal (§ 363).

417. COROLLARY. *If a point lies outside a given circle, the power of the point with respect to the circle is equal to the square of the radius of the circle having the given point for center and orthogonal to the given circle.*

418. Definition. If the common chord of two intersecting circles (A), (B) is a diameter of (B), the circle (B) is said to *be bisected* by (A), and the circle (A) is said to *bisect* (B) (Fig. 109).

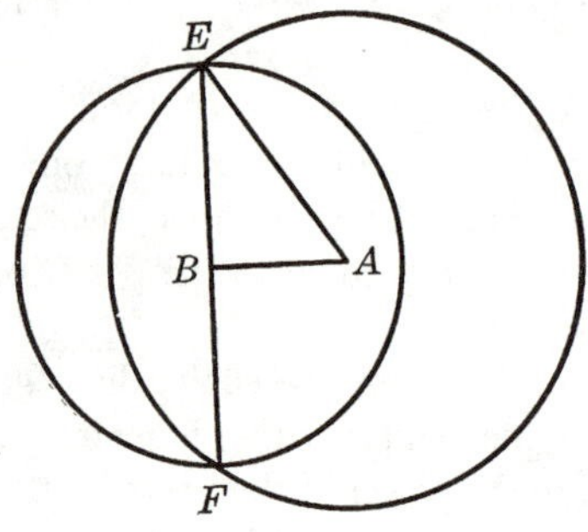

FIG. 109

419. Theorem. *If a circle is bisected by a given circle, the square of the radius of the bisected circle is equal to the negative of the power of the center of the bisected circle with respect to the bisecting circle.*

If the circle (B) is bisected by the given circle (A), and E, F are the ends of their common chord, we have (Fig. 109):

$$AE^2 = BE^2 + AB^2, \quad \text{or} \quad BE^2 = AE^2 - AB^2 = -(AB^2 - AE^2),$$

which proves the proposition.

420. Corollary. *Given a circle (A) and a point B, if B lies outside of (A), B is the center of a circle orthogonal to (A); if B lies inside of (A), B is the center of a circle bisected by (A). In the first case, the square of the radius of the circle (B) is equal to the power of B for (A); in the second case, the square of the radius of (B) is equal to the negative of the power of B for (A).*

EXERCISES

1. Show that (a) the locus of a point whose power for a given circle is constant, is a circle concentric with the given circle; (b) the locus of the center of a circle, with fixed radius, orthogonal to a given circle, is a circle concentric with the given circle.
2. Show that if the sum of the powers of the center of one of two given circles with respect to these two circles is equal to zero, the two circles are orthogonal.
3. Show that (a) the locus of a point the sum of whose powers with respect to two given circles is constant, is a circle having for center the midpoint of the line of centers of the two given circles, and conversely; (b) the locus of a point the sum of whose powers with respect to three given circles is constant, is a circle whose center coincides with the centroid of the triangle formed by the centers of the given circles, and conversely.
4. The square of the distance between a fixed point and a variable point is equal to the sum (or the difference) of the powers of these two points with respect to a fixed circle. Find the locus of the variable point.
5. If two points are inverse with respect to a circle, show that (a) the sum of their powers for the circle is equal to the square of the distance between the points; (b) the product of their powers for the circle is equal to the negative of the product of the square of the radius of the circle and the square of the distance between the points.
6. Show that the sum of the reciprocals of the powers, with respect to a given circle, of any two points inverse with respect to that circle is equal to the negative of the reciprocal of the square of the radius.
7. Show that the power of the orthocenter of a triangle with respect to the circumcircle is equal to four times the power of the same point with respect to the nine-point circle of the triangle.
8. Show that the sum of the powers of the vertices of a triangle for the nine-point circle of the triangle is equal to one-fourth the sum of the squares of the sides of the triangle.
9. If $ABCD$ is a rectangle inscribed in a circle, center O, and if PX, PX', PY, PY' are the perpendiculars from any point P upon the sides AB, CD, AD, BC, prove that $PX \cdot PX' + PY \cdot PY'$ is equal to the power of P with respect to the circle O.

SUPPLEMENTARY EXERCISES

1. A variable circle of fixed radius is orthogonal to a fixed circle. Show that the ends of a diameter of fixed direction of the variable circle move on two fixed circles.

2. Two points M, M' of a fixed line vary so that the product of their distances from a fixed point of the line is constant. Show that the locus of the center of the variable circle determined by M, M', and a fixed point in the plane is a straight line.

3. The escribed circle (I_a) of a triangle ABC meets the circumcircle (O) of ABC in D, and I_aD meets (O) in E. Show that I_aE is equal to the circumdiameter of ABC.

4. Given, in position, the circumcircle of a triangle, the trace of one internal bisector on the opposite side, and the product of the other two sides, to construct the triangle.

5. Show that the square of the radius of the polar circle of a triangle is equal to half the power of the orthocenter of the triangle with respect to the circumcircle of the triangle. (*Hint.* Use §§ 176, 389.)

6. Show that the sum of the powers, with respect to the circumcircle of a triangle, of the symmetrics of the orthocenter with respect to the vertices of the triangle is equal to the sum of the squares of the sides of the triangle.

7. Two unequal circles are tangent internally at A. The tangent to the smaller circle at a point B meets the larger circle in C, D. Show that AB bisects the angle CAD.

8. If the lines joining the vertices A, B, C of a triangle ABC to a point S meet the respectively opposite sides in L, M, N, and the circle (LMN) meets these sides again in L', M', N', show that the lines AL', BM', CN' are concurrent.

F. THE RADICAL AXIS OF TWO CIRCLES

421. Theorem. *The locus of a point which moves so that its powers with respect to two given circles are equal is a straight line perpendicular to the line of centers of the given circles.*

Given the circles (A, a), (B, b), if X is a point of the required locus, we have, by assumption:

$$XA^2 - a^2 = XB^2 - b^2, \quad \text{or} \quad XA^2 - XB^2 = a^2 - b^2;$$

hence the proposition (locus 12, § 11).

422. Definition. The line (§ 421) is called the *radical axis* or, briefly, the *axis* of the two circles.

423. Corollary. The powers, with respect to two given circles, of a point common to these circles are equal, for both are zero (§ 414 a);

hence: *The radical axis of two intersecting circles is the line determined by their points of intersection.*

424. *Remark.* (a) The theorem (§ 421) remains valid, if a or b, or both, are zero (§ 415), i.e., we may speak of the radical axis of a circle and a point, or even of the radical axis of two points, in which case the axis is identical with the mediator of the segment joining the two points.

(b) If two circles are tangent to each other, their radical axis is their common tangent at their common point.

(c) Two concentric circles have no radical axis.

425. Theorem. *The radical axes of three circles, with noncollinear centers, taken in pairs are concurrent.*

If the radical axis of the circles (A), (B) meets the radical axis of the two circles (A), (C) in the point R, and R_a, R_b, R_c are the powers of R for (A), (B), (C), respectively, we have:

$$R_a = R_b \text{ and } R_a = R_c; \quad \text{hence} \quad R_b = R_c,$$

i.e., the point R lies on the radical axis of the circles (B), (C), which proves the proposition.

426. Problem. *Construct the radical axis of two circles.*

(a) If the two circles intersect, the line joining their common points is their radical axis (§ 423).

(b) If the two given circles do not intersect, draw any convenient circle (C) cutting both (A) and (B). Let the common chord of (A) and (C) meet the common chord of (B) and (C) in the point R. The radical axis of (A) and (B) passes through R (§ 425) and is perpendicular to the line of centers AB of these circles.

(c) If the circle (A) is reduced to a point A, draw any circle (C) through A cutting (B). Let the tangent to (C) at A meet the common chord of (C) and (B) in R. The perpendicular from R to the line AB is the required radical axis.

427. Theorem. (*a*) *The locus of a point from which tangents of equal length may be drawn to two circles is the radical axis of the two circles* (§§ 414 a, 421).

(*b*) *If two circles have common tangents, the midpoints of these tangents lie on the radical axis of the two circles.*

428. Theorem. (*a*) *A circle orthogonal to two given circles has its center on the radical axis of the two given circles.*

(*b*) *If a circle has its center on the radical axis of two circles and is orthogonal to one of them, it is also orthogonal to the other circle.*

If a circle (M, m) is orthogonal to two circles (A), (B), the powers of M for these circles are both equal to m^2 (§ 416 a); hence M lies on the radical axis of (A), (B).

If the circle (M, m) is orthogonal to the circle (A), the power of M for (A) is equal to m^2. Now if M lies on the radical axis of the circles (A) and (B), the power of M for (B) is also equal to m^2 (§ 416 a); hence (M, m) is orthogonal to (B) (§ 416 b).

429. Theorem. *The polars, with respect to two given circles, of a point on their radical axis, intersect on the radical axis.*

Let Q be the point of intersection of the polars, with respect to two given circles (A), (B), of a point P of their radical axis u. The two points P, Q are conjugate with respect to both (A) and (B); hence the circle having PQ for diameter is orthogonal to (A), (B) (§ 387). Consequently the midpoint O of PQ lies on u (§ 428). Thus the two points P, O of the three collinear points P, O, Q lie on the line u; hence the same holds for the third point Q, which proves the proposition.

This proof is valid whether or not the given circles intersect. However, when the two circles have two points E, F in common, the proposition is almost obvious, for both polars of P must pass through the harmonic conjugate of P with respect to E, F.

430. Corollary. *If from a point on the radical axis of two circles the four tangents are drawn to these circles, the two chords joining the two pairs of points of contact intersect on the radical axis of the two circles.*

EXERCISES

1. If the squares of the radii of two circles are increased, or diminished, by the same amount, show that the radical axis of the two circles remains unchanged.
2. If from a point on the radical axis of two circles secants are drawn for each of the circles, show that the four points determined on the two circles are cyclic. Conversely.
3. Show that the locus of a point the difference of whose powers with respect to two given circles is constant is a straight line parallel to the radical axis of the two circles.
4. In the plane of a given circle a second circle with a given radius is drawn so that the radical axis of the two circles passes through a given fixed point. Show that the locus of the center of the second circle is a circle.
5. Show that the orthic axis of a triangle is the radical axis of the circumcircle and the nine-point circle of the triangle.
6. If a point describes a line perpendicular to the line of centers of two given circles, show that the point of intersection of the polars of the point for the two circles describes a straight line parallel to the given line.

7. (a) The common chord of a fixed circle and a variable circle passing through a fixed point meets the tangent to the variable circle at the fixed point in a point P. Show that the locus of P is a straight line. (b) Through a given point to draw a line passing through the inaccessible center of a given circle.
8. Through a given point to draw the tangents to a given circle, the center of which is inaccessible.
9. Show that an altitude of a triangle is the radical axis of the two circles having for diameters the medians issued from the other two vertices of the triangle.
10. Show that the radical axis of a point and a circle is the mediator of the segment determined by the given point and its inverse with respect to the circle.
11. If a circle is orthogonal to two given tangent circles, show that it touches the line of centers of the two circles at their common point.
12. Show that the four tritangent circles of a triangle taken in pairs have for their radical axes the bisectors of the angles of the medial triangle of the given triangle.
13. Show that the radical axis of the two circles having for diameters the diagonals AC, BD of a trapezoid $ABCD$ passes through the point of intersection E of the nonparallel sides BC, AD.

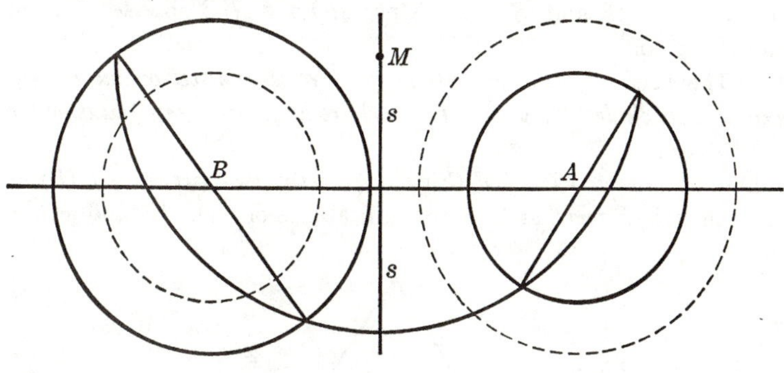

FIG. 110

431. Theorem. *The locus of the center of a circle which bisects two given circles is a straight line perpendicular to the line of centers of the given circles.*

If the circle (M, m) bisects the given circles (A, a), (B, b) (Fig. 110), we have (§ 419):

$$MA^2 + a^2 = m^2 = MB^2 + b^2;$$

hence:

$$MA^2 - MB^2 = b^2 - a^2, \tag{1}$$

which proves the proposition (locus 12, § 11).

432. Definition. The line s determined by (1) (§ 431) is sometimes referred to as the *antiradical axis* of the two given circles.

433. *Remark.* For a point M' of the radical axis r of (A, a), (B, b) we have:

$$M'A^2 - M'B^2 = a^2 - b^2. \tag{2}$$

The formulas (1) and (2) show that the lines s and r are symmetric with respect to the midpoint of the line of centers AB.

It may also be observed that s is the radical axis of the two circles (A, b), (B, a) (Fig. 110).

434. Theorem. *Two antihomologous chords of two circles intersect on the radical axis of the two circles.*

If M, Q are two points of the circle (O), and N', P' their antihomologous points on the circle (O') relative to the same center of similitude of (O), (O'), the four points M, Q, N', P' lie on a circle (§ 401), say (S).

Now the chords MQ and $N'P'$ are the radical axes of the two pairs of circles (S), (O) and (S), (O'); hence MQ and $N'P'$ intersect on the radical axis of (O), (O') (§ 425).

435. Theorem. *If two tangents to two circles intersect on the radical axis of these circles, the points of contact are antihomologous points on the given circles.*

Let the tangents RE, RF (Fig. 111) to the two circles (A), (B) at the points E, F meet in R on the radical axis of (A), (B), and let the

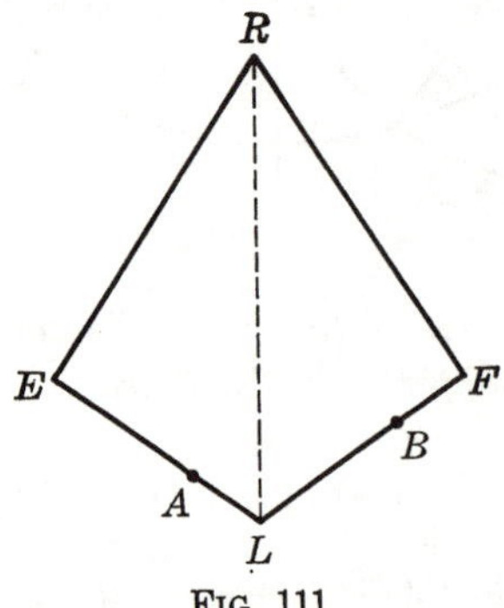

FIG. 111

lines EA, BF meet in L. The tangents RE, RF are equal (§ 427 a); hence from the congruent right triangles REL, RFL we have $LE = LF$. Thus the circle drawn with L as center and $LE = LF$ as radius is tangent to the circles (A), (B) at the points E, F; hence the proposition (§ 409).

436. Converse Theorem. *If two points are antihomologous on two circles, the tangents at these points to the circles intersect on the radical axis of the circles.*

Let the tangent at the point E to the circle (A) (Fig. 112) meet the radical axis of the two circles (A), (B) in R. The circle (R, RE) is

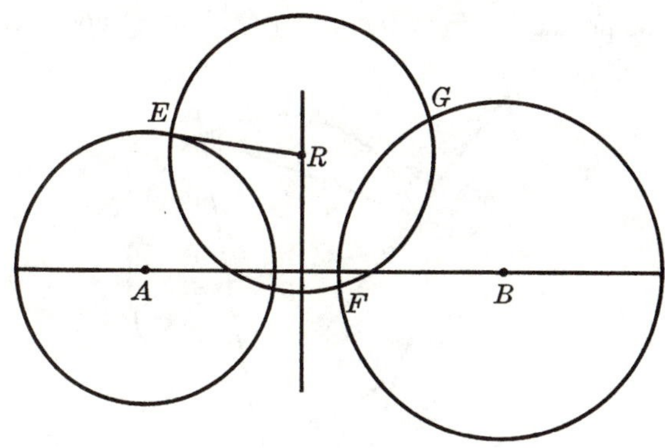

FIG. 112

orthogonal to both (A) and (B) (§§ 364 b, 428 b); hence the tangents to (B) at the points of intersection F, G of (B) with (R, RE) pass through R (§ 366). Thus the pairs of points E, F and E, G are antihomologous points on (A), (B), according to the direct theorem. But the point E of (A) has only two antihomologous points on (B); hence they coincide with F, G, and the proposition to be proved is valid for the pairs of points E, F and E, G, as we have just seen.

437. COROLLARY. *The four points in which two circles are cut by a circle orthogonal to these circles determine three pairs of lines, of which one pair meet on the radical axis, and the other two pairs meet in the two centers of similitude of the two given circles.*

438. Theorem. *The inverses, with respect to two given circles, of a point on their circle of similitude are symmetrical with respect to the radical axis of the two circles.*

If P, Q (Fig. 113) are the inverse points of a point W with respect to the given circles (A, a), (B, b), we have:

$$AP \cdot AW = a^2, \quad BQ \cdot BW = b^2;$$

and if W is a point on the circle of similitude of the given circles, we have (§ 403):

$$WA:WB = a:b.$$

Dividing the first two equalities and taking into account the third, we obtain:

$$AP:BQ = AW:BW,$$

i.e., the lines AB, PQ are parallel.

Now the circle WPQ is orthogonal to both given circles (§ 369); hence the perpendicular from the center O of WPQ upon AB is the

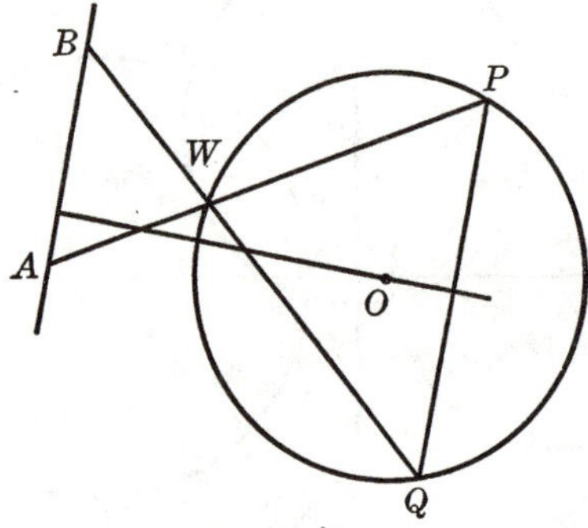

FIG. 113

radical axis of the circles (A, a), (B, b), and since PQ is parallel to AB, this perpendicular is also the mediator of the chord PQ of the circle WPQ; hence the proposition.

439. COROLLARY. *The two polars, with respect to two given circles, of either center of similitude of these circles are symmetrical with respect to the radical axis of the two circles.*

EXERCISES

1. Show that the locus of the center of a circle which passes through a given point and bisects a given circle is a straight line.
2. Through two given points to draw a circle bisecting a given circle.
3. Through a given point to draw a circle bisecting two given circles.
4. Show that the locus of the center of a circle which cuts one given circle orthogonally and bisects a second given circle is a straight line perpendicular to the line of centers of the two given circles.
5. Draw a circle orthogonal to two given circles and bisecting a third given circle.
6. Draw a circle through a given point, bisecting a given circle, and orthogonal to a second given circle.
7. Given two circles (O), (O'), show that the lines joining a point M of (O) to the two centers of similitude of (O), (O') meet (O') in four points, two of which are diametrically opposite on (O'), and the other two are collinear with a fixed point, whatever the position of M.

8. If a circle (L) is orthogonal to two given circles (A), (B), the radical axes of (L), (A) and (L), (B) meet in the pole of the line AB with respect to (L) and cut the line AB in the poles of the radical axis of (A), (B) with respect to these two circles.

SUPPLEMENTARY EXERCISES

1. Show that the radical axis of two circles is equidistant from the radical axes of the two circles with the circle having for diameter the line joining the centers of these two circles.

2. If a line cuts two orthogonal circles in two pairs of harmonic points, show that it passes through the center of one of the circles (or through the centers of both).

3. Two variable circles are tangent to each other and touch a given line in two fixed points. Show that the locus of the midpoint of the second external common tangent of the two circles is a circle.

4. Two variable circles whose radii have a constant ratio touch a given line in two fixed points, on the same side of the line. Find the locus of the point of intersection of the other external common tangent with the radical axis of the two circles.

5. A line antiparallel to the base BC of the triangle ABC meets AB, AC in P, Q, and a parallel to BC meets AB, AC in E, F. Show that the radical axis of the two circles BFQ, CEP remains fixed, when the line EF varies.

6. Given the points A, B on the circles (P), (Q), find a point C on the radical axis r of these circles such that if the lines CA, CB meet the circles (P), (Q) again in E, F, the line EF shall be perpendicular to the line r.

7. Let R, R' be the midpoints of the tangents OP, OP' from O to a given circle. If the chord of contact of the tangents from a point T of RR' to the circle meets RR' in U, show that TOU is a right angle.

G. COAXAL CIRCLES

440. Definition. A group of circles are said to form a *coaxal pencil,* if the same fixed line is the radical axis of any two of the circles of the group.

The fixed line is called the *radical axis,* or, more briefly, the *axis* of the coaxal pencil.

441. Determination of a Coaxal Pencil. In order to form a coaxal pencil one circle (A) and the radical axis r may be taken arbitrarily. Any circle (B) does or does not belong to the pencil depending on whether the radical axis of (A), (B) does or does not coincide with r, for if (C) is another circle such that the radical axis of (A), (C) coincides with r, then the radical axis of (B), (C) coincides with r.

A coaxal pencil may also be determined by taking two circles (A), (B) arbitrarily. The radical axis of these two circles plays the role of radical axis of the pencil.

Thus, summing up: *A coaxal pencil of circles is determined by one circle and the radical axis, or by two circles.*

442. Theorem. (*a*) *A point on the radical axis of a coaxal pencil of circles has the same power with respect to all the circles of the pencil.*

(*b*) *Conversely. If a point has equal powers with respect to the circles of a coaxal pencil, it lies on the axis of the pencil.*

443. Theorem. *The centers of the circles of a coaxal pencil are collinear.*

Let A be the center of a given circle of the pencil, and M the center of any other circle of the pencil. The line AM is perpendicular to the radical axis r of the pencil; hence the center M of any circle of the pencil lies on the perpendicular from the point A to the line r.

444. Definition. The line containing the centers of the circles of a coaxal pencil is called the *line of centers* of the pencil.

445. Discussion. (a) If the radical axis r of a coaxal pencil of circles meets one of the circles in two points E, F, every circle of the pencil must pass through these points, for the power of E (or F) for one of the circles of the pencil being zero (§ 414 a), the same must hold for all the circles of the pencil (§ 442 b).

The points E, F are called the *basic points* of the pencil, and the pencil is said to be *intersecting*, or of the intersecting type.

(b) If r does not cut one circle of the pencil, it cuts none of them. The pencil is said to be *nonintersecting*, or of the nonintersecting type.

(c) If r is tangent to one of the circles of the pencil, say, at T, every circle of the pencil touches r at T, and we have a coaxal pencil of tangent circles.

Note. In what follows the statements concerning coaxal pencils will have reference to types (a) and (b) only.

446. Problem. *With a given point as center to draw a circle belonging to a given coaxal pencil.*

It is assumed that the given center M lies on the line of centers c of the given pencil. Otherwise the problem is impossible.

(a) In the case of an intersecting pencil of circles the required circle (M) has for radius the segment joining M to one of the basic points E, F of the pencil. This radius will be least when M coincides with the trace O of the radical axis r on c, so that EF is a diameter of the circle.

The problem has one and only one solution for any position of M on c.

(b) In the case of a nonintersecting pencil (Fig. 114) the axis r, and therefore also the point O, lies outside the circles of the pencil; hence it is possible to draw a tangent from O to any circle, say (A), of the pencil. Since the point O belongs to r, the length t of this tangent is equal to the length of the tangent from O to any other circle of the pencil, and in particular to the required circle (M). Thus if m is the radius of (M), we have:

$$m^2 = OM^2 - t^2. \tag{1}$$

Thus the problem has one and only one solution if the distance OM of the given center M from the point O is greater than t.

447. Definition. The points of intersection L, L', (Fig. 114) of the line c with the circle (O, t) are called the *limiting points* of the non-intersecting pencil of circles.

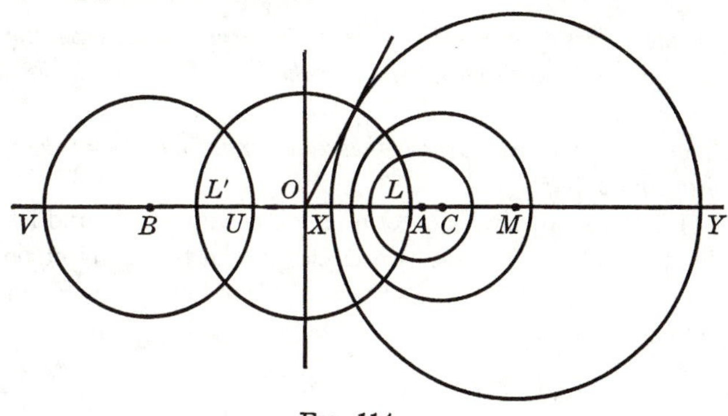

FIG. 114

No circle of the pencil has its center inside the segment LL', and every point M of c outside the segment LL' is the center of such a circle (M).

It follows from the formula (1) (§ 446) that as M moves closer to L (or L') the radius m of (M) approaches zero as a limit. For this reason the points L, L' are sometimes said to be the *point-circles* of the pencil.

448. Theorem. (*a*) *The limiting points of a coaxal pencil of circles are inverse points with respect to every circle of the pencil.*

(*b*) *Conversely. If two points are inverse with respect to a system of circles, the circles form a coaxal pencil.*

If X, Y (Fig. 114) are the points of intersection of the line of centers of the pencil with any circle of the pencil, we have (§ 446 b):

$$OL^2 = t^2 = OX \cdot OY;$$

hence $(LL'XY) = -1$ (§ 347), which proves the proposition.

If two points L, L' are inverse with respect to a system of circles, the centers of the circles lie on the line LL', and if X, Y are the points of intersection of LL' with any circle of the system, we have, by assumption, for the midpoint O of LL':

$$OX \cdot OY = OL^2;$$

hence the point O has equal powers with respect to all the circles of the system. Thus the circles form a coaxal pencil whose radical axis is the perpendicular to the line LL' at the point O.

449. COROLLARY. (*a*) *Any circle through the limiting points of a non-intersecting pencil of coaxal circles is orthogonal to each circle of the pencil.*

(*b*) *All the polars of a limiting point with respect to the circles of the pencil coincide with the perpendicular to the line of centers at the other limiting point of the pencil.*

450. Theorem. *A coaxal pencil of circles cannot have more than one pair of limiting points.*

Let (A), (B) be two circles determining a coaxal pencil, and let the line of centers AB meet these two circles in the two pairs of points X, Y and U, V (Fig. 114). If L, L' are a pair of limiting points of the pencil, we must have:

$$(XYLL') = -1, \quad (UVLL') = -1;$$

hence the pair of points L, L' is unique (§ 350).

451. Problem. *Through a given point to draw a circle belonging to a given coaxal pencil.*

Through an arbitrary point M of the radical axis of the pencil draw a secant meeting one of the circles of the pencil, say (A), in the points E, F, and on the line MP joining M to the given point P construct the point Q, on the same side of M as P, such that $MP \cdot MQ = ME \cdot MF$. If the mediator of PQ meets the line of centers of the pencil in U, the circle (U, UP) satisfies the conditions of the problem. Indeed, the point M has equal powers with respect to the two circles (A) and (U, UP); hence their radical axis coincides with the radical axis of the pencil, and therefore (U, UP) belongs to the pencil.

The problem has one and only one solution.

The method of solution is applicable to a pencil of either type. However, in the case of an intersecting pencil, the given point and the two basic points of the pencil determine the required circle directly.

452. Problem. *Draw a circle tangent to a given line and belonging to a given coaxal pencil of circles.*

From the point of intersection M of the given line s with the radical axis r of the pencil draw a tangent MT to any circle, say (A), of the pencil, and on s lay off the segments $MP = MP' = MT$. If the perpendiculars PU, $P'U'$ to s at the points P, P' meet the line of centers c of the pencil in U, U', the two circles (U, UP), $(U', U'P')$ are the two solutions of the problem.

Consider the case when the line s is parallel to r; when the line s cuts r between the basic points of an intersecting system of circles.

EXERCISES

1. Show that (a) if there is one point whose powers with respect to a group of circles with collinear centers are equal, the circles form a coaxal pencil; (b) if there are two points the powers of each of which with respect to a group of circles are equal, the circles form a coaxal pencil.
2. Show that the circles orthogonal to a given circle and having their centers on a diameter of that circle form a coaxal pencil.
3. If two variable points P, P' are conjugate with respect to a given circle and lie on a fixed line, show that the circle on PP' as diameter describes a coaxal pencil.
4. Show that the circles having their centers on a fixed line and bisecting a given circle form a coaxal pencil.
5. A line touches two circles (A), (B) in the points T, T', and a circle (C) coaxal with (A), (B) cuts TT' in E, F. Show that $(EFTT') = -1$.
6. Draw a circle of given radius and belonging to a given coaxal pencil.
7. Show that (a) the three circles determined by a given point P and the ends of the altitudes of a given triangle have a second point P' in common; (b) if P coincides with the centroid of the triangle, P' lies on the orthic axis; (c) if P lies on the circumcircle, P' lies on the nine-point circle.
8. If a chord of a circle of a coaxal pencil passes through a limiting point of the pencil, show that the projection of the chord upon the line of centers of the pencil is a diameter of a circle of the pencil.

453. Theorem. *If a circle has its center on the radical axis of a coaxal pencil and is orthogonal to one of the circles of the pencil, it is orthogonal to all the circles of the pencil.*

Let (A) be a circle of a given coaxal pencil (U) and (P, p) a circle orthogonal to (A) and having its center P on the radical axis r of (U). The power of P for (A) is equal to p^2 (§ 414 a), and since P lies on r, the power of P for any circle of (U) is equal to p^2 (§ 440 a); hence (P, p) is orthogonal to any other circle of (U) (§ 414 b).

454. Corollary. *If a circle is orthogonal to two circles of a coaxal pencil, it is orthogonal to all the circles of the pencil.*

455. Theorem. *The circles orthogonal to the circles of a coaxal pencil form a coaxal pencil.*

(a) The given pencil (U) is of the intersecting type. The radical axis r of (U) (Fig. 115) passes through the center of any circle (P)

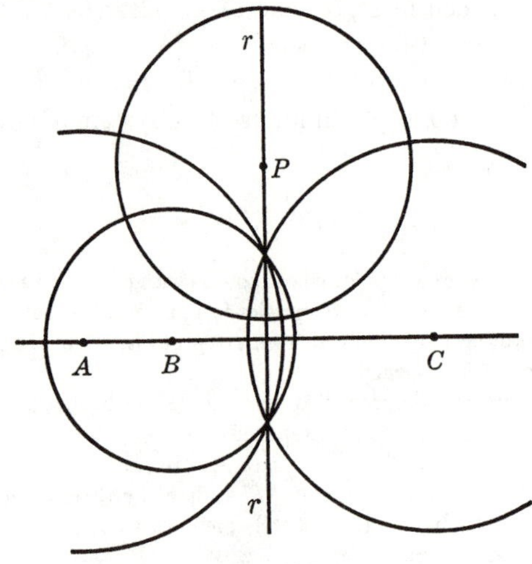

Fig. 115

orthogonal to the circles (A), (B), (C), . . . of (U) (§ 428); hence the basic points of (U) are inverse with respect to (P) (§ 368); hence the proposition (§ 448 b).

(b) The pencil (U) is of the nonintersecting type. If (P) is a circle orthogonal to all the circles of (U), its center P lies on the radical axis r of (U) (§ 428). On the other hand, r is the mediator of the segment LL' determined by the limiting points L, L' of (U) (§ 447); hence a circle (P') may be drawn having P for center and passing through L, L', which circle will be orthogonal to all the circles of (U) (§ 449).

Thus any circle (A) of (U) is orthogonal to both (P) and (P'). But the point P is the center of only one circle orthogonal to (A); hence (P) coincides with (P') and thus passes through L and L', which proves the proposition.

456. Definition. The pencil of circles (W) formed by the circles orthogonal to the circles of the given pencil (U) is said to be *conjugate* to (U). Obviously, the pencil (U) is conjugate to (W).

In two conjugate coaxal pencils every circle of one pencil is orthogonal to every circle of the other pencil.

457. *Remark.* Of two conjugate coaxal pencils one is of the intersecting, the other of the nonintersecting type, and the basic points of one coincide with the limiting points of the other. Furthermore, the radical axis and the line of centers of one pencil are, respectively, the line of centers and the radical axis of the other.

Two conjugate coaxal pencils are determined by a pair of given points.

458. Theorem. *Three, or more, circles orthogonal to the same circle and having their centers on a fixed line form a coaxal pencil.*

Indeed, they form the conjugate pencil (W) of the coaxal pencil (U) determined by the given circle (A) and the given line taken for radical axis.

459. Problem. *Through a given point to draw a circle orthogonal to the circles of a given coaxal pencil.*

The required circle belongs to the conjugate pencil (W) of the given coaxal pencil (U) (§ 456), which reduces the problem to one already solved (§ 451).

460. Theorem. *The polars of a point with respect to the circles of a coaxal pencil are concurrent.*

If the polars of the given point P with respect to any two circles, say (A) and (B), of the given pencil (U) meet in Q, the circle (PQ) having PQ for diameter is orthogonal to both (A) and (B) (§ 387); hence (PQ) is orthogonal to any other circle, say (C), of (U) (§ 454), and therefore P, Q are conjugate with respect to (C); hence the polar of P for (C) will pass through Q, which proves the proposition.

461. *Remark.* The circle (PQ) belongs to the conjugate pencil (W) of (U); hence the point Q is the diametric opposite of the given point P on the circle passing through P and belonging to the pencil (W). This gives a construction of the point Q.

It may also be observed that since the circle (P) passing through P and belonging to the given pencil (U) is orthogonal to (PQ), the line PQ is tangent to (P) at P, and Q is the symmetric of P with respect to the point of intersection of this tangent with the radical axis of (U). This gives another construction of the point Q.

462. Converse Theorem. *If the polars of a given point with respect to three, or more, circles with collinear centers are concurrent, the circles are coaxal.*

If Q is the point common to the polars of the given point P with respect to the circles (A), (B), (C), ... these circles are orthogonal to the circle having PQ for diameter (§ 387); hence the proposition (§ 458).

463. Theorem. *If two points are conjugate with respect to a circle, the square of the distance between them is equal to the sum of their powers with respect to the circle.*

Let A, B be two points conjugate with respect to the given circle (M).

(a) The line AB cuts (M) in two points C, D. Let O be the midpoint of AB. We have, both in magnitude and in sign:

$$\begin{aligned} AC\cdot AD &= (AO + OC)(AO + OD) \\ &= AO^2 + AO(OC + OD) + OC\cdot OD, \\ BC\cdot BD &= (BO + OC)(BO + OD) \\ &= BO^2 + BO(OC + OD) + OC\cdot OD; \end{aligned}$$

hence, adding and observing that:

$$AO^2 = BO^2, \quad \text{and} \quad AO + BO = 0,$$

we obtain:

$$AC\cdot AD + BC\cdot BD = 2\,AO^2 + 2\,OC\cdot OD. \tag{1}$$

Now A, B being conjugate for (M), they are separated harmonically by C, D; hence (1) becomes (§ 346):

$$AC\cdot AD + BC\cdot BD = 2\,AO^2 + 2\,AO^2 = 4\,AO^2 = AB^2, \tag{2}$$

which proves the proposition.

(b) The line AB does not cut (M). Both points A, B lying outside of (M), they may be taken as centers of two circles (A), (B) orthogonal to (M). The points A, B being conjugate to (M), the circle (AB) on AB as diameter is orthogonal to (M) (§ 387); hence the three circles (A), (B), (AB) are coaxal (§ 458) and intersect, for the conjugate pencil determined by (M) and the line AB as radical axis is nonintersecting.

Thus the circle (AB) passes through the points of intersection of the circles (A), (B); hence AB^2 is equal to the sum of the squares of the radii of (A), (B), i.e., to the powers of A, B for (M).

464. Converse Theorem. *If the square of the distance between two points is equal to the sum of the powers of these points with respect to a given circle, the points are conjugate with respect to the circle.*

(a) The line joining the given points A, B cuts the given circle (M) in two points C, D. By assumption, the formula (2) (§ 463) holds for the points A, B, and the formula (1) is valid for any four collinear points; hence we have, from (1) and (2),

$$2\,AO^2 + 2\,OC \cdot OD = AB^2 = 4\,AO^2, \quad \text{or} \quad OC \cdot OD = AO^2,$$

i.e., A, B are separated harmonically by C, D (§ 347); hence the proposition.

(b) The line AB does not cut (M). The square of AB is, by assumption, equal to the sum of the squares of the radii of the circles (A), (B) orthogonal to (M) and having A, B for centers; hence (A), (B) are orthogonal to each other, and therefore the circle (AB) having AB for diameter passes through their points of intersection, i.e., is coaxal with them. Thus (AB) is orthogonal to (M) (§ 454); hence the proposition (§ 386).

465. *Remark.* We have proved incidentally that two circles orthogonal to a third circle and having for their centers two points conjugate with respect to that third circle are orthogonal to each other.

466. COROLLARY. *If three circles are mutually orthogonal, the centers of any two of them are conjugate with respect to the third circle.*

EXERCISES

1. Draw a circle belonging to a given coaxal pencil and bisecting a given circle, not belonging to the pencil.
2. Find the locus of the inverse of a given point with respect to the circles of a coaxal pencil.
3. Show that a common tangent to two circles of a nonintersecting coaxal pencil subtends a right angle at a limiting point of the pencil.
4. P is a point on the radical axis of a coaxal pencil of circles; PH is the tangent from P to a circle of the pencil; prove that $PH = PL$, where L is a limiting point of the given pencil.
5. Show that the tangent from a limiting point to any circle of the coaxal pencil is bisected by the radical axis of the pencil.
6. Show that the tangent from a limiting point to a circle of the coaxal pencil is divided harmonically by every circle of the pencil.
7. If A, B and C, D are the points of intersection of two given nonintersecting circles with their line of centers, L is one of their limiting points, and $AL:LB = CL:LD$, show that the two circles are equal.
8. Through each vertex of a triangle a circle is drawn orthogonal to the circumcircle of the triangle and having its center on the opposite side. Show that the three circles are coaxal.

9. The tangents at the vertices A, B, C to the circumcircle of an acute-angled triangle ABC intersect BC, CA, AB in U, V, W respectively. Prove that the circles on AU, BV, CW as diameters are coaxal, having for radical axis the Euler line of the triangle.

10. If the polar of a given point with respect to a variable circle is fixed, show that the polar with respect to this circle of any other point passes through a fixed point.

11. Given a coaxal pencil of circles, show that the line determined by a fixed point on the radical axis and the pole of that axis with respect to a variable circle of the pencil cuts that circle in two points belonging to a fixed circle.

467. Theorem. *The radical axes of the circles of a coaxal pencil with a circle not belonging to the pencil are concurrent, and their common point lies on the radical axis of the pencil.*

Let (A), (B) be any two circles of the given pencil (U), and (S) a given circle not belonging to (U). The three radical axes of the three pairs of circles:

(1) $(S),\ (A);\ (A),\ (B);\ (S),\ (B)$

have a point, say I, in common (§ 425). Now if (C) is any other circle of (U), the three radical axes of the three pairs of circles:

(2) $(S),\ (A);\ (A),\ (C);\ (S),\ (C)$

have, in turn, a point, say I', in common. But the first two of the axes in (2) are the same as those in (1); hence I' coincides with I, which proves the proposition.

468. *Remark.* (a) If I lies outside of (S), this point is the center of a circle orthogonal to (S) and belonging to the coaxal pencil (W) conjugate to the given pencil (U).

(b) The circle (S) has an analogous point J with respect to the coaxal pencil (W) lying on the radical axis of (W), which line is identical with the line of centers of (U) (§ 457). The points I, J are conjugate with respect to (S) (§ 465).

469. Problem. *Draw a circle belonging to a given coaxal pencil and orthogonal to a given circle not belonging to the pencil.*

Let (S) be the given circle, (U) the given pencil, c the line of centers of (U), and (W) its conjugate pencil. The radical axis of (S) and any circle (K) of (W) cuts the radical axis c of (W) in the point J. The circle belonging to (U) and having J for center solves the problem (§ 468 b).

The problem has one solution, if the point J falls outside of (S).

If (S) belongs to the pencil (W), every circle of (U) is a solution of the problem.

470. Problem. *Draw a circle belonging to a given coaxal pencil and tangent to a given circle not belonging to the pencil.*

Let the radical axis of the given circle (S) and any circle of the given pencil (U) meet the radical axis of (U) in I, and let T, T' be the points of contact of (S) with the tangents from I to (S). If the lines ST, ST' meet the line of centers of (U) in the points C, C', the circles (C, CT), $(C', C'T')$ are two solutions of the problem.

The problem has no solution, if I falls inside of (S).

As an exercise, consider the case when the center of (S) lies on the line of centers of (U); when (S) belongs to (U).

471. Casey's Theorem. *The power of a point with respect to a circle diminished by the power of the same point with respect to a second circle is equal, in magnitude and in sign, to twice the distance of the point from the radical axis of the two circles (the beginning of the segment being on the radical axis), multiplied by the line of centers of the circles (the beginning being the center of the circle first considered).*

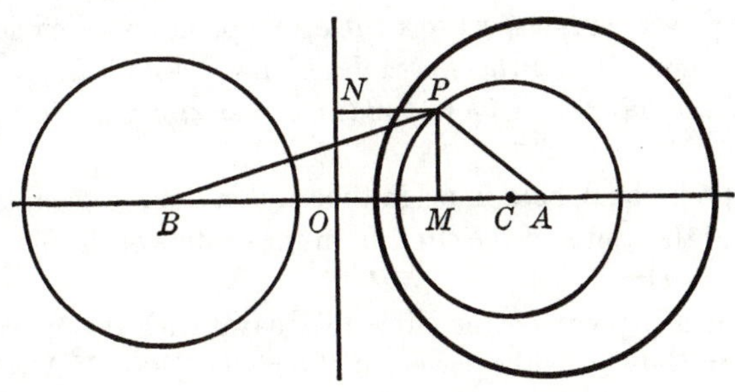

FIG. 116

Let (A, a), (B, b) (Fig. 116) be the given circles, O the trace on AB of their radical axis r, M, N the feet of the perpendiculars PM, PN dropped from the given point P upon AB and r. If d^2 is the required difference, we have:

$$d^2 = (PA^2 - a^2) - (PB^2 - b^2) = (PA^2 - PB^2) - (a^2 - b^2).$$

From the right triangles PMA, PMB we have:

$$PA^2 = PM^2 + AM^2, \quad PB^2 = PM^2 + BM^2;$$

hence:

$$PA^2 - PB^2 = AM^2 - BM^2.$$

The powers of O for the two circles being equal, we have:

$$OA^2 - a^2 = OB^2 - b^2, \quad \text{or} \quad OA^2 - OB^2 = a^2 - b^2.$$

Thus we have:

$$d^2 = (AM^2 - BM^2) - (OA^2 - OB^2).$$

From this relation we obtain successively, both in magnitude and in sign:

$$\begin{aligned} d^2 &= (AM + MB)(AM - MB) - (AO + OB)(AO - OB) \\ &= AB(AM - MB - AO + OB) \\ &= AB[(AO + OM) - (MO + OB) - AO + OB] \\ &= 2\,AB \cdot OM = 2\,AB \cdot NP. \end{aligned}$$

472. Corollary. *The power, with respect to a circle, of a point on a second circle is equal, both in magnitude and in sign, to twice the distance of the point from the radical axis of the two circles (the beginning of the segment being the point on the radical axis), multiplied by the line of centers of the two circles (the beginning being the center of the circle with respect to which the power is taken).*

For the power, with respect to a circle, of a point on the circle is zero.

473. Theorem. *If a circle is coaxal with two given circles, the powers of any point of this circle with respect to the given circles are in a constant ratio.*

This ratio is equal, both in magnitude and in sign, to the ratio of the distances of the center of the circle from the centers of the first and the second given circles.

Let P be any point on the circle (S) coaxal with two given circles (A), (B), and let u, v be the powers of P for (A), (B). If N is the foot of the perpendicular from P upon the radical axis of the pencil (A), (B), (S), we have, both in magnitude and in sign (§ 472):

$$u = 2\,NP \cdot AS, \quad v = 2\,NP \cdot BS;$$

hence:

$$u:v = AS:BS,$$

which ratio is independent of the position of P on (S).

474. *Remark.* The ratio $u:v$ (§ 473) is negative, or positive, according as the center S lies on, or outside, the segment AB.

475. Converse Theorem. *The locus of a point the ratio of whose powers with respect to two given circles is constant, both in magnitude and in sign, is a circle coaxal with the given circles.*

If P is a point of the locus, the circle (S) passing through P and coaxal with the given circles (A), (B) (§ 451) will have for its center the point S which divides the line of centers AB of the given circles in the given ratio, internally, if the ratio is negative, and externally, if that ratio is positive (§ 474). Thus the position of the center S, and therefore of the circle (S), is uniquely determined by the given ratio, independently of the position of P; hence the proposition.

476. *Remark.* If the ratio is given only in magnitude (§ 475), the locus of P consists of two circles coaxal with the given circles and having for centers the two points which divide the line of centers of the two given circles internally and externally in the given ratio.

477. Theorem. *The circle of similitude of two given circles is coaxal with these circles.*

If P is a point of the circle of similitude of the two given circles (A, a), (B, b), we have:

$$PA:PB = a:b, \quad \text{or} \quad (PA^2 - a^2):(PB^2 - b^2) = a^2:b^2.$$

Thus the ratio of the powers of any point of the circle of similitude of the two given circles with respect to these circles is constant; hence the proposition (§ 475).

478. Corollary. *One of the two centers of similitude of two nonintersecting circles lies between the limiting points of the two circles.*

Indeed, the limiting points L, L' of two circles (A), (B) are inverse points with respect to any circle coaxal with (A), (B) (§ 448 a), and in particular with respect to the circle of similitude of (A), (B) (§ 477); hence the centers of similitude S, S' of (A), (B) are separated harmonically by the points L, L'.

479. *Remark I. If from a point on the circle of similitude of two circles tangents can be drawn to these circles, the ratio of these tangents is equal to the ratio of the radii of the circles.*

480. *Remark II. Two circles subtend equal angles at any point of their circle of similitude.*

For if MP, MQ (Fig. 117) are the tangents to the circles (A, a), (B, b) from a point M of their circle of similitude, we have (§ 479):

$$MP:MQ = a:b;$$

hence the two right triangles MPA, MQB are similar.

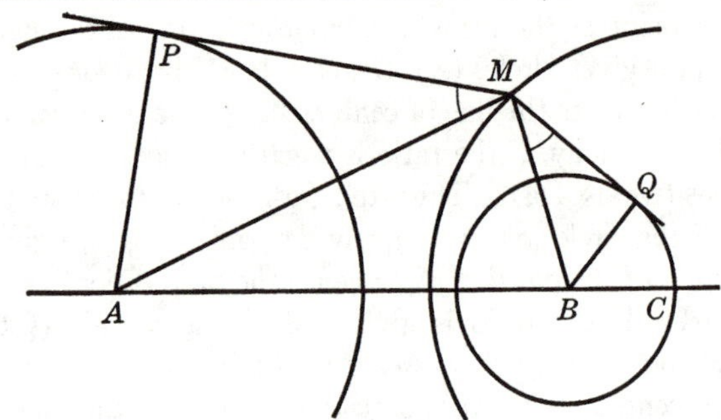

FIG. 117

481. Theorem. *The circle of similitude of two circles is the only circle which is coaxal with these circles and divides their line of centers harmonically.*

The circles having for diameters the segments harmonic to the segment AB determined by the centers A, B of two given circles (A), (B) form a coaxal pencil (X) having the points A, B for limiting points (§ 448 b), and the circle of similitude (S) of (A), (B) belongs to (X).

On the other hand (S) belongs to the coaxal pencil (Z) determined by (A) and (B) (§ 477). Now the two pencils (X) and (Z) cannot have more than one circle in common (§ 441); hence the proposition.

482. Definition. The two circles (S), (S') coaxal with two given circles (A), (B) and having for centers the centers of similitude of the two given circles are called the *circles of antisimilitude* of the given circles.

The external, or direct, center of similitude is the center of the external, or direct, circle of antisimilitude, and the internal, or indirect, center of similitude is the center of the internal, or indirect, circle of antisimilitude.

If the two given circles intersect, both circles of antisimilitude are real. If the given circles do not intersect, only one of the circles of antisimilitude is real (§ 478).

483. Theorem. *The two circles of antisimilitude of two intersecting circles are orthogonal.*

The two circles of antisimilitude (S), (S') and the circle of similitude (SS') of the two given intersecting circles (A), (B) are coaxal, for the three circles belong to the pencil of circles determined by (A), (B) (§§ 482, 477); hence the proposition (§ 365 b).

484. Theorem. *The locus of a point M whose powers with respect to two given circles are equal in magnitude and opposite in sign is a circle coaxal with the given circles.*

The ratio of the powers of the variable point M for the two given circles is constant, namely, equal to -1; hence the proposition (§ 475).

485. Definition. The locus of M (§ 484) is called the *radical circle* of the two given circles.

486. *Remark.* If (A, a), (B, b) are the given circles, we have, for any point M of the radical circle (L):

$$MA^2 - a^2 = -(MB^2 - b^2), \quad \text{or} \quad MA^2 + MB^2 = a^2 + b^2;$$

hence (locus 10, § 11) the center L of (L) is the midpoint of AB, and the square of the radius of (L) is given by the formula:

$$(1) \qquad \tfrac{1}{2}(a^2 + b^2) - \tfrac{1}{4}\,AB^2, \quad \text{or} \quad \tfrac{1}{2}(a^2 + b^2 - \tfrac{1}{2}\,AB^2).$$

487. Discussion. (a) If the given circles intersect, their radical circle is always real, for it is always possible to draw a circle coaxal with them and having for center the midpoint of their line of centers.

(b) If the given circles (A, a), (B, b) do not intersect, the radical circle (L) will be real if the midpoint L of AB falls outside the segment determined by the limiting points of the coaxal pencil determined by the given circles. This will always be the case, if A, B lie on the same side of the radical axis; it may or may not be the case, if A, B lie on opposite sides of the radical axis.

Another test is furnished by the formula (1) (§ 486). The radical circle will be a real circle, a point-circle, or imaginary, depending on whether $a^2 + b^2 - \frac{1}{2}\,AB^2$ is positive, zero, or negative.

(c) If $a^2 + b^2 = AB^2$, the square of the radius of (L) is equal to $\frac{1}{4}\,AB^2$, i.e., if the given circles are orthogonal, their radical circle coincides with the circle having their line of centers for diameter.

488. Theorem. *If the two chords determined on a transversal by two given circles are harmonic, the midpoints of these chords lie on the radical circle of the given circles. Conversely.*

If CD, EF are the two chords determined by the circles (A), (B) on the same transversal, and P is the midpoint of CD, the powers of P for the circles (A), (B) are, respectively:

$$PC\cdot PD = -PC^2, \quad PE\cdot PF.$$

Now if $(CDEF) = -1$, we have (§ 346):

$$(1) \qquad PC^2 = PE \cdot PF.$$

Similarly for the midpoint of EF. Hence the proposition (§ 484).

Conversely, if P lies on the radical circle of (A), (B), we have, by definition:

$$-PC^2 = -PE \cdot PF, \quad \text{or} \quad PC^2 = PE \cdot PF,$$

and the two chords CD, EF are harmonic (§ 347).

489. Corollary. *Through a given point M of the radical circle of two given circles pass, in general, two transversals on which the given circles determine two harmonic chords, namely, the two transversals which are perpendicular at M to the lines joining M to the centers of the given circles.*

It should be observed, however, that, assuming the circle (L) is real, the perpendiculars at M may always be constructed, but they may not cut the given circles in real points.

490. *Remark.* A transversal on which two orthogonal circles determine two harmonic chords passes through the center of, at least, one circle.

For if the given circles are orthogonal, their centers A, B lie on their radical circle (L) (§ 487 c), and if M is any point on (L), the perpendicular to AM will pass through the diametric opposite B of A on (L).

EXERCISES

1. A variable circle APQ passes through the given point A and cuts a given line in P, Q. The power of A with respect to a second variable circle (M) passing through P, Q is constant. Show that the distance between the centers of the two variable circles is constant.

2. A variable circle passes through a fixed point and has its center on a fixed circle. Show that the radical axis of the variable circle and the fixed circle is tangent to a fixed circle.

3. Find the locus of a point such that the length of the tangent from it to a given circle is the mean proportional between a given segment and the distance of the point from a given line.

4. Prove that the circle on an internal bisector of a triangle as diameter is coaxal with the incircle and the excircle whose center lies on the bisector considered. State and prove an analogous proposition for an external bisector.

5. Show that the circle having for ends of a diameter the centroid and the orthocenter of a triangle is coaxal with the circumcircle and the nine-point circle of the triangle.

6. If a line through a point of intersection of two circles meets the circles again in the points P, Q, show that the midpoint of PQ lies on the radical circle of the given circles.
7. Through a given point in the plane to draw a transversal so that the chords determined on it by two given circles shall be harmonic.
8. Draw a line parallel to a given line and cutting two given circles in two pairs of harmonic points.
9. Through the common point O of three circles (P), (Q), (R) to draw a secant meeting the circles in the points A, B, C so that the ratio $AB:AC$ shall have a given value.
10. Given four points in the plane, show that the sum of the inverses of the powers of each of these points with respect to the circle determined by the remaining three points is zero.
11. Show that (a) the power of a point with respect to the radical circle of two given circles is equal to one-half the sum of the powers of the point with respect to the two given circles; (b) the sum of the powers of a point with respect to three given circles is equal to the sum of the powers of the same point with respect to the three radical circles of the given circles taken in pairs.

H. THREE CIRCLES

491. Problem. *Construct a circle orthogonal to three given circles with noncollinear centers.*

The center R of the required circle necessarily lies on the three radical axes of the three given circles (A), (B), (C) taken in pairs (§ 428); hence R is the point common to these three radical axes (§ 425).

The circle (R) having R for center and orthogonal to one of the circles (A), (B), (C) constitutes the solution of the problem.

492. Definition. The point R (§ 491) is called the *radical center* or the *orthogonal center* of the three circles.

The circle (R) is called the *orthogonal circle* of the three given circles.

493. *Remark.* The radical center R has equal powers with respect to the three given circles and is always a real point.

The circle (R) exists only if the power of R for the given circles is positive, or, what is the same thing, if R lies outside the given circles.

494. Theorem. *The orthogonal circle of three given circles is the locus of the point whose polars with respect to the three given circles are concurrent.*

If the polars of the point P with respect to the three given circles meet in Q, the circle (PQ) on PQ as diameter is orthogonal to each of the given circles (§ 387); hence (PQ) coincides with the orthogonal circle (R) of the given circles.

Observe that Q also lies on the orthogonal circle (R), and that the points P, Q are diametrically opposite on (R).

495. Theorem. *If (O) is the circle passing through the centers A, B, C of the three given circles (A, a), (B, b), (C, c), and (R) is the orthogonal circle of these three circles, the distances of A, B, C from the radical axis of the circles (R) and (O) are proportional to a^2, b^2, c^2.*

Indeed, if p is the distance of A from the radical axis m, the quantity $2\,p\cdot RO$ is equal to the difference of the powers of A for (R) and (O) (§ 471). Now these powers are respectively equal to a^2 and zero; hence:

$$a^2 = 2\,p\cdot RO, \quad \text{or} \quad p:a^2 = 1:2\,RO.$$

The right-hand side of this equality being constant, the proposition is proved.

496. Theorem. *The three circles of similitude of three given circles taken in pairs are coaxal.*

Given the circles (A), (B), (C), the circle of similitude (S_{ab}) of the two circles (A), (B) is coaxal with them (§ 477); hence (S_{ab}) is orthogonal to the orthogonal circle (R) of the circles (A), (B), (C) (§ 454).

Again the centers A, B of (A), (B) divide a diameter of (S_{ab}) harmonically (§ 394); hence A, B are inverse points with respect to (S_{ab}), and the circle (O) passing through the centers A, B, C of the given circles is orthogonal to (S_{ab}).

Similarly for the circles of similitude (S_{bc}), (S_{ca}) of the pairs of circles (B), (C) and (C), (A).

Thus the three circles of similitude are orthogonal to the two circles (R) and (O); hence the proposition (§ 455).

497. *Remark.* The proposition remains valid, if the circle (R) does not exist. For the above reasoning shows that the point R has equal powers with respect to the three circles of similitude, and the same holds for the center O of (O); hence the line OR is the radical axis of any two of three circles of similitude considered.

498. COROLLARY. *The centers of the three circles of similitude (S_{ab}), (S_{bc}), (S_{ca}) (§ 496) lie on the radical axis of the circles (R), (O).*

499. Definition. If the coaxal pencil formed by the three circles of similitude of three given circles (§ 496) has two basic points, these points are referred to as the *isodynamic points* of the three given circles.

The isodynamic points of three circles are the limiting points of the coaxal pencil determined by the orthogonal circle of the three given circles and the circle passing through the three centers of the given circles (§ 457).

500. Theorem. *The three inverses, with respect to three given circles, of an isodynamic point of these circles are equidistant from the radical center of the given circles.*

Let W (Fig. 118) be the isodynamic point of the three circles (A), (B), (C), R being their radical center, and O the center of the circle $(O) = ABC$.

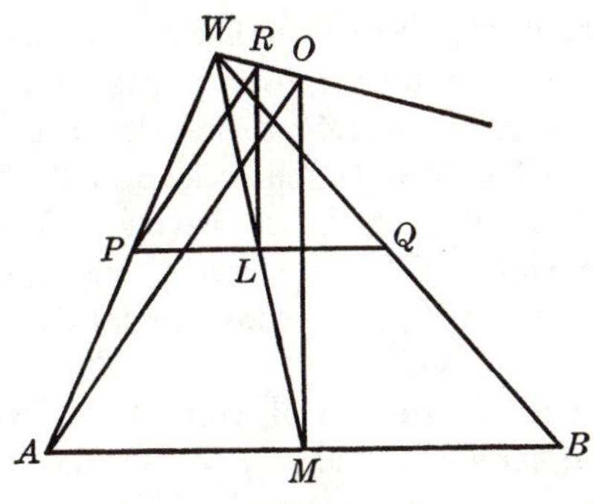

FIG. 118

The inverses P, Q of W for (A), (B) lie on a parallel to AB (§ 438); hence the midpoints L, M of PQ, AB are collinear with W. Now the mediators of PQ, AB pass respectively through R (§ 438) and O; hence:

$$WP:WA = WL:WM = WR:WO;$$

consequently the lines PR, AO are parallel, and we have:

$$RP:OA = WR:WO.$$

The last three terms of this proportion do not depend upon the circle (A) considered; hence the proposition.

501. *Remark.* The point W divides the line of centers RO of the circle (O) and the circle (R') determined by the inverses of W for the three given circles in the ratio of the radii OA, RP of the circles (O), (R'); hence W is a center of similitude of these two circles.

502. Theorem. *If the radii of three variable circles with fixed centers vary so as to remain proportional, the orthogonal circle of the three variable circles describes a coaxal pencil.*

The ratio of the radii of any two of the three variable circles being constant, the centers of similitude of these two circles are fixed; hence the same is true of their circle of similitude. Now the three circles of similitude are coaxal and are orthogonal to the orthogonal circle (R) of the given circles (§ 496); hence (R) describes the coaxal pencil conjugate to the pencil determined by the circles of similitude considered.

503. Theorem. *With three noncollinear given points as centers it is possible to describe three circles so that a given line shall be an axis of similitude of the three circles.*

Let Y, Z be the traces of the given line on the sides AC, AB of the triangle ABC determined by the given points A, B, C; let Y', Z' be the harmonic conjugates of Y, Z with respect to the pairs of points CA, AB; and let (YY'), (ZZ') be the circles on YY', ZZ' as diameters.

The circle (A), of arbitrary radius, having A for center, and the circles (B), (C) having B, C for centers and coaxal, respectively, with (A), (ZZ') and (A), (YY'), are three circles having the desired property.

Indeed, the circle (ZZ') is coaxal with (A), (B) and divides their line of centers AB harmonically; hence (ZZ') is the circle of similitude of (A), (B) (§ 481), and Z is a center of similitude of these two circles.

Similarly Y is a center of similitude of the two circles (A), (C); hence the proposition (§ 407).

It should be observed that the trace X of the given line YZ on BC is a center of similitude of the two circles (B), (C), and that the circle of similitude (XX') of (B), (C) is determined by X and the harmonic conjugate X' of X for the points B, C.

504. *Remark.* The radius of (A) was taken arbitrarily; hence it is possible to describe the three circles in an infinite number of ways, the radii of the various sets of circles being proportional.

505. Theorem. *Given three circles* (A), (B), (C), if (D), (E) *are the radical circles of* (A), (B) *and* (A), (C), *respectively, the radical axis of the circles* (D), (E) *coincides with the radical axis of the circles* (B), (C).

The circles (D), (E) are, respectively, coaxal with the pairs of circles (A), (B) and (A), (C); hence the powers with respect to (D), (E) of the radical center R of the three given circles are equal, for these powers are equal to the power of R for (A). Thus the radical axes of the two pairs of circles (D), (E) and (B), (C) are the perpendiculars from R upon the lines DE, BC. But DE and BC are parallel, for D, E are the midpoints of AB, AC (§ 486); hence the proposition.

EXERCISES

1. If the squares of the radii of three given circles are increased, or diminished, by equal amounts, show that the radical center of the three circles will remain fixed.
2. Show that the radical center of the three circles having for diameters the three sides of a triangle coincides with the orthocenter of the triangle.
3. Show that the orthocenter of a triangle ABC is the radical center of the three circles having for diameters the three cevians AP, BQ, CR.
4. Draw a circle so that the tangents drawn to it from three given points shall have given lengths.
5. Find a point which shall be concyclic with its three inverses with respect to three given circles.
6. Given three nonintersecting circles, show that the three pairs of limiting points determined by the three circles taken two at a time are concyclic.
7. Show that the antiradical axes of three circles taken two at a time are concurrent.
8. If each of three given circles is tangent to the remaining two, show that their orthogonal circle is a tritangent circle of the triangle determined by their centers.
9. Given three circles, to construct a fourth circle such that its radical axes with the given circles shall pass, respectively, through three given points.
10. Given three circles with noncollinear centers, show that any three of their six circles of antisimilitude are coaxal, if their centers are collinear.
11. Show that the distances of an isodynamic point of three circles from the centers of these circles are proportional to the radii of the respective circles.
12. Given the circles (A), (B), (C), a line parallel to BC meets AB, AC in D, E. If (D), (E) are the circles with D, E for centers and coaxal, respectively, with the pairs of circles (A), (B); (A), (C), show that the radical axis of (D), (E) coincides with the radical axis of (B), (C).
13. Show that (a) if on the three sides of a triangle are marked three pairs of points so that any two of these three pairs are concyclic, the three pairs of points lie on the same circle; (b) if the three pairs of points A', A''; B', B''; C', C'' are marked on the three sides BC, CA, AB of the triangle ABC so that:

$$AC' \cdot AC'' = AB' \cdot AB'', \quad BA' \cdot BA'' = BC' \cdot BC'', \quad CB' \cdot CB'' = CA' \cdot CA'',$$

the three pairs of points are concyclic.

SUPPLEMENTARY EXERCISES

1. Show that the foot of the perpendicular from the orthocenter of a triangle upon the line joining a vertex to the point of intersection of the opposite side with the corresponding side of the orthic triangle lies on the circumcircle of the triangle.
2. Construct a triangle given, in position, the orthocenter, one vertex, and the point of intersection of the opposite side with the corresponding side of the orthic triangle.

3. Show that the anticenter of a cyclic quadrilateral has equal powers with respect to the circles having for diameters a pair of opposite sides of the quadrilateral.
4. Given three circles with noncollinear centers, show that the three circles passing through a given point and coaxal with the given circles taken in pairs have a second point in common.
5. Any four points A, B, C, D are taken on a circle; AC, BD intersect in E; AB, CD intersect in F; AD, BC intersect in G. Show that the circumcircles of the triangles EAB, ECD cut the lines AD, BC in four points lying on a fourth circle; if these four circles be taken three at a time, show that the radical centers of the systems so formed will be the vertices of a parallelogram.
6. Given three circles, draw a circle which shall touch one of the given circles, be orthogonal to the second, and bisect the third.
7. A transversal DEF cuts the sides BC, CA, AB of a triangle ABC externally in D, E, F. The three circles having these points for centers and orthogonal to the circumcircle of ABC cut the respective sides externally in P, Q, R and internally in P', Q', R'. Show that the triads of points P, Q, R; P, Q', R'; P', Q, R'; P', Q', R are collinear, and that the triads of lines AP', BQ', CR'; AP', BQ, CR; AP, BQ', CR; AP, BQ, CR' are concurrent.

I. THE PROBLEM OF APOLLONIUS

506. Ten Problems. Consider the problem of determining a circle subject to three conditions taken from among the following: The circle is to pass through one or more points, P; to be tangent to one or more lines, L; to be tangent to one or more circles, (C). This question gives rise to the following ten problems, symbolically represented by (1) PPP, (2) PPL, (3) PLL, (4) LLL, (5) PPC, (6) PLC, (7) PCC, (8) LLC, (9) LCC, (10) CCC.

507. Problem I. *Construct a circle passing through three given points* (PPP).

508. Problem II. *Construct a circle passing through two given points and tangent to a given line* (PPL) (§ 452).

OTHERWISE. Let the line AB joining the given points A, B meet the given line t in I, and let t touch the required circle ABT in T (Fig. 119). If B' is the symmetric of B with respect to t, we have the following equalities among angles:

$$\begin{aligned}\angle ATB' &= ATB + BTB' = ATB + BTI + BTI \\ &= ATB + BAT + BTI = IBT + BTI \\ &= 180° - BIT.\end{aligned}$$

Thus the known segment AB' subtends at the point T a known angle; hence T may be determined (locus 7, § 11).

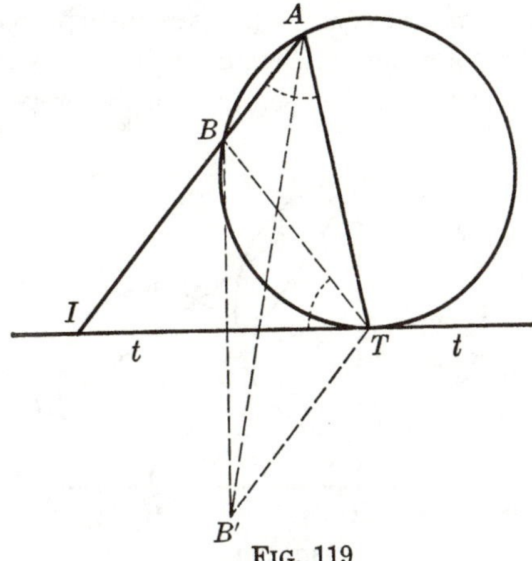

FIG. 119

The symmetric of T with respect to I yields another solution of the problem.

509. Problem III. *Construct a circle tangent to two given lines and passing through a given point* (PLL).

The required circle also passes through the symmetric P' of P with respect to the bisector of the angle formed by the given lines, and the problem reduces to the preceding one.

510. Problem IV. *Construct a circle tangent to three given lines* (LLL) (§ 117).

Consider the case when two of the lines, or all three, are parallel.

511. Problem V. *Draw a circle passing through two given points and tangent to a given circle* (PPC) (§ 470).

512. Problem VI. *Draw a circle passing through a given point, tangent to a given line and to a given circle* (PLC).

Let the required circle (I) (Fig. 120) passing through the given point B touch the given circle (A) in L and the given line s in D. The point L being a center of similitude of the two circles, the line LD passes through one end, say E, of the diameter EF of (A) parallel to the radius ID of (I). The point E is thus known, for EF is the diameter of (A) perpendicular to the line s.

Let EF meet s in G. The diameter EF subtends a right angle at L; hence DF subtends right angles at both L and G, and therefore $LFGD$ is a cyclic quadrilateral. Thus we have:

$$EF \cdot EG = EL \cdot ED = EB \cdot EC,$$

and the segment EC may be constructed as a fourth proportional, which in turn locates the point C on the line EB and reduces the problem to the case (PPL), which gives two solutions of the proposed problem.

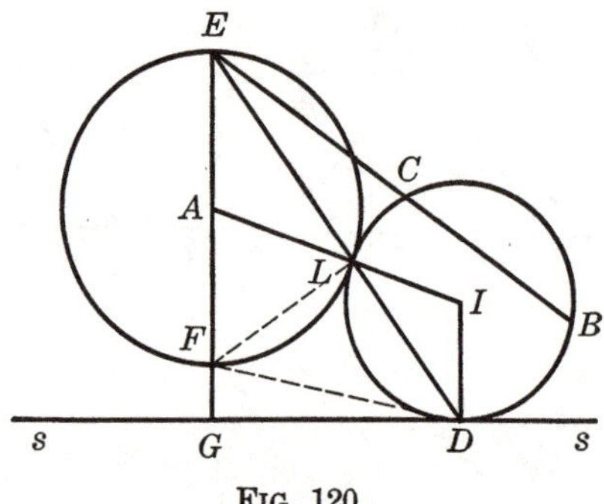

FIG. 120

The point F may be made to play the same role as E, and we obtain two other solutions — in all, four solutions.

513. Problem VII. *Draw a circle passing through a given point and tangent to two given circles* (PCC).

The points of contact M, N of the required circle (O) (Fig. 121) with the given circles (C), (D) are collinear with a center of similitude S of

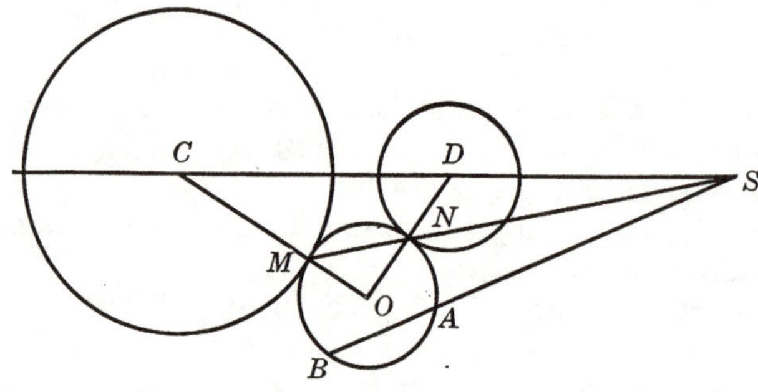

FIG. 121

(C), (D) (§ 409); hence if the line SA joining S to the given point A on (O) meets (O) again in B, we have:

$$SA \cdot SB = SM \cdot SN.$$

Now the value of the right-hand side of this equality is known (§ 400); hence B may be determined on the given line SA, and the problem reduced to the case (PPC), which gives two solutions of the proposed problem.

By making use of the other center of similitude S' of (C), (D) we obtain two other solutions — in all, four solutions.

514. Problem VIII. *Draw a circle tangent to two given lines and to a given circle* (LLC).

Let a, b be the given lines, (C, c) the given circle, and (O, r) the required circle. The circle $(O, r + c)$ (Fig. 122) will pass through C and through the symmetric C' of C with respect to the bisector t of the angle formed by a, b, for O lies on t. Moreover, $(O, r + c)$ will be

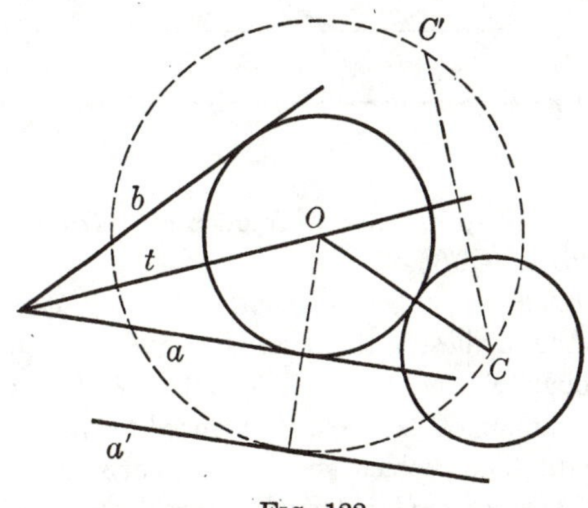

FIG. 122

tangent to the line a' drawn parallel to a at a distance c from a. Thus the construction of $(O, r + c)$ is reduced to the case (PPL), and the center O of (O, r) is determined.

The problem (PPL) has two solutions. Now the symmetric C' of C may be taken with respect to either of the bisectors t, t' of the angles formed by a, b, and the parallel a' to a may be drawn on either side of a; hence (LLC) may have eight solutions.

515. Problem IX. *Draw a circle tangent to a given line and to two given circles* (LCC).

Let p be the given line, (A, a), (B, b) the given circles, and (O, r) the required circle (Fig. 123). If a is smaller than b, the circle $(O, r + a)$ will pass through A, touch the circle $(B, b - a)$, and be tangent to the line p' drawn parallel to p at a distance a. Thus the problem of drawing $(O, r + a)$ is reduced to the case (PLC), and the center O of (O, r) is determined.

Since (PLC) has four solutions, and the line p' may be drawn on either side of p, the problem (LCC) may have eight solutions.

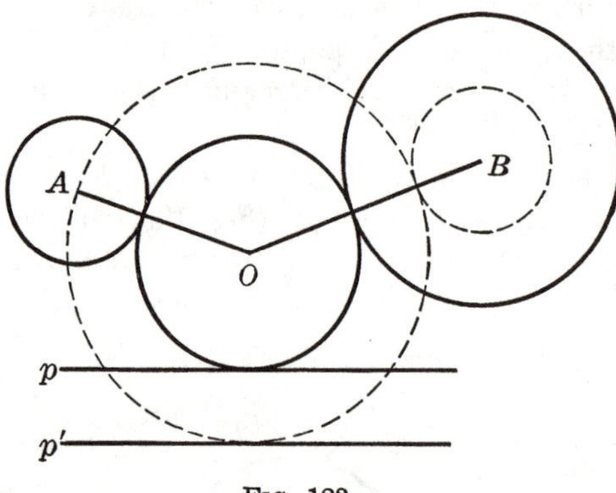

FIG. 123

516. Problem X. The Problem of Apollonius. *Draw a circle tangent to three given circles (CCC).*

Let (A, a), (B, b), (C, c) be the given circles, and (O, r) the required circle. If a is the smallest of the three given radii, the circle $(O, r + a)$ will pass through A and touch the circles $(B, b - a)$, $(C, c - a)$. Thus the construction of $(O, r + a)$ is reduced to the case (PCC), and O is determined. The problem may have eight solutions.

517. Problem. *On a given line to find a point the sum (or the difference) of whose distances from two given points shall have a given length.*

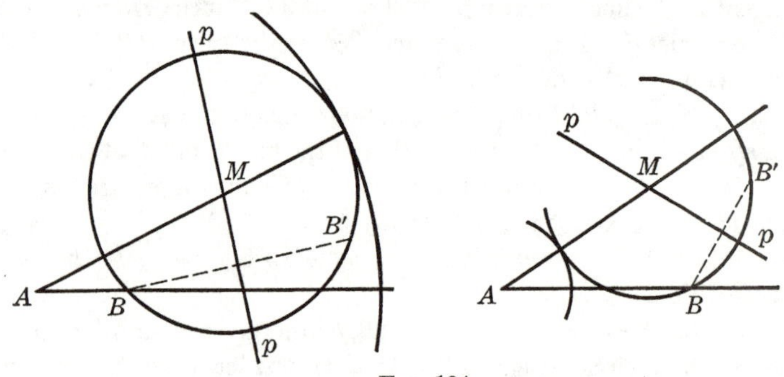

FIG. 124

Let p be the given line, A, B the given points, m the given length, and M the required point on p. The circle (M, MB) (Fig. 124) passes through the symmetric B' of B with respect to p and is tangent to the circle (A, m); hence (M, MB) may be constructed (§ 511), and M is thus determined.

EXERCISES

1. Given a point A outside a given circle (O), construct a circle touching the line OA and the circle (O) and having its center on a tangent from A to (O).
2. Draw a circle so that the chords which it determines in three given circles shall have given lengths.
3. Construct a triangle given the base, the altitude to the base, and the sum (or the difference) of the other two sides.
4. Construct a triangle given the base, the sum (or the difference) of the other two sides, and the angle which the median makes with the base.
5. Construct a triangle given the base, the corresponding altitude, and the sum of the two medians to the other two sides $(a, h_a, m_b + m_c)$.
6. Given a point P on the side AB of the triangle ABC, find a point Q on AC such that the length PQ shall be equal to the sum of the distances of the points P, Q from BC.
7. On a given line to find a point such that the sum of the tangents from that point to two given circles shall be equal to a given length.

SUPPLEMENTARY EXERCISES

1. If G is the centroid of a triangle ABC, show that the powers of the vertices A, B, C for the circles GBC, GCA, GAB, respectively, are equal.
2. On the circumcircle of the triangle ABC to find a point M such that $MA^2 = MB \cdot MC$.
3. ABC is a triangle, A' the midpoint of BC, P the pole of BC for the circumcircle; through the midpoint of $A'P$ a line is drawn parallel to BC meeting AB, AC, produced, in Q, R; show that A, P, Q, R are cyclic.
4. Through two given points P, Q on the circumcircle (O) of a triangle ABC the two circles are drawn touching the side BC, in A', A''. Let B', B'' and C', C'' be the analogous points on CA and AB. Assuming that A', B', C' are external, and A'', B'', C'' are internal to (O), show that (A', B', C'), (A', B'', C''), (A'', B', C''), (A'', B'', C') are collinear triads of points, and (AA'', BB'', CC''), (AA'', BB', CC'), (AA', BB'', CC'), (AA', BB', CC'') are triads of concurrent lines.
5. Show that the ratio of the distances of the centers of two circles from the radical axis of these circles is equal to the ratio of the distances of this line from its poles with respect to the circles considered.
6. Three circles, centers A, B, C, have a point D in common and intersect two-by-two in the points A', B', C'. The common chord DC' of the first two circles meets the third circle in C''. Let A'', B'' be the analogous points on the other two circles. Prove that the segments $A'A''$, $B'B''$, $C'C''$ are twice as long as the altitudes of the triangle ABC.

7. Two circles (O, R), (O', R') are such that quadrilaterals may be constructed that are inscribed in the first and circumscribed about the second. If $OO' = c$, prove that (a) $(R^2 - c^2)^2 = 2R'^2(R^2 + c^2)$; (b) R^2 cannot be smaller than $2R'^2$; (c) if $R^2 = 2R'^2$, the circles are concentric and all the quadrilaterals are squares.

8. Two variable points are marked on a tangent to a given circle, at equal distances from a fixed point on that tangent. Show that the locus of the point of intersection of the two tangents from the two points to the circle is a straight line.

9. In what direction should a billiard ball placed on a circular billiard table be shot so as to pass through the initial position after two successive reflexions?

10. Three circles are drawn each touching a different side of a triangle and all touching the circumcircle externally, in the same point. Prove that (a) the points of contact with the sides of the triangle are collinear; (b) if the contacts with the circumcircle are internal, the points of contact with the sides are the feet of three concurrent cevians.

11. Construct a triangle given, in position, the incircle, the midpoint of a side, and a point on the external bisector of the angle opposite that side.

12. Two given circles (O), (O') are touched by a common tangent in A, A'. Through a fixed point of AA' a variable secant is drawn meeting (O) in B, C, and (O') in B', C'. Show that the locus of the points of intersection of the lines AB, AC with the lines $A'B'$, $A'C'$ is a circle coaxal with the given circles.

13. A variable circle (M) touches a given circle (O) in I and cuts a second given circle (O') orthogonally in H, K. If the lines HK, IO' meet in P, and the line IO' meets (M) again in N, show that (a) the lines MN, MP meet the line OO' in two fixed points; (b) the variable circle (M) is tangent to a second fixed circle.

14. Show that the three circles having for diameters three concurrent cevians are met by the perpendiculars dropped from the orthocenter upon the respective cevians in three pairs of points belonging to the same circle.

15. Given three concurrent cevians, show that the perpendiculars dropped upon them from the orthocenter meet the circles having for diameters the corresponding sides of the triangle in three pairs of points lying on a circle.

16. Show that the circles having for diameters three concurrent cevians meet the circles having for diameters the corresponding sides of the triangle in three pairs of points lying on a circle.

17. A straight line through the circumcenter of a triangle ABC meets BC, CA, AB in P, Q, R; prove that the circles having for diameters AP, BQ, CR have two points in common, one on the circumcircle, the other on the nine-point circle, and that the common chord of the circles passes through the orthocenter. If P', Q', R' are the symmetrics of P, Q, R with respect to the circumcenter O, prove that the lines AP', BQ', CR' meet in a point on the circumcircle.

18. Draw the three circles each passing through two vertices of a triangle and orthogonal to its incircle. Show that (a) the triangle determined by the centers of these three circles is homothetic to the triangle formed by the points of contact of the incircle; (b) the circumcenters of these two triangles coincide, and their circumradii differ by half the inradius of the given triangle. Consider the excircles.

19. Given two fixed points A, B and a variable line passing through a fixed point. The two circles are constructed passing through A, B and tangent to the line at M and N. Prove that the circle through M, N, and A (or B) passes through a second fixed point.
20. If S is a center of similitude of two orthogonal circles intersecting in A and C, prove that if B, D are the traces on the line of centers of the polars of S for these circles, $ABCD$ is a square.
21. Construct two circles given their two centers of similitude, the length of one common tangent, and the direction of another.
22. Find the locus of a point P such that the tangents PA, PB, PC drawn from P to three given coaxal circles shall satisfy the relation $PA \cdot PB : PC^2 = k$.
23. From two variable points C, D lying on a fixed line s and equidistant from the center O of a given circle (O) the tangents CE, DF are drawn to (O) intersecting in M. (a) Find the locus of M. (b) Find the locus of the incenter and of the orthocenter of the triangle MEF. (c) Find the locus of the circumcenter of the triangle PEF, where P is a given point.
24. If two circles intersect in A, B and touch a third circle in C, D, show that $AC:AD = BC:BD$.
25. Show that the polar of a point common to two intersecting circles with respect to a variable circle tangent to the given circles passes through a fixed point. (It is assumed that the variable circle touches the given circles in a fixed way, i.e., either both circles internally, or both circles externally, or one circle internally and the other externally.) If the given circles are tangent to each other, show that the proposition is valid for their point of contact.
26. Show that the three chords of intersection of the circumcircle of a triangle with the escribed circles meet the corresponding sides of the triangle in three collinear points.
27. If the tangents p, q, r drawn to a circle (S) from the vertices of a triangle, sides a, b, c, are such that one of the products ap, bq, cr is equal to the sum of the other two, prove that (S) touches the circumcircle.
28. Four circles may be drawn tangent to the sides AB, AC of a triangle ABC and to its circumcircle (O). Show that the four polars of A for these four circles pass each through a tritangent center of ABC.

IX

INVERSION

518. Definitions. Given a point O, the *center*, or the *pole*, *of inversion*, and a constant k, the *constant of inversion*, to any point P of the plane we make correspond a point P', on the line OP, such that $OP \cdot OP' = k$. The point P' is said to be the *inverse* of P. It is clear that P is the inverse of P'.

If the constant k is positive, the two points P, P' are taken on the same side of the center of inversion O; if k is negative, the points P, P' lie on opposite sides of O.

If the constant of inversion k is positive, the circle (O) having O for center and k for the square of its radius is called the *circle of inversion.* Any point on the circumference of (O) coincides with its inverse, and any two inverse points P, P' are also inverse with respect to (O), in the sense in which inverse points with respect to a circle were defined before (§ 358). The radius of (O) is called the *radius of inversion.*

The line OP joining a point P to the center of inversion O is called the *radius vector* of P.

The inversion having O for center and k for constant of inversion is sometimes denoted by (O, k).

If the points P, P' correspond to each other in the inversion (O, k), and the point P describes a given curve (C), the curve (C') described by P' is said to be the *inverse of the curve* (C).

519. Theorem. *Two points P, Q and their corresponding points P', Q' are collinear, or concyclic.*

In the latter case the lines PQ, $P'Q'$ are antiparallel with respect to the radii vectors OPP', OQQ'.

520. Problem. *Given the lengths of the radii vectors of two given points and the length of the segment joining those points, to find the length of the segment joining their inverse points.*

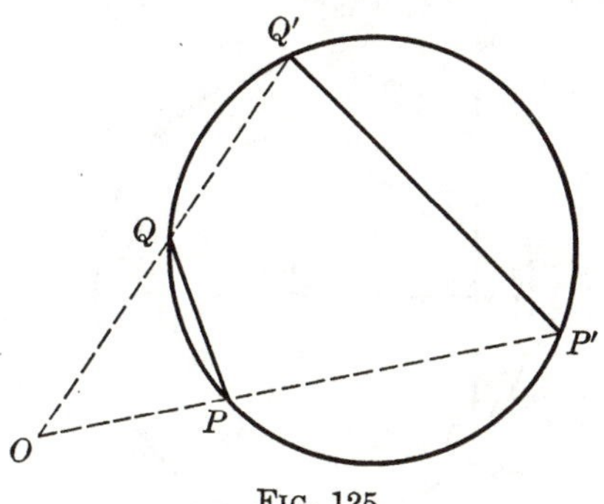

Fig. 125

Let P', Q' be the points corresponding to the two given points P, Q in the inversion (O, k) (Fig. 125). We have $OP \cdot OP' = k$, and from the similar triangles $OPQ, OP'Q'$ we have:

$$P'Q' : PQ = OP' : OQ = OP \cdot OP' : OP \cdot OQ = k : OP \cdot OQ;$$

hence:

(f) $$P'Q' = PQ(k : OP \cdot OQ).$$

521. Theorem. *If a point describes a straight line not passing through the center of inversion, the inverse point describes a circle passing through the center of inversion.*

Let A be the foot of the perpendicular from the center of inversion O upon the given line x, P any point on x, and A', P' the inverses of A, P (Fig. 126). The lines AP, $A'P'$ are antiparallel with respect to the lines OAA', OPP' (§ 519); hence angle $A'P'P = PAO = 90°$, and therefore the locus of P' is the circle (OA') having OA' for diameter.

Otherwise. Let B be the symmetric of O with respect to x, and B' the inverse of B (Fig. 126). The four points B, B', P, P' being cyclic, the two triangles $OBP, OB'P'$ are similar, and since OPB is isosceles, we have $B'P' = B'O$. Thus the variable point P' is at a fixed distance $B'O$ from the fixed point B', which proves the proposition.

522. *Remark.* (a) The proposition is often stated, more succinctly: *The inverse of a straight line is a circle through the center of inversion.*

(b) The proofs suggest two methods for constructing the circle into which a given line is inverted.

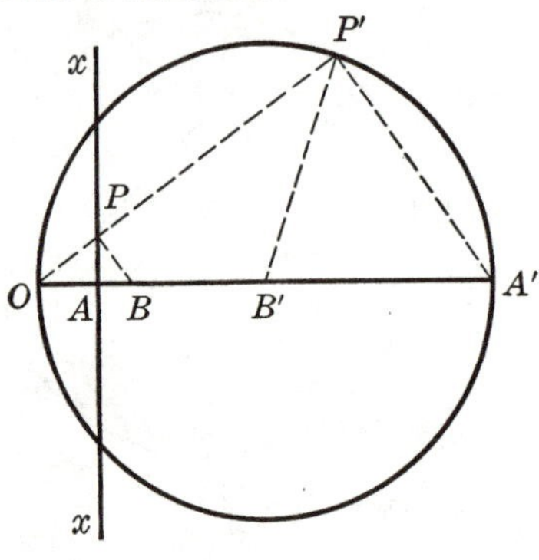

FIG. 126

(c) In the figure the pairs of inverse points lie on the same side of the center of inversion O, which shows that the constant of inversion k is positive and the circle of inversion (O) is real. However, the proofs given are valid whether k is positive, or negative. The reader may draw the figure for the latter case.

(d) If the circle of inversion is real, the given line is the radical axis of this circle and the circle inverse to the line.

Indeed, we have (Fig. 126):

$$AO \cdot AA' = AO(AO + OA') = AO^2 + AO \cdot OA' = AO^2 - OA \cdot OA'.$$

Now $AO \cdot AA'$ is the power of A for the circle (OA'), and $AO^2 - OA \cdot OA'$ is the power of A with respect to the circle of inversion (O), for $OA \cdot OA'$ is the square of the radius of (O); hence the proposition.

(e) We have:

$$k = OB \cdot OB' = 2OA \cdot OB';$$

hence the radius OB' of the inverse circle (OA') is equal to $(k:2OA)$.

523. Converse Theorem. *If a point describes a circle passing through the center of inversion, the inverse point describes a straight line perpendicular to the line joining the center of the given circle to the center of inversion.*

The proof is readily obtained by reversing the steps in either of the proofs of the direct proposition.

The proposition is often stated more briefly: *The inverse of a circle passing through the center of inversion is a straight line.*

524. Theorem. *A straight line and a circle may be considered inverse figures in two different ways.*

From the preceding two propositions (§§ 521, 523) it follows that if a circle and a line are two inverse figures, the center of inversion must be an end of the diameter d of the given circle perpendicular to the given line x. On the other hand, if we take an end O of d as center of inversion and for constant of inversion k the product of the distances of O from x and from the diametric opposite of O on the given circle, the line x will invert into the given circle (§ 521); hence the proposition.

If x is tangent to the circle, the diametric opposite of the point of contact is the only center of inversion available.

525. Theorem. *If a point describes a circle not passing through the center of inversion, the inverse point describes a circle not passing through the center of inversion.*

Let the line OP joining the center of inversion O to any point P of the given circle (C) meet (C) again in Q (Fig. 127), and let P' be the

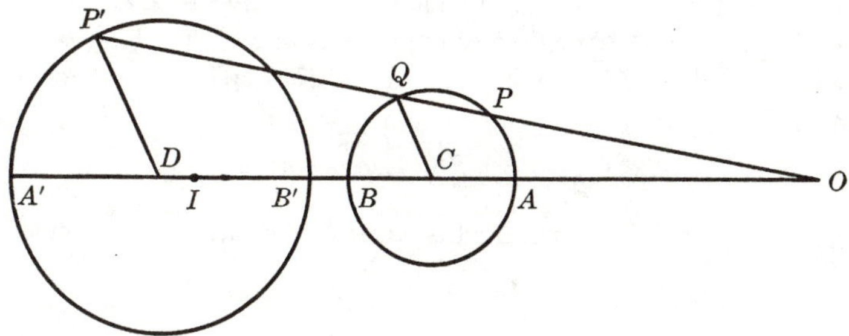

FIG. 127

inverse of P in the inversion (O, k). We have, denoting by p the power of O for (C):

$$OP \cdot OP' = k, \quad OP \cdot OQ = p; \quad \text{hence} \quad OP' : OQ = k : p.$$

Thus as Q varies on (C) with the point P, the point P' describes a circle (D) homothetic to (C), the point O and the constant $k:p$ being the center and the ratio of homothecy.

526. *Remark.* (a) The proposition is sometimes stated more succinctly: *The inverse of a circle is a circle.*

(b) The center of inversion is a center of similitude of the given circle and its inverse.

(c) The two inverse points P, P' are two antihomologous points on the two inverse circles. The point P' describes the circle (D) in the sense opposite to the sense in which the point P describes the given circle (C).

(d) If R, R' are the radii of the circles (C), (D), we have (Fig. 127):

$$R':R = OP':OQ = OP\cdot OP':OP\cdot OQ = k:p;$$

hence:

$$R' = R(k:p).$$

(e) If the constant of inversion is equal to the power of the center of inversion with respect to the given circle, the circle inverts into itself, for in that case the point P' coincides with the point Q.

When the constant of inversion is positive, and therefore the circle of inversion real, the result thus obtained may be stated: *A circle orthogonal to the circle of inversion inverts into itself.*

527. *Observation.* It is essential to observe that the center D of the circle (D) (§ 526) is not the inverse of the center C of the given circle (C). The inverse I of C may be found in the following manner.

Let A, B be the diametrically opposite points of (C) lying on the line OCD, and A', B' their inverses (Fig. 127). We have, in magnitude and in sign:

$$OA + OB = 2\,OC, \quad OA\cdot OA' = OB\cdot OB' = OC\cdot OI = k;$$

hence, dividing the terms of the first equality by these equal products:

$$\frac{1}{OA'} + \frac{1}{OB'} = \frac{2}{OI},$$

i.e., the point I is the harmonic conjugate of O with respect to A', B' (§ 348), or, in other words, *the center of a circle inverts into the inverse, with respect to the inverse circle, of the center of inversion.*

528. Theorem. *Any two circles may be considered as the inverse to one another, in two different ways.*

If two circles are the inverse to one another, the center of inversion is a center of similitude of the two circles (§ 526 b), and the constant of inversion is the product of the distances of this center of similitude from two points on the two circles antihomologous with respect to the center of similitude considered (§ 526).

On the other hand, given the two circles, if the center and constant of inversion be chosen in the manner just indicated, one circle will invert into the other (§ 525). Hence the proposition.

529. Definitions. If two curves have a point P in common, the angle formed by the tangents at P to the curves is said to be the *angle of intersection* of the curves.

The angle of intersection of a curve and a straight line is the angle formed by the straight line and the tangent to the curve at a point common to the curve and the given line.

In particular, the angle of intersection of a line and a circle is a right angle if, and only if, the line passes through the center of the circle.

530. Theorem. *The tangents to two inverse curves at two corresponding points are symmetrical with respect to the mediator of the segment determined by the points of contact.*

Let A, A' and B, B' be two pairs of corresponding points on two inverse curves $(C), (C')$ (Fig. 128). In the cyclic quadrilateral $ABB'A'$ (§ 519) angle $A'AB = A'B'D$.

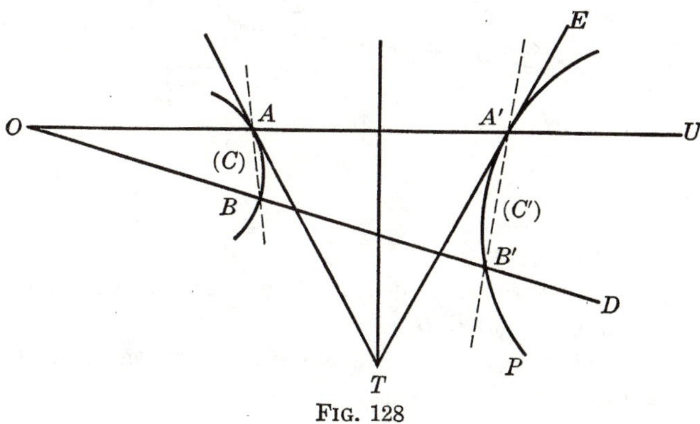

FIG. 128

Now if B approaches A, the limiting positions of the chords AB, $A'B'$ are the tangents AT, $A'T$ to (C), (C') at A, A', and the limiting position of the line OBB' is the line OAA'. Thus the limiting values of the angles $A'AB$, $A'B'D$ are the angles $A'AT$, $EA'U = AA'T$; hence the proposition.

Observe that while the angles TAA', $TA'A$ are equal in magnitude, they are opposite in sign, if the radius vector OAA' is taken for the initial line.

531. Theorem. *If two curves intersect, their angle of intersection is equal to the angle of intersection of the inverse curves at the corresponding point.*

Indeed, if AT, AR are the tangents to the two given curves intersecting at A, and $A'T$, $A'R$ are the tangents to the respective inverse curves at the corresponding point A', the points T, R lie on the mediator of the segment AA', and the angles TAR, $TA'R$ of the two congruent triangles TAR, $TA'R$ are equal.

This important proposition is often stated thus: *Inversion preserves angles.*

532. COROLLARY. (*a*) *Inversion preserves orthogonality.* (*b*) *Inversion preserves contact.*

533. Peaucellier's Cell. Let $ABCD$ (Fig. 129) be a rhombus formed by four rigid bars of equal length hinged together, and let thc joints

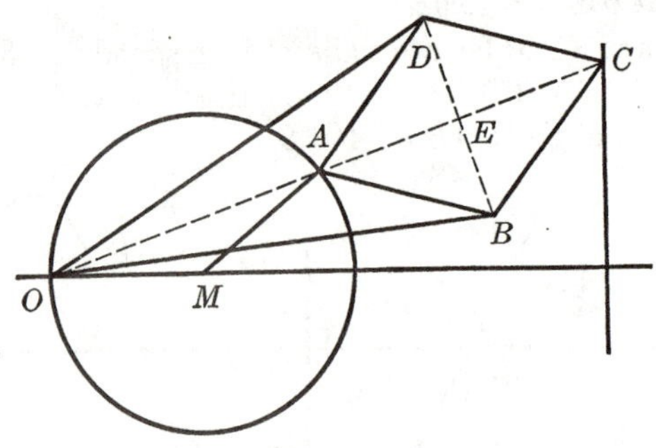

FIG. 129

B, D be connected with a fixed point O by means of two equal rigid bars hinged at O, and at B, D.

The points O, A, C and the midpoint E of BD lie on the mediator of BD, and we have:

$$\begin{aligned} OA \cdot OC &= (OE - AE)(OE + AE) = OE^2 - AE^2 \\ &= (OD^2 - DE^2) - (AD^2 - DE^2) = OD^2 - AD^2; \end{aligned}$$

hence A, C are inverse points in the inversion whose circle of inversion has for center the point O and for radius the leg of a right triangle, the hypotenuse and the other leg of which are the fixed lengths OD and AD. Thus if the point A is made to describe any curve, the point C will describe the inverse curve.

In particular, if A is connected by a rigid rod to a fixed point M, and $MA = MO$, the point A will describe a circle passing through O;

hence the point C will describe a straight line perpendicular to MO (§ 523).

This linkage, known as Peaucellier's Cell, transforms circular motion into rectilinear motion.

534. *Observation.* When a figure (F) is transformed by an inversion, the relations existing within the figure (F) appear in the inverse figure (F') in a more or less changed form. The preceding basic properties of inversion enable us to infer what property (P') the figure (F') will have, if the figure (F) has the property (P), and vice versa.

This relation between the properties of the two figures (F), (F') is often made use of, in the following manner. In order to prove that the figure (F) has a certain property, we transform (F) by a suitable inversion into a figure (F'). It often happens that in the new figure we readily notice a property the corresponding property of which in (F) is precisely the property of (F) to be proved. This correspondence between the two properties is the required proof.

Similar considerations apply to construction problems.

The following propositions illustrate the method.

535. Theorem. *If two of the four circles determined by four points taken three at a time are orthogonal, the remaining two circles are also orthogonal.*

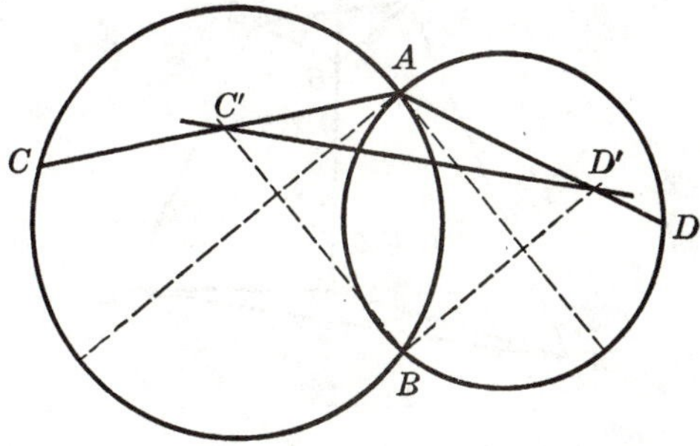

FIG. 130

Let A, B, C, D (Fig. 130) be the four given points, and suppose that the two circles ABC, ABD are orthogonal. The inversion having (A, AB) for circle of inversion transforms the point B into itself, and the two circles ABC, ABD into two straight lines (§ 523) forming a right angle (§ 532 a) at B and passing, respectively, through the inverses C', D' of C, D.

The circle CDA transforms into the line $C'D'$ (§ 523) and the circle CDB into the circle $C'D'B$ (§ 525). Now since $C'BD'$ is a right angle, the line $C'D'$ is a diameter of the circle $C'D'B$ and is therefore orthogonal to that circle. From this we infer that the circles (CDA), (CDB) are orthogonal (§ 532 a), which proves the proposition.

OTHERWISE. The inversion (C, p), where p is the power of C for the circle ABD, transforms this circle into itself (§ 526 e), and the lines CA, CB, CD meet this circle again in the inverses A', B', D' of the points A, B, D. The circle CAB inverts into the straight line $A'B'$, and this line is a diameter of the circle ABD, for the circles ABC, ABD are orthogonal, by assumption.

The circles CDA, CDB invert into the lines $D'A', D'B'$. Now these two lines are perpendicular; hence the circles CDA, CDB are orthogonal.

536. Ptolemy's Theorem and Its Extension. *In a quadrilateral the sum of the products of the two pairs of opposite sides is equal to or greater than the product of the diagonals depending on whether the quadrilateral is cyclic or not cyclic.*

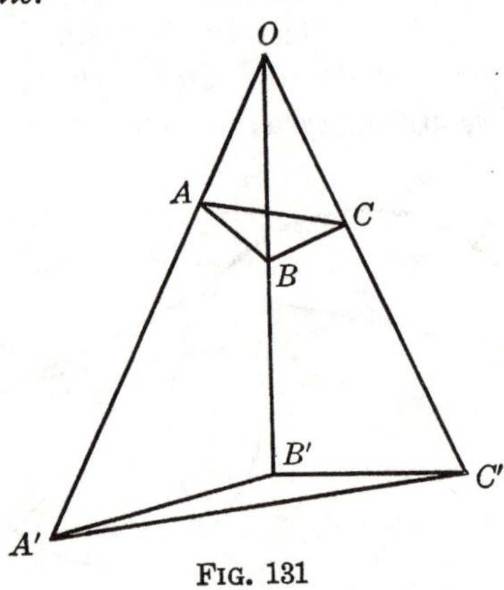

FIG. 131

Let $OABC$ be the given quadrilateral (Fig. 131) and A', B', C' the inverse points of A, B, C in the inversion (O, k). We have (§ 520):

$$\frac{AB}{OA\cdot OB} = \frac{A'B'}{k}, \quad \frac{BC}{OB\cdot OC} = \frac{B'C'}{k}, \quad \frac{AC}{OA\cdot OC} = \frac{A'C'}{k}.$$

Now if $OABC$ is not cyclic, the points A', B', C' are not collinear (§ 525), so that:

$$(1) \qquad A'B' + B'C' > A'C';$$

hence:

(2) $$\frac{AB}{OA \cdot OB} + \frac{BC}{OB \cdot OC} > \frac{AC}{OA \cdot OC},$$

or:

(3) $$AB \cdot OC + BC \cdot OA > AC \cdot OB.$$

But if $OABC$ is cyclic, the points A', B', C' are collinear (§ 523); hence in (1), and therefore also in (2) and (3), the sign of inequality is to be replaced by the sign of equality.

537. Theorem. *If a circle is tangent internally (or externally) to the circumcircle of a triangle and to two sides of the triangle, the line joining its points of contact with the sides passes through the incenter (or corresponding excenter) of the triangle.*

Let the circle PQT touch the circumcircle (O) of the triangle ABC internally in T (Fig. 132) and the sides AB, AC in P, Q. Let I be

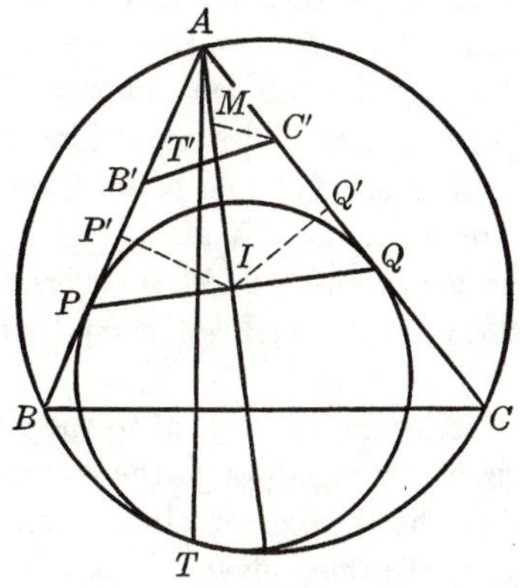

FIG. 132

the midpoint of PQ. If B', C' are the inverses of the points B, C in the inversion (A, AI^2), the line $B'C'$ is the inverse of (O), and the line AT meets $B'C'$ in the inverse T' of T.

Let P', Q' be the inverses of P, Q. We have $AP \cdot AP' = AI^2$, and since the triangle AIP is right-angled at I, the point P' coincides with the foot of the perpendicular from I upon the hypotenuse AP. Similarly for the point Q'. Now the circle $P'Q'T'$ is the inverse of PQT and is therefore tangent to $B'C'$ at T' (§ 532 b); hence the point I is the excenter of the triangle $AB'C'$ relative to A.

The circle $IB'C'$ meets AI in the incenter M of the triangle $AB'C'$ (§ 122 a), and in the cyclic quadrilateral $IB'MC'$ we have:

$$\angle B'IM = B'C'M = \tfrac{1}{2}\, B'C'A = \tfrac{1}{2}\, ABC,$$

for the lines BC, $B'C'$ are antiparallel with respect to AB, AC. Now the lines BI and $B'I$ are also antiparallel with respect to AB, AI, for the points B', I are the inverses of B, I; hence:

$$\angle ABI = B'IA = B'IM = \tfrac{1}{2}\, ABC,$$

and I is the incenter of ABC.

The proof in case of external contact is analogous.

538. Problem. *Find an inversion which transforms three given circles into themselves.*

The power p of the required center of inversion S with respect to each of the given circles must be equal to the constant of inversion k (§ 526 e); hence the three powers must be equal, i.e., S must coincide with the radical center of the three given circles. Furthermore, the constant of inversion must be equal to the power of S with respect to the given circles. The problem has one and only one solution (§ 493).

If the centers of three given circles are collinear, the problem has either an infinite number of solutions, or no solution, depending on whether the circles are or are not coaxal.

539. Problem. *Find the inversion which transforms three given circles with noncollinear centers into three circles having their centers on a given straight line.*

The three inverse circles are orthogonal to the given line s; hence s is the inverse of a circle passing through the required center of inversion and orthogonal to the three given circles, i.e., O is a point of the orthogonal circle (R) of the three given circles. Thus if (R) is real, O is either end of the diameter of (R) perpendicular to s.

540. Theorem. *A coaxal pencil of circles inverts into a coaxal pencil of circles.*

Let (P), (Q) be any two circles of the coaxal pencil (E) conjugate to the given coaxal pencil of circles (F). Any circle (C) of (F) inverts into a circle (C') orthogonal to the inverses (P'), (Q') of (P), (Q) (§ 532 a); hence (C') describes a coaxal pencil (F') (§ 455).

541. Discussion. One of the two given conjugate coaxal pencils (E), (F), say (F), has two basic points A, B, which points are the limiting points of (E). The inverse circles (F') of (F) pass through the inverses A', B' of A, B.

The circles of (E) invert into a coaxal pencil (E') conjugate to (F'); hence the points A', B' are the limiting points of (E').

If one of the points A, B, say A, is taken for center of inversion, the circles of (F) will invert into straight lines passing through the inverse B' of B; every circle of (E') will be orthogonal to the lines passing through B', i.e., B' will be the common center of all the circles of (E').

542. Corollary. *In an inversion two points inverse with respect to a circle invert into two points inverse with respect to the inverse circle of the given circle.*

Indeed, the limiting points A, B (§ 541) are inverse with respect to any given circle, say (M), of (E), and the points A', B' are inverse with respect to the inverse (M') of (M).

543. Theorem. *Two inverse circles and the circle of inversion are coaxal.*

Let (C) and (C') be two inverse circles and (S) the circle of inversion.

If (S) and (C) have two points in common, the circle (C') passes through these points, since the points of (S) invert into themselves.

If (S) and (C) have no common points, they determine a nonintersecting coaxal pencil of circles having two limiting points L, L'. These points being inverse with respect to (S) invert into each other, and since they are inverse with respect to (C), they are also inverse with respect to (C') (§ 542); hence (S), (C), (C') belong to the same coaxal pencil (§ 448 b).

544. Corollary. *If two circles are inverse with respect to a third circle as circle of inversion, this circle is a circle of antisimilitude of the first two circles.*

Indeed, the center S of the third circle (S) is a center of similitude of the first two circles (C), (C') (§ 526 b), and (S) is coaxal with (C), (C'); hence the proposition.

545. Theorem. *In a triangle the circumcircle, the nine-point circle, the polar circle, and the circumcircle of the tangential triangle are coaxal.*

The vertex A and the foot D of the altitude AD of the triangle ABC are inverse points with respect to the polar circle (H) of ABC; similarly for the analogous pairs of points B, E and C, F. Thus the circumcircle $(O) = ABC$ and the nine-point circle $(N) = DEF$ of ABC are inverse with respect to (H); hence (O), (N), (H) are coaxal (§ 543).

The point of intersection L of the tangents to (O) at B, C is the pole of BC for (O); hence L and the midpoint A' of BC are inverse with respect to (O). Similarly for the analogous pairs of points M, B' and

N, C'. Thus the circumcircle $(T) = LMN$ of the tangential triangle of ABC and $(N) \equiv A'B'C'$ are inverse with respect to (O), and therefore (O), (T), (N) are coaxal.

Consequently the four circles (O), (N), (H), (T) are coaxal (§ 441).

546. Corollary. *The circumcenter of the tangential triangle lies on the Euler line of the given triangle* (§ 443).

EXERCISES

1. Show that the tangents to a curve from the center of inversion are also tangent to the inverse curve.

2. If two circles are inverted with respect to any center, into what does their line of centers invert?

3. Prove that the formula (f) (§ 520) remains valid, if the points P, Q are collinear with the center of inversion.

4. Show that the inverse of a harmonic range, with respect to a point on the line of the range, is a harmonic range.

5. Given three noncollinear points P, A, B, and a variable point M collinear with A, B, prove that the angle of intersection of the two circles PAM, PBM is constant.

6. The tangent LM to a given circle (O) at a variable point M meets a given line s in L. Show that the locus of the projection P of M upon OL is a circle.

7. Given a circle (O) and the two fixed points A, B, a variable line through B meets (O) in C, D, and the lines AC, AD meet (O) again in E, F. Show that the center of the circle AEF describes a straight line.

8. If the line joining the ends A, B of a diameter AB of a given circle (O) to a given point P meets (O) again in A', B', show that the circle $PA'B'$ is orthogonal to (O).

9. Given two circles (A), (B) intersecting in E, F, show that the chord $E'F'$ determined in (A) by the lines MEE', MFF' joining E, F to any point M of (B) is perpendicular to MB.

10. Two rectangular chords BC, DE of a given circle intersect within the circle, in the point A. Show that the altitude of the triangle ABD issued from A is a median of the triangle ACE.

11. The points A', B', C' are marked on the altitudes AH, BH, CH of the triangle ABC so that we have, both in magnitude and in sign:

$$HA \cdot HA' = HB \cdot HB' = HC \cdot HC'.$$

Show that H is a tritangent center of the triangle $A'B'C'$.

12. A variable tangent to a given circle, center A, cuts a second given circle, center B, in the points C, D. Prove that the circle BCD is tangent to a fixed circle. *Hint.* Consider the inversion (B, BC^2).

13. An angle BAC of fixed magnitude revolves about its fixed vertex A and meets a fixed line s in B, C. Show that the circle ABC is tangent to a fixed circle.

14. Show that the product of the powers of a center of similitude of two circles with respect to these circles is equal to the fourth power of the radius of the corresponding circle of antisimilitude.

15. If the circle (B) passes through the center A of the circle (A), show that the radical axis of (A), (B) is the inverse of (B) with respect to (A) taken as circle of inversion.

16. Show that the circle of similitude and the radical axis of two circles are inverse with respect to a circle of antisimilitude of the two given circles.

17. Of the four lines joining the ends of two rectangular diameters of two orthogonal circles, two lines pass through one point of intersection of the two circles, and the remaining two lines pass through the other point of intersection of the two circles.

18. (a) Prove that the inverse of the circumcircle (O) of a triangle ABC with respect to the incircle (I), as circle of inversion, is the nine-point circle of the triangle XYZ determined by the points of contact of (I) with the sides of ABC; (b) prove, by inversion, Euler's formula $d^2 = R^2 - 2\,Rr$; (c) state and prove by inversion the analogous propositions regarding the excircles.

19. Given four points in a plane, each set of three is inverted with respect to the fourth; show that the four inverse triangles are similar.

20. Through two given points to draw a circle intersecting a given circle at a given angle.

21. Invert the theorem: The altitudes of a triangle meet in a point. *Hint.* Use the orthocenter as center of inversion.

RECENT GEOMETRY OF THE TRIANGLE

A. POLES AND POLARS WITH RESPECT TO A TRIANGLE

547. Harmonic Relations. (a) Let P be a point in the plane of the triangle ABC (Fig. 133), L, M, N the traces of the lines AP, BP, CP on the sides BC, CA, AB, and L', M', N' the harmonic conjugates of L, M, N, for the respective pairs of vertices of ABC.

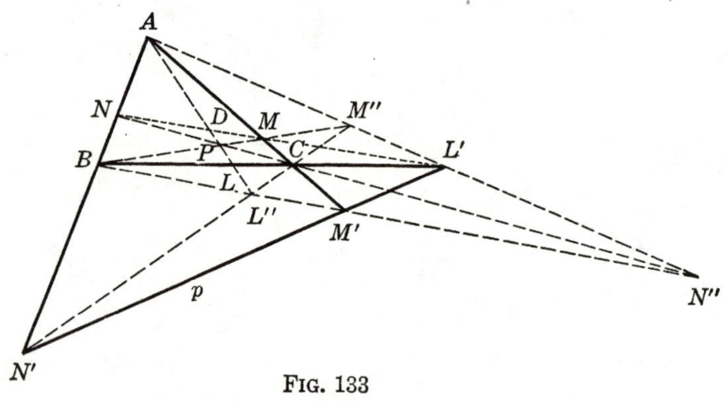

FIG. 133

Theorem. *The points L', M', N' are collinear.*

(b) Let p be a line in the plane of the triangle ABC meeting the sides BC, CA, AB in the points L', M', N', and let L, M, N be the harmonic conjugates of L', M', N' for the respective pairs of vertices of ABC.

Theorem. *The lines AL, BM, CN are concurrent.*

In both cases (a) and (b) we have, by construction, both in magnitude and in sign:

$$\frac{BL'}{L'C} = -\frac{BL}{LC}, \quad \frac{CM'}{M'A} = -\frac{CM}{MA}, \quad \frac{AN'}{N'B} = -\frac{AN}{NB};$$

hence, multiplying:

$$\text{(F)} \qquad -\frac{BL' \cdot CM' \cdot AN'}{L'C \cdot M'A \cdot N'B} = \frac{BL \cdot CM \cdot AN}{LC \cdot MA \cdot NB}.$$

In case (a) the right-hand side of the formula (F) is equal to unity, by Ceva's theorem, and it follows from the resulting equality that the points L', M', N' are collinear, by Menelaus' theorem.

In case (b) the left-hand side of (F) is equal to unity, by Menelaus' theorem, and it follows from the resulting equality that the lines AL, BM, CN are concurrent, by Ceva's theorem.

548. Definitions. The line $p = L'M'N'$ (§ 547) is said to be the *trilinear polar*, or the *harmonic line* of the point P for the triangle ABC, and P is the *trilinear pole*, or the *harmonic pole* of the line p for ABC.

549. Theorem. *The trilinear polar of a point for a triangle is the axis of perspectivity of this triangle and the cevian triangle of the point for the given triangle.*

We have (Fig. 133):

$$(ABNN') = -1, \quad (ACMM') = -1,$$

and the two harmonic ranges have the point A in common; hence (§ 357) the lines BC, MN, $M'N'$ are concurrent, i.e., the point L' is collinear with M, N. Similarly M' is collinear with L, N, and N' with L, M. Hence the proposition.

550. *Remark.* Given the triangle ABC and the point P, the above proposition (§ 549) gives a simple method for the construction of the trilinear polar p of P for ABC, with ruler alone.

551. Theorem. *The trilinear polar of a point for a triangle is also its trilinear polar for the cevian triangle of this point with respect to the given triangle.*

We have $A(BCLL') = -1$ (Fig. 133); hence the point L' is the harmonic conjugate, with respect to M, N, of the trace D of the line APL on the line $L'MN$ (§ 353). Thus L' is a point on the trilinear polar of P for the triangle LMN. Similarly for M' and N'. Hence the proposition.

552. Definition. The vertices $L'' = (BM', CN')$, $M'' = (CN', AL')$, $N'' = (AL', BM')$ of the triangle formed by the lines AL', BM', CN' (§ 547) are said to be the *harmonic associates* of P for the triangle ABC.

553. Theorem. *The harmonic associates of the point P (§ 552) lie on the lines AP, BP, CP and are separated harmonically from P by the pairs of points A and L, B and M, C and N, respectively.*

From the two harmonic pencils (Fig. 133):

$$B(ACMM') = -1, \quad C(ABNN') = -1$$

cut by the line APL, it follows that both BM' and CN' pass through the harmonic conjugate of P for A and L; hence that harmonic conjugate is the point L'' common to the lines BM', CN'. Similarly for M'' and N''.

554. Corollary. *The harmonic associates of a point P with respect to a triangle ABC form a triangle perspective to ABC, the point P being the center of perspectivity, and the axis of perspectivity being the trilinear polar of P for ABC.*

555. *Remark.* Given the triangle ABC and the line $p = L'M'N'$, the lines AL', BM', CN' determine the points L'', M'', N'' and the lines AL'', BM'', CN'' meet in the harmonic pole P of the line p for the triangle ABC.

556. Theorem. *The point P and its three harmonic associates with respect to a triangle ABC form a group of four points, any three of which determine a triangle circumscribed about and perspective to ABC, the center of perspectivity being the fourth point of the group.*

The axes of perspectivity are the lines $L'MN, LM'N, LMN'$.

The proof is readily obtained from the figure (Fig. 133).

557. *Remark.* The figure 133 was constructed given the triangle ABC and the point P. The same figure, however, would be obtained, if instead of P one of its harmonic associates, say L'', with respect to ABC were given. For, given L'', its harmonic conjugate with respect to A, L is the point P, and the rest of the figure follows.

EXERCISES

1. Show that the orthic axis of a triangle is the trilinear polar of the orthocenter with respect to the given triangle and with respect to the orthic triangle; the vertices of the given triangle are the harmonic associates of the orthocenter with respect to the orthic triangle.
2. Show that the trilinear polar of the incenter of a triangle passes through the feet of the external bisectors, and this line is perpendicular to the line joining the

incenter to the circumcenter; the excenters of the triangle are the harmonic associates of the incenter.

3. Construct the harmonic associates of the centroid. Does the centroid have a trilinear polar?
4. If L, M, N are the traces of the lines AP, BP, CP on the sides BC, CA, AB of the triangle ABC, and L', M', N' the traces, on the same sides, of the trilinear polar of P for ABC, show that the midpoints of the segments LL', MM', NN' are collinear.
5. If A' is the point of intersection of the side BC of the triangle ABC with the trilinear polar p of a point P on the circumcircle of ABC, show that the circle APA' passes through the midpoint of BC.
6. With the vertices of a given triangle as centers, it is possible to describe three circles so that the feet of three given concurrent cevians shall be centers of similitude for the respective pairs of circles.
7. Let L, M, N, be the feet of the cevians AP, BP, CP of the triangle ABC, and let P' be a point on the trilinear polar of P for ABC. If the lines AP', BP', CP' meet MN, NL, LM in X, Y, Z, show that the triangle XYZ is circumscribed about the triangle ABC.

B. LEMOINE GEOMETRY

a. SYMMEDIANS

558. Definition. The symmetric of a median of a triangle with respect to the internal bisector issued from the same vertex is a *symmedian* of the triangle.

A triangle has three symmedians.

Observe that a median and the corresponding symmedian of a triangle are also symmetric with respect to the external bisector issued from the same vertex.

559. Theorem. *The traces, on the circumcircle of a triangle, of a median and the corresponding symmedian determine a line parallel to the side of the triangle opposite the vertex considered.*

Indeed, the arc PQ (Fig. 134) intercepted on the circumcircle of the triangle ABC by the median AP and the symmedian AQ has the same midpoint M as the arc subtended by the side BC.

This proposition affords a simple construction of the symmedians of a triangle.

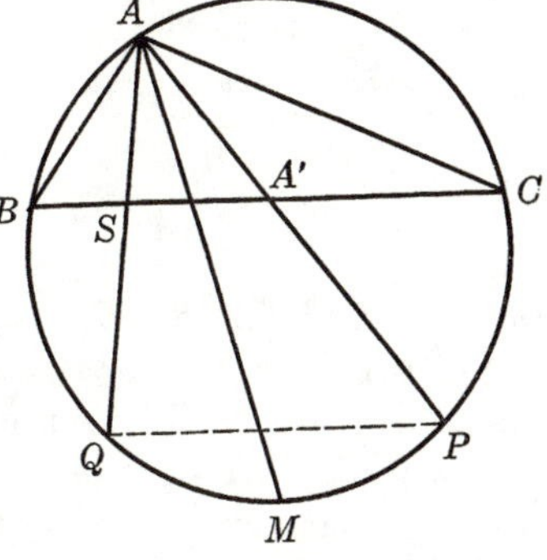

FIG. 134

560. Theorem. *The symmedian issued from a vertex of a triangle passes through the point of intersection of the tangents to the circumcircle at the other two vertices of the triangle.*

The point of intersection D of the tangents DB, DC to the circumcircle (O) of the triangle ABC at B, C is the pole of BC for (O); hence the diameter II' of (O) (Fig. 135) passing through D is the mediator of BC. Thus $(DA'II') = -1$, and therefore $A(DA'II') = -1$.

Now IAI' is a right angle; hence AI, AI' are the bisectors of the angle DAA' (§ 355). But AI is the internal bisector of the angle A, and AA' is the median of ABC; hence AD is the symmedian of ABC issued from A, by definition.

561. Theorem. *The segments into which a side of a triangle is divided by the corresponding symmedian are proportional to the squares of the adjacent sides of the triangle.*

The polar BC of the point D (Fig. 135) meets the tangent AT_a to the circumcircle (O) at A in the pole T_a of the symmedian AD; hence

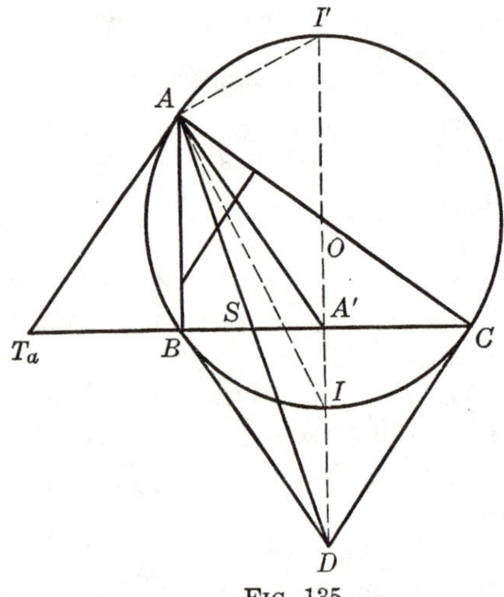

FIG. 135

the trace S of BC on AD is separated harmonically from T_a by the points B, C of (O). Now T_a divides the segment BC externally in the ratio $AB^2 : AC^2$ (§ 319); hence the proposition.

562. *Remark.* We have proved also that *a symmedian of a triangle is the harmonic conjugate of the tangent to the circumcircle at the vertex considered with respect to the two sides passing through that vertex.*

563. Definition. The tangents to the circumcircle of a triangle at the vertices are sometimes referred to as the *external symmedians* of the triangle.

564. Theorem. (*a*) *A symmedian of a triangle bisects any antiparallel to the side of the triangle relative to the symmedian considered.*

Indeed, any antiparallel to BC with respect to AB, AC (Fig. 135) is parallel to the tangent AT_a (§§ 185, 191); hence the proposition (§ 351).

(*b*) *Conversely. If a segment limited by two sides of a triangle is bisected by the corresponding symmedian, it is antiparallel, with respect to those two sides, to the third side of the triangle.*

Indeed, the problem of § 41 has only one solution.

565. Theorem. *The distances from a point on a symmedian of a triangle to the two including sides are proportional to these sides.*

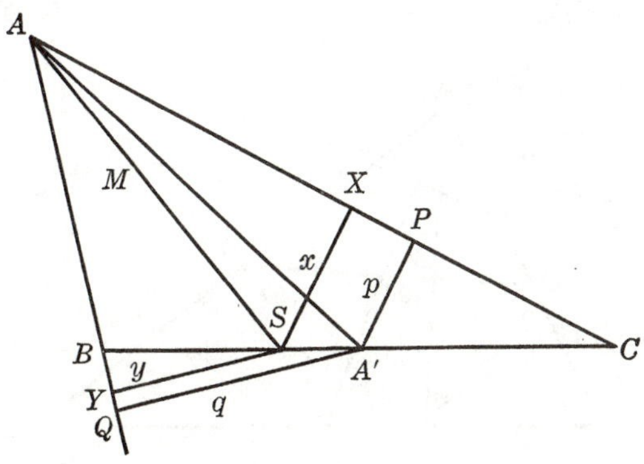

FIG. 136

Let x, y be the distances from the foot S of the symmedian AS (Fig. 136) to the sides AC, AB of ABC. Since the triangles ASC, ASB have a common altitude issued from A, we have:

$$bx:cy = \text{area } ASC:\text{area } ASB = SC:SB = b^2:c^2 \text{ (§ 561)};$$

hence:

$$x:y = b:c.$$

Now the distances from any point of AS to AC, AB are proportional to x, y; hence the proposition.

566. *Remark.* The preceding proposition (§ 565) is valid for the points of the external symmedian of the triangle. The proof is analogous.

567. Theorem. *If the distances from a point to two sides of a triangle are proportional to these sides, the point lies on the symmedian relative to the third side.*

It is assumed that the point lies within the angle of the two sides considered.

If M is a point whose distances from b, c are proportional to b, c, and S is the trace of AM on BC (Fig. 136), then the distances from S to b, c are also proportional to b, c. Hence S divides BC into two segments whose ratio is equal to $b^2:c^2$, as is seen by the proof of the preceding theorem (§ 565). Now there is only one point which divides BC internally in this ratio; hence the proposition.

568. Theorem. *If from a point on the symmedian (median) perpendiculars are drawn to the including sides of the triangle, the line joining the feet of these perpendiculars is perpendicular to the corresponding median (symmedian) of the triangle.*

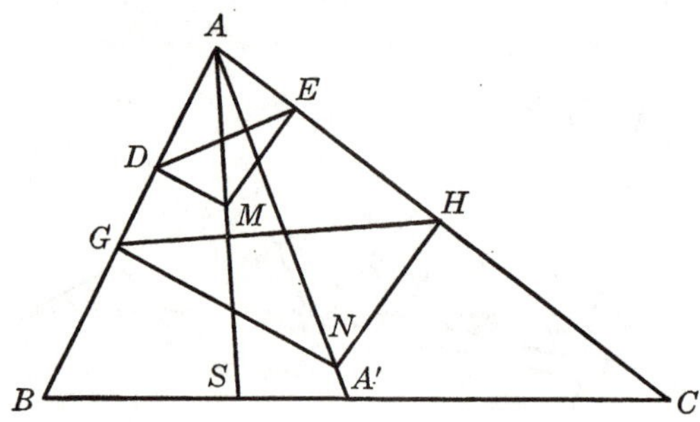

FIG. 137

In the triangle ABC let AA' and AS be the median and the symmedian issued from A (Fig. 137). If MD, ME are the perpendiculars to AB, AC from a point M of AS, the quadrilateral $ADME$ is cyclic; hence:

$$\angle DEM = DAM = EAA'.$$

Now ME is perpendicular to AE; hence ED is perpendicular to AA'.

In a similar way it may be shown that GH is perpendicular to AS.

569. Theorem. *If two antiparallels to two sides of a triangle are of equal length, they intersect on the symmedian relative to the third side of the triangle.*

Let DE, FH (Fig. 138) be antiparallel to the sides AC, AB, respectively, of the triangle ABC, and let M be their point of intersection. We have:

$$\angle HFC = DEB = A;$$

hence the triangle FME is isosceles. Thus:

$$FM = EM, \quad ED = FH; \text{ hence } DM = HM.$$

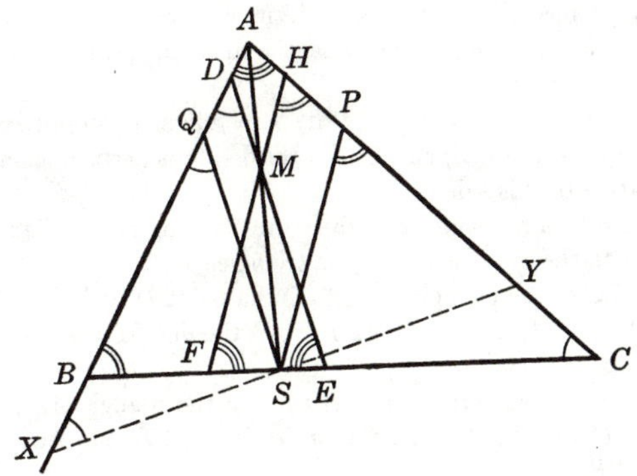

Fig. 138

Let S be the trace of AM on BC, and SQ, SP the parallels through S to DE, FH. From similar triangles we have:

$$SQ:MD = AS:AM = SP:MH,$$

and since $MD = MH$, we have $SQ = SP$.

Now the triangles BSQ, CSP are both similar to the triangle ABC; hence:

$$BS:SQ = c:b, \quad SP:SC = c:b;$$

hence, multiplying:

$$BS:SC = c^2:b^2,$$

since $SQ = SP$. Thus S is the foot of the symmedian (§ 561).

570. Converse Theorem. *The two antiparallels to two sides of a triangle drawn through a point on the symmedian relative to the third side are of equal length.*

The proposition is proved by taking the steps of the direct proof in reverse order.

571. Corollary. *The symmedian relative to a given side of a triangle is the locus of the points through which antiparallels of equal length may be drawn to the other two sides of the triangle.*

EXERCISES

1. On the side AB of the triangle ABC lay off $AB' = AC$, and on AC lay off $AC' = AB$. Show that the midpoint of $B'C'$ lies on the symmedian of ABC issued from A.

2. On the sides AB, AC of the triangle ABC squares are constructed external to ABC. Show that the sides of these squares opposite the sides AB, AC intersect on the symmedian of ABC issued from A.

3. Construct a triangle given the altitude, the median, and the symmedian issued from the same vertex.

4. Show that the line joining the feet of the perpendiculars dropped upon two sides of a triangle from the midpoint of the third side is perpendicular to the symmedian relative to this side.

5. Show that a side of a triangle and the corresponding symmedian are conjugate with respect to the circumcircle of the triangle.

6. On the side BC of the triangle ABC to find a point D such that if the parallels through D to AB, AC meet AC, AB in E, F, the line EF shall be antiparallel to BC.

7. If the symmedian issued from the vertex A of the triangle ABC meets the circumcircle in D, and P, Q, R are the projections of D upon BC, CA, AB, show that $PQ = PR$.

8. Show that the parallel to the side BC through the vertex A of ABC, the perpendicular from the circumcenter O upon the symmedian issued from A, and the perpendicular upon AO from the point of intersection T of the tangents to the circumcircle at B and C are three concurrent lines.

9. If A', B' are the inverse points of A, B in an inversion of center O, show that the median and the symmedian of the triangle AOB issued from O are respectively the symmedian and the median of the triangle $A'OB'$.

10. Show that the three points of intersection of the symmedians of a triangle with the circumcircle determine a triangle having the same symmedians as the given triangle.

11. The median and the symmedian of a triangle ABC issued from A meet the circumcircle in M', N'. Show that the Simson lines of M', N' are respectively perpendicular to AN', AM'.

b. THE LEMOINE POINT

572. Theorem. *The three symmedians of a triangle are concurrent.*

Let K be the point common to the two symmedians BK, AK (Fig. 139) of the triangle ABC, and p, q, r the distances of K from the sides a, b, c of ABC. We have (§ 565):

$$p:r = a:c, \quad q:r = b:c; \quad \text{hence} \quad p:q = a:b,$$

and therefore CK is the third symmedian of the triangle ABC (§ 567).

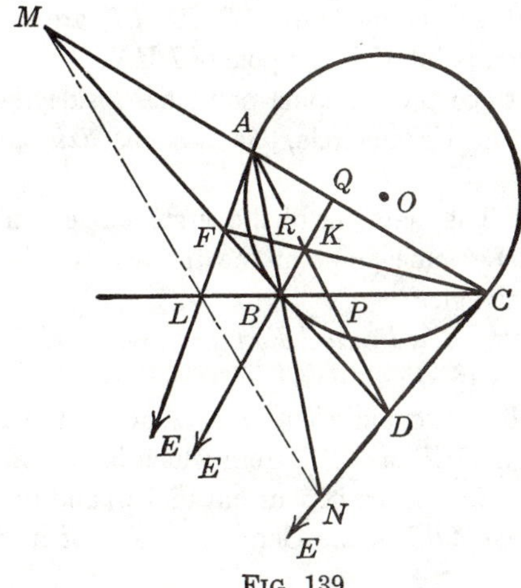

FIG. 139

573. Definition. The point common to the three symmedians of a triangle is called the *symmedian point*, or the *Lemoine point*, of the triangle. The point is usually denoted by K.

574. COROLLARY. *The distances of the Lemoine point K from the sides of the triangle are proportional to the respective sides, and K is the only point, within the triangle, having this property. However, the vertices of the tangential triangle of the given triangle also possess this property* (§§ 565, 560).

575. *Remark.* The proposition (§ 572) may be proved by making use of other properties of the symmedians. The reader may do so by utilizing § 561 and Ceva's theorem. Other ways will be pointed out in what follows.

576. Definition. The line containing the traces, on the sides of a given triangle, of the tangents to the circumcircle at the respectively opposite vertices (§ 319) is called the *Lemoine axis* of the given triangle.

577. Theorem. *The Lemoine point of a triangle is the pole of the Lemoine axis with respect to the circumcircle of the triangle.*

If the tangent AL to the circumcircle (O) of the triangle ABC at the vertex A meets BC in L (Fig. 139), the polar of L for (O) passes through A and through the pole D of BC (§ 379); hence this polar coincides with the symmedian AD (§ 560). Similarly for the analogous points M, N. Now the points L, M, N being collinear, it follows both that the three symmedians AD, BE, CF are concurrent, and that their common point K is the pole of LMN.

Observe that since the Lemoine point lies inside the triangle and therefore inside the circumcircle, *the Lemoine axis does not cut the circumcircle.*

578. Definition. The diameter of the circumcircle of a triangle passing through the Lemoine point of the triangle is called the *Brocard diameter* of the triangle.

579. COROLLARY. *The Brocard diameter of a triangle is perpendicular to the Lemoine axis* (§ 375 a).

580. *Remark.* The circumcircle (O) of ABC is the incircle of the tangential triangle DEF of ABC, from which it follows both that the symmedians DA, EB, FC are concurrent (§ 330) and that the Lemoine point of a triangle ABC is the Gergonne point of its tangential triangle DEF.

581. Theorem. *The Lemoine point of a triangle is the trilinear pole, for the triangle, of the Lemoine axis of the triangle.*

The trace L (Fig. 139) of the Lemoine axis LMN on the side BC has for harmonic conjugate, with respect to B and C, the foot P of the symmedian AP of ABC. Similarly for the analogous pairs of points M and Q, N and R. It follows therefore both that the symmedians AP, BQ, CR are concurrent, and that their common point K is the trilinear pole of LMN for ABC.

582. COROLLARY. *The vertices of the tangential triangle of a given triangle are the harmonic associates of the Lemoine point of the given triangle.*

583. Theorem. *The pedal triangle of the Lemoine point has this point for its centroid.*

Let X, Y, Z (Fig. 140) be the projections of the symmedian point K of the triangle ABC upon the sides BC, CA, AB, and let the parallel to KZ through Y meet KX, extended, in X'.

The triangles KYX', ABC are similar, for their sides are respectively perpendicular; hence:

$$KY:AC = YX':AB.$$

But (§ 565):

$$KY:AC = KZ:AB;$$

hence $X'YKZ$ is a parallelogram, and the line KX' bisects YZ, i.e., XK is a median of the triangle XYZ. Similarly for YK and ZK. Hence the proposition.

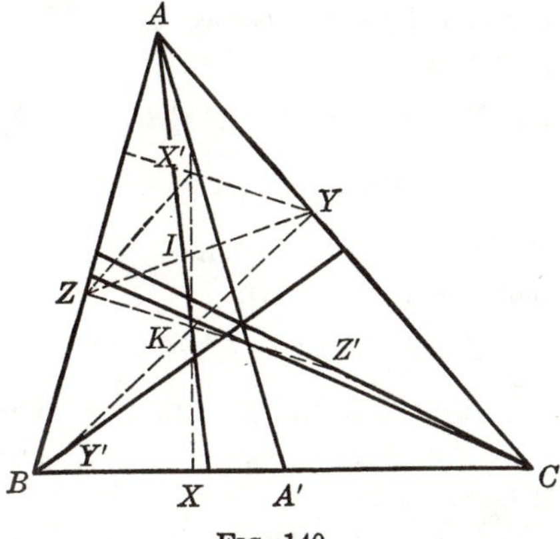

FIG. 140

584. COROLLARY. The lines YX', ZX' are parallel to KZ, KY and therefore perpendicular to AB, AC; hence X' is the orthocenter of the triangle AYZ, and therefore the line AX' is perpendicular to YZ, i.e., AX' is the median of ABC issued from A (§ 568).

On the other hand, if I is the midpoint of YZ, we have:

$$KX' = 2\,KI = KX;$$

hence: *The symmetric, with respect to the Lemoine point, of the projection of this point upon a side of the triangle lies on the median of the triangle relative to the side considered.*

585. Problem. *Construct a triangle, given, in position, two vertices and the Lemoine point.*

FIRST SOLUTION. Let X be the projection of the given symmedian point K (Fig. 140) upon the side joining the two given vertices B, C. The midpoint A' of BC and the symmetric X' of X with respect to K lie on the median of the required triangle ABC passing through A (§ 584).

The line joining the projections Y, Z of K upon AC, AB is perpendicular to $AX'A'$ (§ 565) and passes through the midpoint I of KX' (§ 584). Thus the point Z may be determined as the intersection of the perpendicular from the point I to the line $A'X'$ with the circle having KB for diameter. The lines $A'X'$ and BZ determine the third required vertex A. The problem may have two, one, or no solutions.

SECOND SOLUTION. Let B, C be the given vertices and K the given symmedian point of the required triangle ABC, and G its centroid.

If the parallels through A to BG, CG meet BC in D, E, and A' is the midpoint of BC, we have:

$$A'B:BD = A'G:GA = A'C:CE = 1:2;$$

hence:

$$BD = CE = BC = a,$$

and the points D, E may be constructed.

On the other hand we have (§ 558):

$$\angle DAB = GBA = KBA', \quad \angle CAE = GCA = KCA'.$$

Thus the known segments BD, CE subtend at A two known angles; hence A lies on two known arcs (locus 7, § 11). The problem may have two, one, or no solutions.

586. Theorem. *The Lemoine point of a triangle is the point of intersection of the lines joining the midpoints of the sides of the triangle to the midpoints of the corresponding altitudes.*

The vertex A of the triangle and the foot P of the symmedian AP separate harmonically the symmedian point K from the pole D of BC with respect to the circumcircle (O) of ABC (§ 582); hence $A'(APKD) = -1$, where A' is the midpoint of BC. Now the altitude AA_h is parallel to $A'D$; hence $A'K$ passes through the midpoint of AA_h (§ 351). Similarly for the other altitudes.

587. Theorem. *The isotomic of the orthocenter of a triangle is the Lemoine point of the anticomplementary triangle.*

The altitude $A'D'$ of the anticomplementary triangle $A'B'C'$ of ABC meets BC in the isotomic P of the foot D of the altitude AD of ABC, and P bisects the altitude $A'D'$. On the other hand A is the midpoint of $B'C'$. Thus in the triangle ABC the line AP is the isotomic of AD, and in the triangle $A'B'C'$ the line AP joins the midpoint A of the side $B'C'$ to the midpoint P of the altitude $A'D'$. Similarly for BQ and CR. Hence the proposition (§ 586).

EXERCISES

1. If DEF is the orthic triangle of ABC, show that the symmedian points of the triangles AEF, BFD, CDE lie on the medians of ABC.
2. Given the triangle ABC, show that there are three other triangles $A'BC$, $B'CA$, $C'AB$ having a side in common with ABC and the same Lemoine point K as ABC. Prove that the lines AA', BB', CC' are concurrent.
3. Show that the distances of the vertices of a triangle from the Lemoine axis are proportional to the squares of the respective altitudes.

4. Construct a triangle given, in position, the projections of its Lemoine point upon the three sides.
5. Construct a triangle given, in position, a vertex, the centroid, and the symmedian point.
6. Construct a triangle ABC given, in position, the indefinite sides AB, AC and the symmedian point K.
7. Show that the sides of the pedal triangle of the symmedian point of a given triangle are proportional to the medians of the given triangle, and the angles of the pedal triangle are equal to the angles between the medians of the given triangle.
8. If through the vertices of a triangle perpendiculars are drawn to the medians of the triangle, show that the symmedian point of the triangle thus formed coincides with the centroid of the given triangle.
9. Through the vertices B, C of the triangle ABC parallels are drawn to the tangent to the circumcircle (O) at A, meeting AC, AB in A', A''. The line $A'A''$ meets BC in U. Show that AU and its analogues BV, CW are concurrent.
10. If on each side of a triangle as base a square is constructed external to the triangle, the lines joining the vertices of the triangle formed by the external sides of the three squares, extended, to the corresponding vertices of the basic triangle meet in a point. (This point coincides with the Lemoine point of the basic triangle.)

c. THE LEMOINE CIRCLES

588. Theorem. *The three parallels to the sides of a triangle through the Lemoine point determine on these sides six concyclic points.*

Let the three parallels EKF', $D'KF$, DKE' to the sides BC, CA, AB of the triangle ABC (Fig. 141) through the symmedian point K inter-

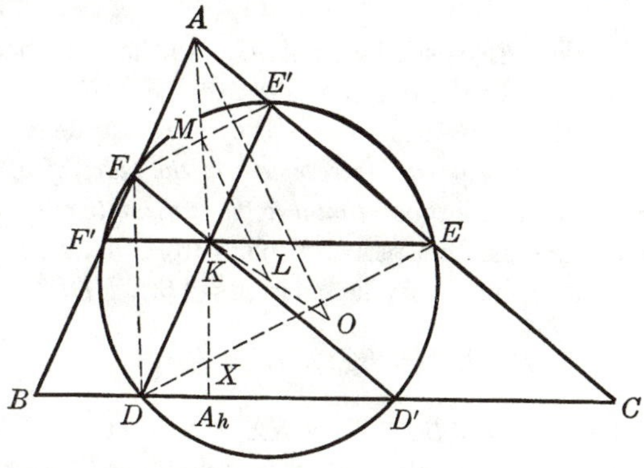

FIG. 141

sect these sides in the points D, D'; E, E'; F, F'. Since $AFKE'$ is a parallelogram, the line $E'F$ is bisected by the symmedian AK and is therefore antiparallel to BC (§ 564 b), hence also to EF', i.e., the four points E, E', F, F' lie on a circle.

For analogous reasons the same holds for the groups of points F, F', D, D' and D, D', E, E'. Now if these three circles are distinct, their radical axes $DD' = BC, EE' = CA, FF' = AB$ must meet in a point (§ 425), which is clearly not the case. On the other hand, if two of these circles coincide, the third obviously coincides with them. We conclude that the points D, D', E, E', F, F' are concyclic.

589. Definitions. The circle of these six points (§ 588) is known as the *first Lemoine circle* of the triangle.

The parallels considered are often called the *Lemoine parallels.*

590. Theorem. *The center of the first Lemoine circle of a triangle lies midway between the circumcenter and the Lemoine point of the triangle.*

Let O be the circumcenter of triangle ABC (Fig. 141), L the midpoint of OK, and M the common midpoint of AK and $E'F$. The line ML joins the midpoints of two sides of the triangle KOA and is therefore parallel to the third side OA; but the circumradius OA is perpendicular to $E'F$, for $E'F$ is antiparallel to BC (§§ 185, 188, 588); hence ML is the mediator of $E'F$. Similarly the mediators of DF' and $D'E$ pass through L. Hence the proposition.

591. *Remark. The triangles DEF, $D'E'F'$ are congruent.*

Indeed, in the first Lemoine circle we have (Fig. 141):

$$\angle FDE = FF'E = B, \quad \angle DEF = DD'F = C;$$

hence the triangle DEF is similar to ABC, and the same holds for the triangle $F'D'E'$. Thus the two triangles DEF, $F'D'E'$ are similar and are inscribed in the same circle; hence they are congruent.

592. Theorem. *The segments intercepted on the sides of a triangle by the first Lemoine circle are proportional to the cubes of the respective sides.*

If AA_h, KX are the two perpendiculars from A and K upon BC (Fig. 141), we have, in the similar triangles ABC, KDD':

$$DD':KX = BC:AA_h = BC^2:2S,$$

where S is the area of ABC. Now KX and the two analogous distances are proportional to the respective sides of ABC (§ 574); hence the proposition.

593. Theorem. *The three antiparallels to the sides of a triangle through the Lemoine point of the triangle determine on the respective pairs of noncorresponding sides six points lying on a circle whose center coincides with the Lemoine point.*

The three antiparallels are equal (§ 570) and are bisected by the symmedian point K (§ 564); hence K is equidistant from the ends of these antiparallels and is therefore the center of a circle passing through the six points considered (Fig. 142).

594. Definitions. The three antiparallels considered (§ 593) are often referred to as the *Lemoine antiparallels.*

The circle is known as the *second Lemoine circle*, or the *cosine circle*, of the triangle. The latter name is due to the following property of this circle.

595. Theorem. *The segments intercepted on the sides of a triangle by the second Lemoine circle are proportional to the cosines of the respectively opposite angles of the triangle.*

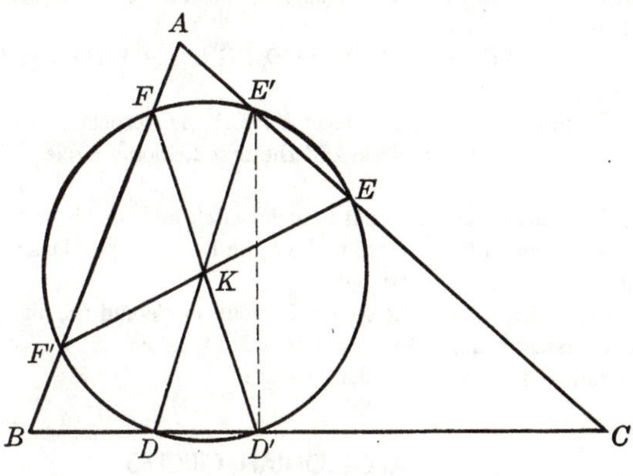

FIG. 142

Let the three antiparallels EKF', $D'KF$, DKE' (Fig. 142) to the sides BC, CA, AB of the triangle ABC through the symmedian point K meet these sides in the points D, D'; E, E'; F, F'. The antiparallel DE' being a diameter of the circle (K) (§ 593), we have in the right triangle $D'DE'$:

$$DD' = DE' \cos D'DE' = DE' \cos A.$$

Similarly:

$$EE' = EF' \cos B, \quad FF' = FD' \cos C.$$

But $DE' = EF' = FD'$ (§ 593); hence the proposition.

596. Theorem. *The second Lemoine circle is bisected by the first Lemoine circle.*

Through the symmedian point K of ABC draw the parallel $B'C'$ and the antiparallel $B''C''$ to BC, in the angle A. The four points B', C', B'', C'' being cyclic, we have:

$$KB' \cdot KC' = KB'' \cdot KC'',$$

i.e., the point K belongs to the radical axis of the two Lemoine circles. Now K is the center of the second Lemoine circle; hence the proposition.

EXERCISES

1. Show that the segments $E'F$, $F'D$, $D'E$ (Fig. 141) are equal.
2. Show that the lines $E'F$, $F'D$, $D'E$ produced (Fig. 141) form a triangle whose incircle is concentric with the first Lemoine circle of ABC and equal to the nine-point circle of ABC.
3. Show that the sum of the areas $AE'F$, $BF'D$, $CD'E$ (Fig. 141) is equal to the area DEF.
4. Show that the lines $E'F$, $F'D$, $D'E$ meet BC, CA, AB respectively (Fig. 141) on the radical axis of the circumcircle and the first Lemoine circle of the triangle ABC.
5. Show that the radical axis of the first Lemoine circle and the nine-point circle of a triangle passes through the points in which the Lemoine parallels meet the corresponding sides of the orthic triangle.
6. Show that the radical axis of the second Lemoine circle and the nine-point circle of a triangle passes through the points in which the Lemoine antiparallels meet the corresponding sides of the medial triangle.

C. THE APOLLONIAN CIRCLES

597. Definition. The interior and exterior bisectors of the angles A, B, C of a triangle ABC meet the opposite sides BC, CA, AB in the points U, U'; V, V'; W, W', respectively. The circles on UU', VV', WW' as diameters are called the *Apollonian circles*, or the *circles of Apollonius*, of the triangle ABC (Fig. 143).

598. Theorem. *The Apollonian circles pass through the respective vertices of the triangle.*

The circle (L) on UU' as diameter is the locus of points the ratio of whose distances from the points B, C is equal to the ratio (locus 11, § 11):

$$BU:CU = BU':CU'.$$

Now $AB:AC = BU:CU$; hence A is a point on the locus. Similarly for the other two Apollonian circles (M), (N).

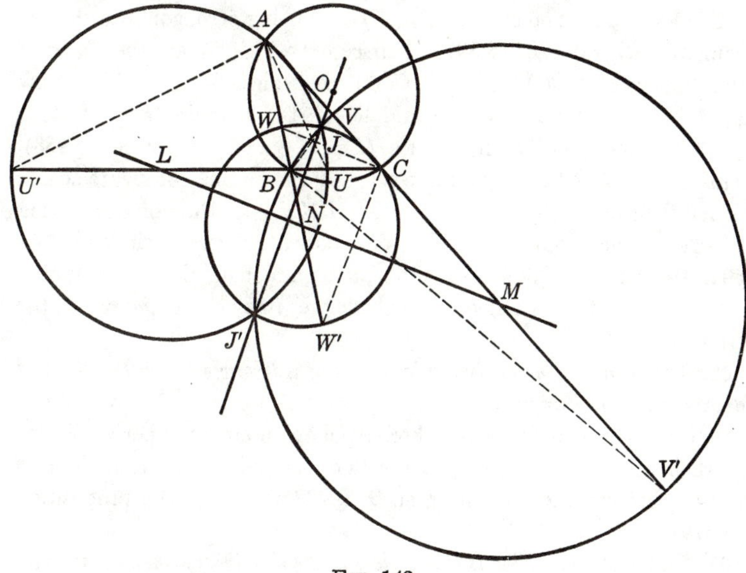

FIG. 143

OTHERWISE. The segment UU' subtends a right angle at the point A and therefore A lies on the circle (L).

Similarly for the other two Apollonian circles (M), (N).

599. Theorem. *The circumcircle of a triangle is orthogonal to the three Apollonian circles of the triangle.*

We have $(BCUU') = -1$; therefore the points B, C are inverse with respect to the circle (L) (Fig. 143); hence (L) is orthogonal to the circumcircle (O) of ABC (§ 369). Similarly for the circles (M), (N).

600. Theorem. *The three Apollonian circles of a triangle have two points in common.*

The Apollonian circle (M) passes through the vertex B, which point lies inside the circle (L); hence the two circles (L), (M) intersect (Fig. 143). Similarly for the pairs of circles (L) and (N), (M) and (N).

Now if J is one of the two points common to (L), (M), we have (§ 598):

$$JA:JC = BA:BC, \quad JB:JC = BA:CA;$$

hence:

$$JA:JB = CA:CB,$$

i.e., the point J also belongs to the circle (N).

OTHERWISE. The circles (L), (M), (N) being orthogonal to the circumcircle (O) of ABC (§ 599), their centers L, M, N are the traces of the tangents to (O) at A, B, C on the sides BC, CA, AB. Now L, M, N are collinear (§ 319) and lie on the Lemoine axis of ABC (§ 576); consequently the circles (L), (M), (N) are coaxal (§ 458). This coaxal pencil is of the intersecting type, for the conjugate coaxal pencil determined by (O) and the line LMN is of the nonintersecting type, since the Lemoine axis LMN does not cut the circumcircle (§ 577).

601. Definition. The two points J, J' common to the three Apollonian circles of a triangle are called the *isodynamic points* of the triangle.

602. Theorem. *The isodynamic points of a triangle lie on the Brocard diameter of the triangle.*

The radical axis JJ' of the three Apollonian circles is perpendicular to their line of centers, i.e., to the Lemoine axis of the triangle, and passes through the circumcenter O (§ 428 a); hence the proposition (§ 579).

603. COROLLARY. *The isodynamic points are inverse with respect to the circumcircle and divide the Brocard diameter harmonically* (§ 448).

604. Theorem. *The Apollonian circles of a triangle are the circles of similitude of the three circles, taken in pairs, whose centers are the vertices of the given triangle and whose radii are proportional to the respective altitudes of the triangle.*

We have (§ 598) (Fig. 143):

$$BU:CU = AB:AC = h_b:h_c = kh_b:kh_c,$$

where k is an arbitrary factor. Thus U is the internal center of similitude of the two circles (B, kh_b), (C, kh_c), and U' is therefore their external center of similitude.

Similarly V and V', W and W' are the centers of similitude of the pairs of circles (C, kh_c) and (A, kh_a), (A, kh_a) and (B, kh_b).

605. *Remark.* The properties of §§ 599, 600 are immediate consequences of this proposition (§§ 496 ff., § 604).

606. Theorem. *The common chord of the circumcircle and an Apollonian circle of a triangle coincides with the corresponding symmedian of the triangle.*

The circumcircle (O) (Fig. 144) and an Apollonian circle, say, (L) are orthogonal (§ 599); hence their common chord is the polar of the center L of (L) with respect to (O); hence the proposition (§ 561).

607. Corollary. *The traces of the symmedians of a triangle on the circumcircle lie also on the respective Apollonian circles of the triangle.*

608. Theorem. *The polar of the circumcenter of a triangle with respect to an Apollonian circle coincides with the corresponding symmedian of the triangle.*

Indeed, the common chord of the two orthogonal circles (O) and (L) (§ 606) is also the polar of O with respect to (L).

609. Corollary. *The isodynamic points of a triangle are separated harmonically by the circumcenter and the Lemoine point of the triangle.*

610. *Remark. The pole of the Brocard diameter with respect to an Apollonian circle of the triangle is the trace on the Lemoine axis of the symmedian through the vertex considered.*

Indeed, the pole of the Brocard diameter OK with respect to the circle (L) lies on the diameter LMN of (L) perpendicular to OK and on the polar AK of the point O of OK with respect to (L).

611. *Observation.* Let A', B', C' (Fig. 144) be the points in which the symmedians AKA', BKB', CKC' of the triangle ABC meet the circumcircle (O). Thus (O) cuts the Apollonian circles (M), (N) in the four points B, B', C, C' and is orthogonal to these circles (§ 599); hence of the three pairs of lines BB' and CC', BC and $B'C'$, BC' and $B'C$ one pair, namely the first, intersects on the radical axis OK of the circles (M), (N) (§ 437), in the point K, and the remaining two pairs intersect in the centers of similitude of (M), (N), which centers of similitude lie on the line of centers of the two circles, i.e., on the Lemoine axis LMN of ABC.

612. Theorem. *The center of an Apollonian circle of a triangle is a center of similitude of the other two Apollonian circles of the triangle.*

Indeed, one of the centers of similitude of the two circles (M), (N) (Fig. 144) lies on the line BC and on the Lemoine axis LMN of ABC (§ 611), and these two lines have in common the center L of the third Apollonian circle (L) of ABC.

613. Corollary. *A symmedian of a triangle meets the Lemoine axis in a center of similitude of the two Apollonian circles relative to the other two vertices of the triangle.*

The second center of similitude L' (Fig. 144) of the circles (M), (N) is harmonically separated from L by the centers M, N; hence the line AL' is harmonically separated from AL by the lines ABN, ACM, i.e., AL' is the symmedian issued from A (§ 562).

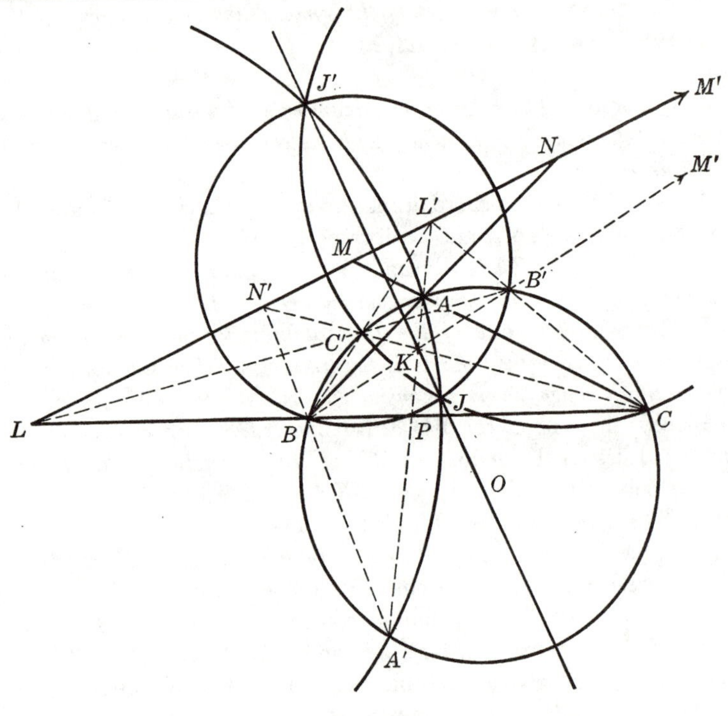

FIG. 144

614. Theorem. *The traces of two symmedians of a triangle on the circumcircle are collinear with the center of the Apollonian circle passing through the third vertex.*

Indeed, the lines $BC, B'C'$ (§ 611) (Fig. 144) meet the Lemoine axis in the same point L (§ 612).

615. Theorem. *The two lines joining two vertices of a triangle to the traces of their symmedians on the circumcircle meet at the trace of the third symmedian on the Lemoine axis of the triangle.*

The lines $BC', B'C$ meet in the center of similitude L' of the two Apollonian circles (M), (N) (§ 611), and this point is also the trace of the symmedian AKA' on the Lemoine axis LMN (§ 613).

616. Theorem. *The three traces of the symmedians of a triangle on its circumcircle determine a triangle having the same Apollonian circles as the given triangle.*

The Apollonian circle (L_o) of $A'B'C'$ passing through A' must be orthogonal to (O) (Fig. 144) and have its center on the tangent to (O) at A' and on the line $B'C'$. Now both these lines pass through the point L (§§ 606, 614); hence (L_o) is identical with the circle (L) of ABC. Similarly for the circles (M), (N).

617. COROLLARY. *The two triangles considered have the same symmedians, the same Lemoine point, the same Lemoine axis, and the same isodynamic points.*

618. Definition. A triangle and the triangle having for vertices the traces of the symmedians of the first on its circumcircle are called *cosymmedian triangles.* The name is justified by their properties (§ 617).

Two cosymmedian triangles are perspective, the center and axis of perspectivity being the Lemoine point and the Lemoine axis of both triangles.

619. Theorem. *A symmedian meets the corresponding side of the triangle in the Lemoine point of the triangle having for one vertex the corresponding vertex of the cosymmedian triangle, and for the remaining two vertices the isodynamic points of the two cosymmedian triangles considered.*

The circumcircle (O) of ABC (Fig. 144) is the Apollonian circle of the triangle $A'JJ'$ relative to the vertex A', for (O) passes through A', is orthogonal to the circumcircle (L) of $A'JJ'$, and has its center on the side JJ' opposite A'. The common chord $A'A$ of (O) and (L) is therefore a symmedian of the triangle $A'JJ'$ (§ 606).

If P is the trace of the symmedian AA' on BC, we have $(LPBC) = -1$; hence $N'(LPBC) = -1$. But the lines $N'B$ and $N'C$ are identical with $N'BA'$ and $N'CC'K$ (§ 615); hence $(A'KL'P) = -1$. Now L' is the pole of the side JJ' of the triangle $A'JJ'$ for the circumcircle (L) of $A'JJ'$ (§ 610); hence the harmonic conjugate P of L' with respect to the vertex A' and the trace K of $A'L'$ on the opposite side JJ' is the Lemoine point of $A'JJ'$ (§§ 581, 582), which proves the proposition.

620. *Remark.* The traces Q, R of the symmedians BK, CK of ABC on CA, AB are the Lemoine points of the triangles $B'JJ'$, $C'JJ'$. The traces P', Q', R' of the symmedians $AA'K$, $BB'K$, $CC'K$ on the sides $B'C'$, $C'A'$, $A'B'$ of the cosymmedian triangle are the Lemoine points of the triangles AJJ', BJJ', CJJ'.

621. Theorem. *The isodynamic points of a triangle and one vertex of that triangle determine a second triangle whose isodynamic points are* *those vertices of the cosymmedian triangle of the given triangle which do not correspond to the vertex of the given triangle considered.*

The line joining the circumcenter L of the triangle $A'JJ'$ (Fig. 144) to the Lemoine point P of $A'JJ'$ (§ 619) is the Brocard diameter of $A'JJ'$; hence LP meets the Apollonian circle (O) of $A'JJ'$ in the isodynamic points B, C of $A'JJ'$, which proves the proposition.

622. *Remark.* The Lemoine axis of the triangle $A'JJ'$ (§ 621) is perpendicular to the Brocard diameter LP and passes through the center O of the Apollonian circle (O) of $A'JJ'$; hence the Lemoine axis of $A'JJ'$ is the mediator of BC.

Thus the Lemoine axes of the six triangles $JJ'A', JJ'B', JJ'C', JJ'A, JJ'B, JJ'C$ *have a point in common.*

623. **Problem.** *Construct a triangle given, in position, the symmedian point, the circumcenter, and one vertex* (K, O, A).

Let the line AK meet the circle (O, OA) again in A' (Fig. 144). If L' is the harmonic conjugate of K with respect to A, A' (§ 577) and P the harmonic conjugate of L' with respect to K, A' (§ 619), the line joining P to the pole L of AK with respect to the circle (O, OA) meets this circle in the two remaining vertices of the required triangle ABC.

The problem has one and only one solution, assuming that K lies inside the circle (O, OA).

624. **Theorem.** *An Apollonian circle of a triangle makes an angle of* $120°$ *with each of the remaining two circles.*

The point L (Fig. 144) is a center of similitude of the two Apollonian circles (M), (N) (§ 612), and the circle (L) is coaxal with (M), (N); hence these two circles are inverse with respect to (L) (§§ 527, 543), and therefore the angle between (L), (M) is equal to the angle between (L), (N), i.e., (L) bisects the angle between the circles (M), (N). Similarly for the circle (M) with respect to (L), (N), and again for the circle (N) with respect to (L), (M). Thus each of the Apollonian circles bisects the angle between the remaining two (§ 531); hence the proposition.

EXERCISES

1. Show that the mediators of the internal bisectors of the angles of a triangle meet the respective sides of the triangle in three collinear points.
2. If in the triangle ABC we have $BC > CA > AB$, show that the sum of the reciprocals of the diameters of the Apollonian circles relative to BC and AB is equal to the reciprocal of the diameter of the Apollonian circle relative to CA.

3. Show that the distances from the vertices of a triangle to an isodynamic point are proportional to the respectively opposite sides.
4. Construct three circles so that the center of each shall lie on the circle of similitude of the other two.
5. Show that the median through a given vertex of a triangle meets the Apollonian circle passing through the vertex considered and the circumcircle in two points which, with the other two vertices of the triangle, determine a parallelogram.
6. Show that the symmetrics of the Apollonian circles of a triangle with respect to the mediators of the corresponding sides of the triangle are coaxal, their radical axis passing through the circumcenter of the triangle.

D. ISOGONAL LINES

625. Definition. Two lines passing through the vertex of a given angle and making equal angles with the bisector of the given angle are said to be *isogonal* or *isogonal conjugates.*

Thus the altitude and the circumdiameter issued from the same vertex of a triangle are a pair of isogonal lines (§ 73 b). Another example of two isogonal conjugate lines is furnished by the median and the symmedian issued from the same vertex of a triangle (§ 558).

A bisector of an angle coincides with its own isogonal conjugate for that angle.

626. Theorem. *The line joining the traces, on the circumcircle of a triangle, of two isogonal lines of an angle of the triangle is parallel to the side opposite the vertex considered.*

627. Theorem. *The line joining the two projections of a given point upon the sides of an angle is perpendicular to the isogonal conjugate of the line joining the given point to the vertex of the angle.*

Those are generalizations of the propositions relative to symmedians (§§ 559, 568). The proofs remain the same.

628. Theorem. *The distances, from the sides of an angle, of two points on two isogonal lines are inversely proportional.*

Let M, N be two points on the two isogonal lines AM, AN (Fig. 145), MQ, MP, NR, NS the four perpendiculars from M, N to the sides AB, AC of the angle BAC.

From the two pairs of similar right triangles AMQ and ANS, AMP and ANR we have:

$$\text{(a)} \qquad MQ:NS = AQ:AS = AM:AN = AP:AR = MP:NR,$$

and the proposition follows from the equality of the first and last ratios.

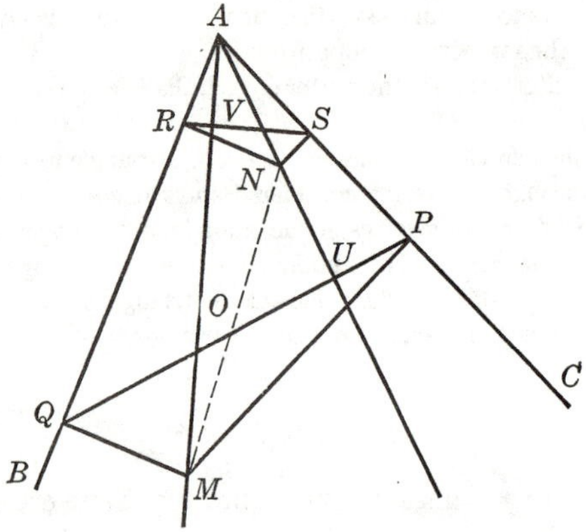

FIG. 145

629. COROLLARY. *The product of the distances of two points on two isogonal lines from one side of the angle is equal to the product of their distances from the other side of the angle.*

Indeed, we have (§ 628):

$$MP \cdot NS = MQ \cdot NR.$$

630. Converse Theorem. *If the distances of two points from the sides of a given angle are inversely proportional, the points lie on two isogonal conjugate lines.*

We have, by hypothesis (Fig. 145):

$$MP:MQ = NR:NS,$$

and the angles M, N are supplementary to the angle A; hence the two triangles MPQ, NRS are similar, and therefore the angles QPM, NRS are equal. Now from the two cyclic quadrilaterals $APMQ, ARNS$ we have:

$$\angle QAM = QPM, \quad \angle NAS = NRS;$$

hence the proposition (§ 625).

631. Theorem. *The four projections upon the two sides of an angle of two points on two isogonal conjugate lines are concyclic.*

Indeed, we have, from the formula (a) (§ 628):

$$AQ \cdot AR = AP \cdot AS;$$

hence P, Q, R, S are concyclic.

632. Corollary. *The two lines PQ, RS are antiparallel with respect to the sides of the angle PAQ.*

633. *Remark. The circle $PQRS$* (§ 631) *has for its center the midpoint of MN (Fig. 145).*

The center of the circle considered is the point common to the mediators of the segments QR and PS, and each of these mediators passes through the midpoint O of MN.

634. Theorem. *If two lines are antiparallel with respect to an angle, the perpendiculars dropped upon them from the vertex are isogonal in the angle considered.*

If PQ, RS are two antiparallels in the angle PAQ (Fig. 145), and U, V the feet of the perpendiculars from A upon PQ, RS, the right triangles AUP, ARV have their angles P, R equal; hence the angles PAU, RAV are equal, which proves the proposition.

EXERCISES

1. The isogonal conjugate of the parallel through A to the side BC of the triangle ABC meets BC in M, and the symmedian AS meets BC in S. Show that $BM:CM = BS:CS$.
2. (a) Show that if M, N are the traces on BC of the two isogonal lines AM, AN, we have $CM \cdot CN : BM \cdot BN = b^2 : c^2$; (b) the isogonal conjugates of the three lines joining the vertices of a triangle to the points of intersection of the respectively opposite sides with a transversal meet these sides in three collinear points.
3. Show that the line joining a given point to the vertex of a given angle has for its isogonal line the mediator of the segment determined by the symmetrics of the given point with respect to the sides of the angle.
4. Two triangles ABC, $A'B'C'$ are perspective and inscribed in the same circle (O); if A'', B'', C'' are the traces, on BC, CA, AB, of the isogonal conjugates of the lines $A'A$, $B'B$, $C'C$, show that the lines AA'', BB'', CC'' are concurrent.
5. If P is the point common to the circumcircles of the four triangles determined by four given lines (§ 297), show that the isogonal conjugates of the six lines joining P to the points of intersection of the given four lines are parallel.
6. Given two isogonal conjugate lines, if the segment determined on one of them by the vertex considered and the opposite side of the triangle is multiplied by the chord intercepted on the other line by the circumcircle, show that the product is equal to the product of the two sides of the triangle passing through the vertex considered.
7. Show that the four perpendiculars to the sides of an angle at four concyclic points form a parallelogram whose opposite vertices lie on isogonal conjugate lines with respect to the given angle.

635. Theorem. *The isogonal conjugates of the three lines joining a given point to the vertices of a given triangle are concurrent.*

Let M' (Fig. 146) be the point of intersection of the isogonals AM', BM' of the lines AM, BM. We have (§ 628):

$$ME:MF = M'F':M'E', \quad MF:MD = M'D':M'F';$$

hence, multiplying:

$$ME:MD = M'D':M'E',$$

i.e., the lines CM, CM' are isogonal conjugates (§ 630); hence the proposition.

636. Definition. The points M, M' are said to be *isogonal conjugate points*, or *isogonal points* with respect to the triangle ABC.

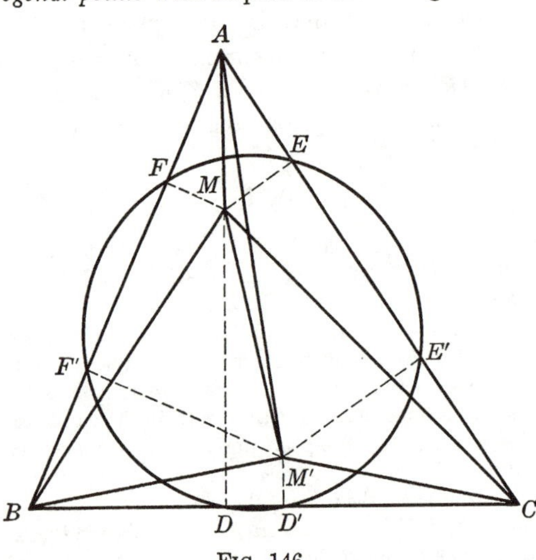

FIG. 146

Thus the orthocenter and the circumcenter are a pair of isogonal conjugate points with respect to the triangle; the centroid and the symmedian point are another.

A tritangent center of a triangle coincides with its own isogonal conjugate for that triangle.

637. COROLLARY. *The perpendiculars dropped from the vertices of a triangle upon the corresponding sides of the pedal triangle of a given point are concurrent.*

Indeed, the lines AM', BM', CM' (Fig. 146) are perpendicular to EF, FD, DE (§ 627).

638. *Remark. If the point M* (§ 635) *lies on the circumcircle of the triangle, the isogonal lines of AM, BM, CM are parallel.* Conversely, *if the isogonal conjugates of the lines AM, BM, CM are parallel, the point M lies on the circumcircle of the triangle.*

If X, Y, Z are the projections of M upon the sides of ABC, the isogonal conjugates p, q, r of AM, BM, CM are perpendicular to the lines YZ, ZX, XY (§ 627). Now if M lies on the circumcircle (O) of ABC, the points X, Y, Z are collinear (§ 282 a); hence p, q, r are parallel.

Conversely, if p, q, r are parallel, the same holds for YZ, ZX, XY (§ 627), i.e., the points X, Y, Z are collinear, and therefore M lies on (O) (§ 282 b).

639. Definition. The perpendiculars erected at the vertices of a triangle to the lines joining these vertices to a given point form a triangle called the *antipedal triangle* of the given point with respect to the given triangle.

640. Theorem. *If two points are isogonal for a triangle, the pedal triangle of one is homothetic to the antipedal triangle of the other.*

If E, F are the feet of the perpendiculars from the given point M upon the sides AC, AB of the triangle ABC, the line EF is perpendicular to the isogonal conjugate of AM in the angle A (§ 627), which line passes through the isogonal conjugate M' of M for ABC; hence the line EF is parallel to the side of the antipedal triangle of M' passing through A.

Similarly for the other pairs of sides of the pedal triangle of M and the antipedal triangle of M'. Hence the proposition.

641. Theorem. *The six projections of two isogonal conjugate points upon the sides of a triangle are concyclic.*

The six projections lie by fours on circles having for centers the midpoint of the segment determined by the given isogonal points (§§ 631, 633); hence the three circles are identical.

642. Definition. This circle (§ 641) is often referred to as the *pedal circle* of the two isogonal conjugate points.

643. Converse Theorem. *The circumcircle of the pedal triangle of a point for a given triangle cuts the sides of the given triangle again in the vertices of the pedal triangle of a second point, which point is the isogonal conjugate of the first point with respect to the given triangle.*

Let (L) be the circle circumscribing the pedal triangle XYZ of the point M for the triangle ABC. If $X'Y'Z'$ is the pedal triangle for ABC of the isogonal conjugate M' of M for ABC, the six points X, Y, Z, X', Y', Z' lie on a circle (§ 641) which is identical with (L), for the two circles have the three points X, Y, Z in common. Hence the proposition.

644. Theorem. *If two points are isogonal with respect to a triangle, each is the center of the circle determined by the symmetrics of the other with respect to the sides of the triangle.*

Let D, E, F be the projections of the point M upon the sides of the triangle ABC, and D', E', F' the symmetrics of M with respect to the sides of ABC.

The point M is the homothetic center of the two triangles DEF, $D'E'F'$, the homothetic ratio being $1:2$; hence if L, L' are the circumcenters of DEF, $D'E'F'$, we have $ML:ML' = 1:2$, i.e., L' is the symmetric of M with respect to L and therefore L' coincides with the isogonal conjugate M' of M for ABC (§ 641).

645. Theorem. *The radical center of the three circles having for diameters the segments intercepted by a given circle on the sides of a triangle is the isogonal conjugate of the center of the given circle with respect to the triangle.*

Let D, E, F be the midpoints of the segments intercepted by a circle (M) on the sides BC, CA, AB of the triangle ABC, and let (D), (E), (F) be the circles having the intercepted segments for diameters.

The lines AB, AC are the radical axes of the circles (M) and (F), (M) and (E); hence the point A lies on the radical axis of the circles (E), (F) (§ 425). Thus the radical axis of (E), (F) is the perpendicular dropped from A upon the side EF of the pedal triangle DEF of M for ABC. Similarly for the radical axes of (F) and (D), (D) and (E). Hence the proposition (§ 637).

646. Theorem. *If two isogonal conjugate points for a triangle are collinear with the circumcenter, their pedal circle is tangent to the nine-point circle of the triangle.*

If X, Y are the ends of a circumdiameter of a triangle ABC, the simsons x, y of X, Y are perpendicular and the point $L = xy$ lies on the nine-point circle (N) of ABC (§ 295). The projections X', Y' of X, Y upon BC lie on x, y, respectively, and are symmetrical with respect to the midpoint A' of BC; hence the circle $(X'Y')$ having $X'Y'$ for diameter passes through L.

Let P, P' be *any* two points of the diameter XY, and D, D' their projections upon BC. If O is the circumcenter and R the circumradius of ABC, we have:

$$A'D : A'X' = OP : R, \quad A'D' : A'X' = OP' : R;$$

hence, multiplying:

$$A'D \cdot A'D' : A'X'^2 = OP \cdot OP' : R^2,$$

or, since L lies on $(X'Y')$:

$$A'D \cdot A'D' : A'L^2 = OP \cdot OP' : R^2.$$

The right-hand side of this proportion does not depend upon the side BC considered; neither does the point $L = xy$; hence we have, denoting by B', C' the midpoints of AC, AB, and by E, E'; F, F' the projections upon these sides of the points P, P':

(a) $$A'D \cdot A'D' : A'L^2 = B'E \cdot B'E' : B'L^2 = C'F \cdot C'F' : C'L^2.$$

Now suppose that P, P' *are isogonal conjugates* for ABC. The six projections $D, D', \ldots$ of P, P' lie on a circle (S) (§ 641), and the products $A'D \cdot A'D', \ldots$ represent the powers of the points A', B', C' for (S). If we consider $A'L^2, \ldots$ as the powers of the points A', B', C' for the point-circle (L), it follows from the formula (a) that the circle (N) is coaxal with the circles (S), (L) (§ 475); hence the center N of (N) is collinear with L and the center S of (S). Moreover, since the point L lies on (N), the radical axis of the coaxal system is the tangent to (N) at L; therefore the circle (S) passes through L and is thus tangent to (N) at L. Hence the proposition.

647. Corollary. If the isogonal points P, P' coincide in a tritangent center of ABC (§ 634), the circle (S) coincides with the corresponding tritangent circle; hence: *Each tritangent circle of a triangle touches the nine-point circle of that triangle.*

This is Feuerbach's theorem (§ 215).

648. *Remark.* It follows from the proof of the preceding theorem (§ 644) that the Feuerbach point (§§ 214, 215) relative to a given tritangent circle is the point of intersection of the simsons of the ends of the circumdiameter passing through the corresponding tritangent center. The Feuerbach points may thus be located without making use of either the nine-point circle or the tritangent circles.

EXERCISES

1. Given two isogonal points for a triangle, show that the product of their distances from a side of the triangle is equal to the analogous product for each of the other two sides.

2. Show that the sum of the angles subtended by a side of a triangle at two isogonal points is equal to a straight angle increased by the angle of the triangle opposite the side considered.

3. Show that the vertices of the tangential triangle of ABC are the isogonal conjugates of the vertices of the anticomplementary triangle of ABC.
4. Show that the lines joining the vertices of a triangle to the projections of the incenter upon the mediators of the respectively opposite sides meet in a point — the isotomic conjugate of the Gergonne point of the triangle.
5. Show that the three symmetrics of a given circle with respect to the sides of a triangle have for radical center the isogonal point of the center of the given circle with respect to the triangle.

E. BROCARD GEOMETRY

a. THE BROCARD POINTS

649. Definition. Consider the circle (AB) passing through the vertices A, B of the triangle ABC (Fig. 147) and tangent at B to the side BC. Similarly a circle (BC) may be described on the side BC

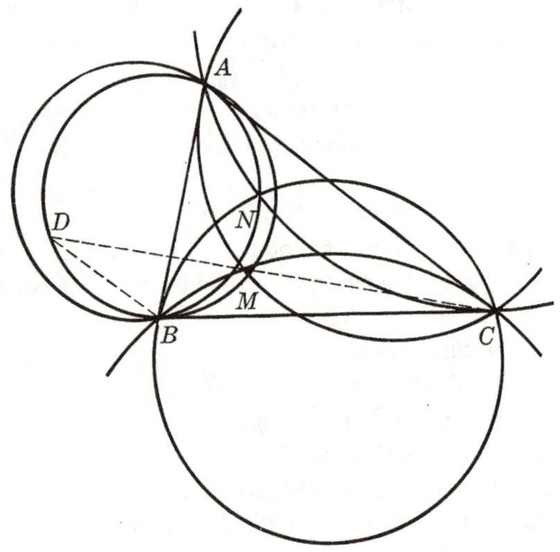

FIG. 147

tangent at C to CA, and again a circle (CA) on CA tangent at A to AB. These three circles will be referred to as the *direct group of adjoint circles.*

Taking the vertices A, B, C by twos in the circular permutation BAC, we obtain the *indirect group of adjoint circles* (BA), (AC), (CB), the circle (BA) being described on the side BA, tangent at A to the side AC, and similarly for the other two circles.

650. Theorem. *The three adjoint circles of the direct group have a point in common.*

The two circles (AB), (BC) have the point B in common (Fig. 147); hence they meet in another point M, inside the triangle ABC. Now the circle (AB) being tangent to BC at B, angle $AMB = 180° - B$, and similarly angle $BMC = 180° - C$, hence:

$$\angle AMC = 360° - (180° - B) - (180° - C) = B + C = 180° - A.$$

Thus the point M belongs also to the arc of the circle (CA) which lies inside of ABC; hence the proposition.

651. Theorem. *The three adjoint circles of the indirect group have a point N in common.*

The proof is similar to the proof of the preceding proposition (§ 650).

652. Definition. The two points M, N (§§ 650, 651) are known as the *Brocard points* of the triangle. They are frequently denoted by the letters Ω, Ω'.

653. Theorem. (*a*) *Angle MAB = angle MBC = angle MCA, and the point M is the only one having this property.*

(*b*) *Angle NAC = angle NCB = angle NBA, and the point N is the only one having this property* (Fig. 147).

The angle MAB inscribed in the circle (AB) is measured by half the arc BM, and the angle MBC formed by the chord BM and the tangent BC to the same circle is also measured by half the arc BM; hence these angles are equal. Similarly for the angles MBC and MCA.

Now if M' is any other point such that angle $M'AB = M'BC$, the circle through the points M', A, B will be tangent to BC at B, i.e., M' will be a point of the circle (AB). Similarly, in order to satisfy the condition angle $M'BC = M'CA$, the point M' will have to lie on (BC); hence M' is identical with M. Similarly for part (b)

654. Theorem. *The Brocard points are a pair of isogonal points of the triangle.*

If N' is the isogonal conjugate of the Brocard point M (Fig. 147), we have

$$\angle MAB = N'AC, \quad \angle MBC = N'BA, \quad \angle MCA = N'CB,$$

and

$$\angle MAB = MBC = MCA;$$

hence:

$$\angle N'AC = N'BA = N'CB,$$

i.e., N' coincides with N (§ 653).

655. Definition. The angle $MAB = NAC$ (§ 654) is called the *Brocard angle* of the triangle and is frequently denoted by ω.

656. Construction. Given the triangle ABC (Fig. 148) and one circle of the direct group of adjoint circles, say (CA), the Brocard point M and the Brocard angle may be constructed as follows.

Through the point of contact A of (CA) draw a parallel to the opposite side BC meeting (CA) again in I. The line BI joining I to the third vertex B meets (CA) again in the required point M, and IBC is the Brocard angle of ABC.

Indeed, we have:

$$\angle MAB = MCA = MIA = MBC.$$

Similarly for the second Brocard point.

657. *Remark I.* In the triangles ACI, ACB (Fig. 148) we have angle $AIC = BAC$, angle $IAC = ACB$; hence angle $ACI = ABC$; the line CI is therefore tangent to the circumcircle (O) of ABC. Consequently the point I may be determined by the parallel AI to BC and

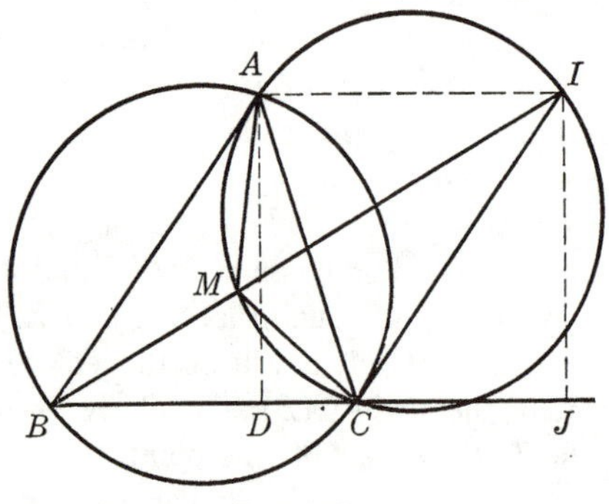

FIG. 148

the tangent CI at C to the circumcircle (O) of ABC. Thus the Brocard angle may be constructed without the aid of an adjoint circle; by repeating this construction twice we may determine the Brocard point.

658. *Remark II.* Let D, J be the feet of the perpendiculars from A, I upon BC (Fig. 148). We have:

$$\frac{BJ}{IJ} = \frac{BD}{IJ} + \frac{DC}{IJ} + \frac{CJ}{IJ} = \frac{CJ}{IJ} + \frac{BD}{AD} + \frac{DC}{AD}$$

or:

$$\cot \omega = \cot A + \cot B + \cot C.$$

659. Theorem. *The shape of a triangle is determined by one of its angles and its Brocard angle.*

Through the vertex A of the given angle BAC (Fig. 148) and through an arbitrary point C of its side AC draw two lines AM, CM, inside the angle A, making with the given sides AB, AC, respectively, angles equal to the given Brocard angle ω, and meeting in the point M. If B is a point on the side AB at which the segment CM subtends an angle equal to ω, the triangle ABC has the given angle A and the given Brocard angle ω.

There are, in general, two points, B and B', on the side AB at which CM subtends an angle ω. The two triangles ABC and $AB'C$ are inversely similar.

660. Problem. *Construct a triangle given a side, an angle, and the Brocard angle.*

The two given angles determine a triangle similar to the required triangle (§ 659); the required triangle is then readily obtained.

661. Theorem. *The traces on the circumcircle of a triangle of the lines joining the vertices to a Brocard point form a triangle congruent to the given triangle.*

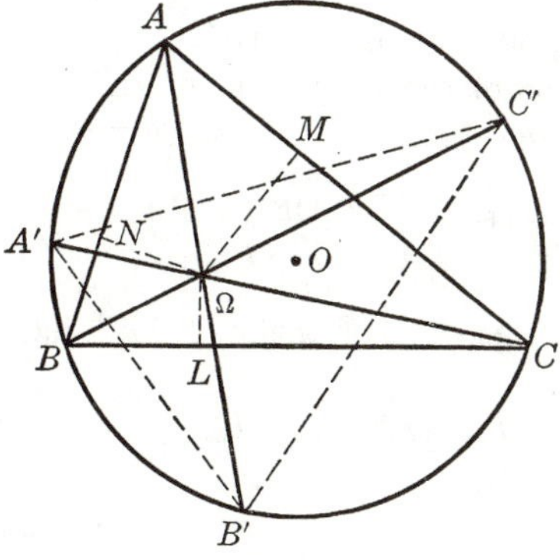

FIG. 149

Let the line $A\Omega$, $B\Omega$, $C\Omega$, meet the circumcircle (O) of ABC in B', C', A'. In the triangle $A'B'C'$ (Fig. 149) we have:

$$\angle A' = B'A'C + C'A'C = B'AC + C'BC = B'AC + BAB' = A.$$

Likewise for the angles B' and C'. Thus the two triangles ABC, $A'B'C'$ are similar and inscribed in the same circle; hence they are congruent.

Similarly for the second Brocard point.

Observe that the triangle $A'B'C'$ may be obtained from the triangle ABC by a counterclockwise rotation of angle $2\,\omega$ about the circumcenter O of ABC, for:

$$\angle AOA' = 2\,ACA' = 2\,\omega.$$

662. *Remark.* The point Ω is the second Brocard point of the triangle $A'B'C'$, for:

$$\angle \Omega A'C' = CA'C' = CBC' = \omega;$$

likewise

$$\angle \Omega C'B' = BC'B' = BAB' = \omega.$$

663. Corollary. *The two Brocard points of a triangle are equidistant from the circumcenter of the triangle.*

In the two congruent triangles ABC, $A'B'C'$, to the point Ω' of ABC corresponds the point Ω of $A'B'C'$, hence the two points Ω, Ω' are equidistant from the common circumcenter of the two triangles.

664. Theorem. *The pedal triangle of a Brocard point is similar to the given triangle.*

Let L, M, N be the projections of the Brocard point Ω upon the sides BC, CA, AB of the triangle ABC (Fig. 149). Taking into account the definition of the Brocard point and the two cyclic quadrilaterals ΩMAN and ΩNBL, we have, successively:

$$\begin{aligned}\angle A &= \Omega AN + \Omega AM = \Omega BL + \Omega AM = \Omega NL + \Omega AM \\ &= \Omega NL + \Omega NM = MNL.\end{aligned}$$

Similarly for the other pairs of angles of ABC and MNL.

Likewise for Ω'.

665. Corollary. *The pedal triangles of the two Brocard points are congruent.*

For they are similar (§ 664) and have the same circumcircle (§ 654).

EXERCISES

1. Show that the antipedal triangles of the Brocard points are similar to the given triangle.
2. Construct a triangle, given, in position, two indefinite sides and a Brocard point.
3. Show that (a) the triangles ΩCA, $\Omega' BA$ are similar (point out other pairs of similar triangles); (b) $A\Omega : A\Omega' = b : c$ (state two analogous formulas); (c) $B\Omega \cdot C\Omega' = C\Omega \cdot A\Omega' = A\Omega \cdot B\Omega'$; (d) $c \cdot C\Omega = b \cdot B\Omega'$ (state two analogous formulas); (e) $A\Omega : B\Omega : C\Omega = (b:a):(c:b):(a:c)$, and similarly for Ω'.

4. Let A' and A'', B' and B'', C' and C'' be the points in which the lines joining the vertices of a triangle ABC to the Brocard points meet the circumcircle of the triangle, and K, K', K'' the Lemoine points of the triangles ABC, $A'B'C'$, $A''B''C''$. Prove that the lines KK', KK'' pass each through a Brocard point of ABC.

b. THE BROCARD CIRCLE

666. Definition. The circle (OK) having for diameter the segment determined by the circumcenter O and the symmedian point K of a triangle ABC is called the *Brocard circle* of ABC (Fig. 150).

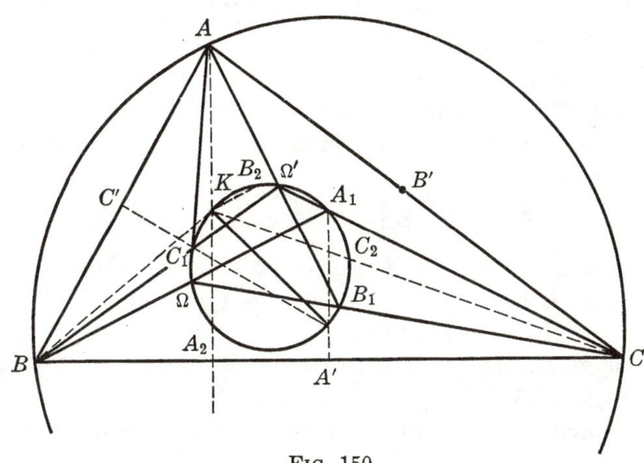

FIG. 150

The mediators of the sides BC, CA, AB meet the Brocard circle (OK) again in the vertices A_1, B_1, C_1 of *Brocard's first triangle* $A_1B_1C_1$ of ABC.

The symmedians AK, BK, CK of ABC meet (OK) again in the vertices A_2, B_2, C_2 of *Brocard's second triangle* $A_2B_2C_2$ of ABC.

667. Theorem. *The Brocard points of a triangle lie on the Brocard circle.*

The angle KA_1O is a right angle (Fig. 150); hence KA_1 is parallel to BC, and therefore A_1A' is equal to the distance of K from BC. Thus (§ 574):

(1) $$A_1A':BC = B_1B':CA = C_1C':AB.$$

Let us show that the point of intersection P of AC_1 with BA_1 lies on the circle (OK).

From (1) it follows that the two isosceles triangles ABC_1, BCA_1 are similar; hence:

$$\angle BA_1A' = AC_1C' = PC_1O.$$

Thus the points A_1, C_1 are concyclic with O, P; hence P lies on (OK). Again, considering the triangles BCA_1, CAB_1, it may be shown that A_1B and B_1C meet on (OK). Hence the three lines A_1B, B_1C, C_1A meet in the same point P on (OK). Now from the similitude of the three isosceles triangles A_1BC, B_1CA, C_1AB we have:

$$\angle PAB = PBC = PCA;$$

hence P coincides with the Brocard point Ω.

In a similar way we may show that the three lines B_1A, A_1C, C_1B meet on (OK) in the second Brocard point Ω'.

668. *Remark.* From the three similar isosceles triangles considered (§ 667) we have:

$$AC_1{:}AB_1 = c{:}b, \quad BA_1{:}BC_1 = a{:}c, \quad CA_1{:}CB_1 = a{:}b.$$

669. Theorem. *The Brocard circle is concentric with the first Lemoine circle of the triangle.*

Indeed, both circles have for their center the midpoint of the segment determined by the circumcenter and the symmedian point of the triangle (§§ 590, 666).

670. Theorem. (*a*) *The Brocard circle of a triangle is orthogonal to the Apollonian circles.*

(*b*) *The radical axis of the Brocard circle and the circumcircle coincides with the Lemoine axis.*

(a) The Lemoine point K is conjugate to the circumcenter O with respect to each of the Apollonian circles of a triangle (§ 608); hence the proposition (§§ 387, 666).

(b) The Brocard circle (OK) and the circumcircle (O) are both orthogonal to the Apollonian circles of the triangle (§ 599); hence (OK) and (O) determine the pencil (U) conjugate to the pencil (W) formed by the Apollonian circles (§ 456). Thus the line of centers of (W), i.e., the Lemoine axis (§ 600), is the radical axis of (U) (§ 455), which proves the proposition.

671. Theorem. *The Brocard circle is tangent to the radical axis of the two Lemoine circles.*

The line of centers of the two Lemoine circles lies on the Brocard diameter of the triangle, and the radical axis of these two circles passes through the symmedian point of the triangle (§ 596); hence the proposition.

672. Theorem. *The lines joining the vertices of a triangle to the corresponding vertices of the first Brocard triangle concur in the isotomic of the Lemoine point with respect to the triangle.*

The line A_1OA' (Fig. 150) passes through the pole U of BC with respect to the circumcircle (O) of ABC, and so does the symmedian AK (§ 560). Now if K' is the trace of AK on BC, the parallel KA_1 to the ray $A'K'$ of the harmonic pencil $A'(AK'KU)$ meets $A'A$ in the midpoint of the segment KA_1, i.e., the lines AA_1, AK are isotomic. Similarly for the pairs of lines BB_1 and BK, CC_1 and CK. Hence the proposition.

673. *Remark.* The first Brocard triangle is perspective to the basic triangle in three different ways, the centers of perspectivity being the two Brocard points (§ 667) and the isotomic of the Lemoine point (§ 672).

674. Theorem. *Brocard's first triangle is inversely similar to the given triangle.*

Indeed, we have (Fig. 150):

$$\angle A = B_1KC_1 = B_1A_1C_1,$$

for the sides of these angles are parallel; similarly for the other pairs of angles of the two triangles.

675. Theorem. *The centroid of a triangle is also the centroid of the first Brocard triangle.*

Let X be the symmetric of A_1 with respect to BC (Fig. 150). It is readily seen that angle $C_1BX = B$ and we have (§ 668):

$$BC_1:BX = BC_1:BA_1 = c:a;$$

hence the triangles BC_1X, ABC are similar. In like manner it may be shown that the triangle CB_1X is similar to ABC. Hence the two triangles BC_1X and CB_1X are similar. But $BX = CX$; hence the two triangles are congruent, and we have:

$$B_1X = BC_1 = AC_1, \quad C_1X = CB_1 = AB_1.$$

Thus AC_1XB_1 is a parallelogram. Let U be the point of intersection of its diagonals B_1C_1, AX. The lines AA' and A_1U are medians of the triangle AA_1X; hence they trisect each other, say in G. On the other hand the lines AA', A_1U are medians of the triangles ABC, $A_1B_1C_1$, respectively; hence G is the centroid of these two triangles.

676. Problem. *Construct a triangle given, in position, its first Brocard triangle.*

The required triangle, ABC is inversely similar to its given first Brocard triangle $A_1B_1C_1$ (§ 674), and the first Brocard triangle $A'B'C'$ of $A_1B_1C_1$ is inversely similar to this triangle; hence the two triangles ABC, $A'B'C'$ are directly similar. Moreover, the angles $(B_1C_1, B'C')$, (B_1C_1, BC) are equal; hence the lines BC, $B'C'$ are parallel. Thus the triangles ABC, $A'B'C'$ are homothetic.

The triangles ABC, $A_1B_1C_1$ have the same centroid (§ 675), and so do the triangles $A_1B_1C_1$, $A'B'C'$; hence the two triangles ABC, $A'B'C'$ have the same centroid G, which is therefore their homothetic center. Furthermore, we have:

$$GA':GA_1 = GA_1:GA.$$

The segment GA may thus be determined and the point A located on the line GA'. The points B, C may then be constructed as the corresponding points of B', C' in the homothecy $(G, GA:GA')$.

677. Theorem. *The perpendiculars from the midpoints of the sides of the first Brocard triangle upon the respective sides of the given triangle meet in the nine-point center of the given triangle.*

The perpendiculars from the midpoints of the sides of the triangle $A_1B_1C_1$ (Fig. 150) upon the sides of ABC are parallel to the lines OA_1, OB_1, OC_1; hence these perpendiculars meet in the complementary point of O for the triangle $A_1B_1C_1$, i.e., in a point O' such that $GO':GO = -1:2$, where G is the centroid of $A_1B_1C_1$. But G is also the centroid of ABC (§ 675); hence O' coincides with the nine-point center of ABC.

678. Theorem. *The parallels through the vertices of a triangle to the corresponding sides of the first Brocard triangle meet on the circumcircle of the given triangle.*

If the parallels through the vertices B, C of ABC to the sides A_1C_1, A_1B_1 of the first Brocard triangle $A_1B_1C_1$ (Fig. 150) meet in R, we have:

$$\angle BRC = B_1A_1C_1 = BAC;$$

hence R lies on the circumcircle (O) of ABC, the points R, A lying on the same side of BC.

It is thus seen that the line BR is met by the parallels through C and A in the same point R of (O).

679. Definition. The point R (§ 678) is called the *Steiner point* of the triangle.

680. Corollary. *The perpendiculars dropped from the vertices of a triangle upon the corresponding sides of the first Brocard triangle meet in a point.*

Indeed, these perpendiculars pass through the diametric opposite N of the point R on (O).

681. Definition. The point N (§ 680) is called the *Tarry point* of the given triangle.

682. Theorem. *The vertices of the second Brocard triangle are the midpoints of the segments determined by the circumcircle of the given triangle on the symmedians, produced, of the given triangle.*

The angle OA_2K inscribed in the circle (OK) (Fig. 150) is a right angle; hence OA_2 is the perpendicular dropped from the circumcenter O upon the symmedian AK of ABC; hence the proposition.

683. Theorem. *The direct and the indirect adjoint circles tangent to the sides of the same angle of a triangle meet in the corresponding vertex of the second Brocard triangle.*

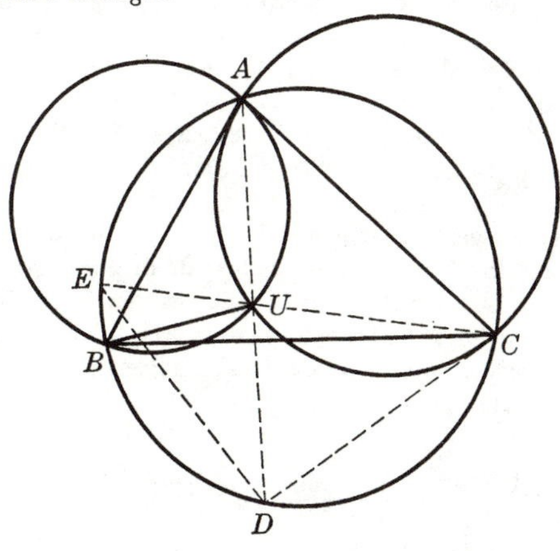

Fig. 151

Let U be the point of intersection of the two adjoint circles (CA), (BA) (Fig. 151) of the triangle ABC, and let the lines AU, CU meet the circumcircle (O) in D, E. We have:

(1) $$\angle UAB = ACU = ACE = ADE,$$
$$\angle UBA = CAU = CAD = CED.$$

Thus the three triangles UAB, UAC, UDE are similar. Now:

$$\angle BAD = BCD; \quad \text{hence} \quad \angle BCA = ECD,$$

i.e., the chords AB, DE are equal, and therefore the triangles UAB, UDE are congruent, and $AU = UD$.

In the similar triangles UAB, UAC the altitudes issued from U are proportional to the sides AB, AC; hence U is a point on the symmedian of ABC (§ 567) passing through A. Thus the proposition is proved.

EXERCISES

1. Show that the polar of the circumcenter of a triangle for the second Lemoine circle is the radical axis of this circle and the Brocard circle.

2. Show that the lines joining the vertices of a triangle to the corresponding vertices of the first Brocard triangle divide the respective sides of the given triangle in the inverse ratio of the squares of the other two sides.

3. Show that the Lemoine axis and the Brocard circle are inverse with respect to the circumcircle of the triangle.

4. Show that the segments joining the symmedian point to the vertices of the first Brocard triangle are bisected by the corresponding medians of the given triangle.

5. Show that the line joining the symmedian point K of a triangle to the nine-point center O' passes through the center Z' of the Brocard circle of the anticomplementary triangle, and $KO' = O'Z'$.

6. Show that the circumcenter O of ABC, the Lemoine point K_1 of the medial triangle $A_1B_1C_1$, and the circumcenter Z' of the anticomplementary triangle $A'B'C'$ are collinear, and $OK_1 = K_1Z'$.

7. Show that the symmedian point and the circumcenter of a triangle are the Steiner point and the Tarry point of the first Brocard triangle.

8. Show that the Simson line of one point of intersection of the Brocard diameter with the circumcircle of the triangle is either parallel or perpendicular to the bisector of the angle formed by a side of the triangle and the corresponding side of Brocard's first triangle.

F. TUCKER CIRCLES

684. Concyclic Points. With the Lemoine point K of the triangle ABC (Fig. 152) as homothetic center a triangle $A'B'C'$ is constructed directly homothetic to ABC. Let the sides $B'C'$, $C'A'$, $A'B'$ meet the sides AC and AB, BA and BC, CB and CA in the points E and F', F and D', D and E'.

Theorem. *The six points D, E, F, D', E', F' are concyclic.*

The diagonal $E'F$ of the parallelogram $AE'A'F$ is bisected, in U, by the diagonal AA', i.e., by the symmedian $AA'K$ of ABC; hence $E'F$ is antiparallel to BC. Similarly, $F'D$ is antiparallel to AC, and $D'E$ to AB.

The two parallel lines DD', EF' and the two antiparallels DF', $D'E$ to AC, AB, respectively, form an isosceles trapezoid; hence the four points D, D', E, F' lie on a circle. Similarly, the points E, E', F, D' lie on a circle, and also F, F', D, E'. Now these three circles cannot

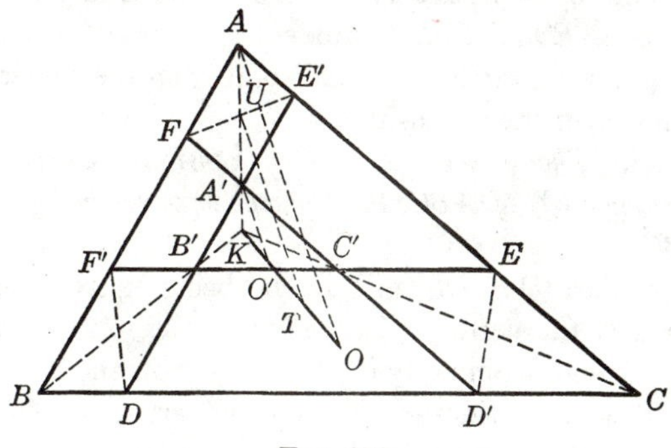

FIG. 152

all be distinct, for, if they were, their radical axes would be the lines ED', FE', DF' which form a triangle, whereas the radical axes of three distinct circles are concurrent (§ 425); hence at least two of these circles coincide. But then the third circle coincides with them, and the proposition is proved.

685. Definition. The circle (§ 684) is called a *Tucker circle* of the triangle.

The homothetic ratio of the triangles ABC, $A'B'C'$ may be taken arbitrarily; hence the triangle ABC has an infinite number of Tucker circles.

The value of the homothetic ratio is determined, if one of the vertices of $A'B'C'$, say A', is chosen on the corresponding symmedian AK. If A' coincides with A, the corresponding Tucker circle coincides with the circumcircle of ABC. When A' coincides with K, i.e., when the antiparallels $D'F, \ldots$ are drawn through K, the corresponding Tucker circle coincides with the first Lemoine circle of ABC.

686. Theorem. *The centers of the Tucker circles lie on the Brocard diameter of the given triangle.*

The circumradius $O'A'$ (Fig. 152) of the triangle $A'B'C'$ is parallel to the circumradius OA of ABC, and O' is collinear with K and O (§ 47). Thus the parallel UT to OA through the midpoint U of AA' passes through the midpoint T of OO'. Now OA is perpendicular to

$E'F$ (§§ 185, 188); hence UT is the mediator of the chord $E'F$ of the Tucker circle. Similarly the mediators of DF' and $D'E$ pass through T. Hence the proposition.

687. Theorem. *The center T of the Tucker circle* (§ 686) *is the incenter of the triangle $A''B''C''$ formed by the lines $E'F$, $F'D$, $D'E$, produced.*

The segments $E'F, F'D, D'E$ are equal (§ 682); therefore these chords of the Tucker circle are equidistant from the center T of this circle, which proves the proposition.

688. Theorem. *The triangle $A''B''C''$* (§ 687) *is homothetic to the tangential triangle* (Q) *of ABC, the homothetic center being the Lemoine point of ABC.*

The antiparallels $D'E$, DF' to AB, AC being equal, their point of intersection, i.e., the vertex A'' of $A''B''C''$, lies on the symmedian AK of ABC (§ 569). Similarly for B'', C''. Now the lines $D'E$, $E'F$, $F'D$ are parallel to the sides of (Q), and the vertices of (Q) lie on the symmedians of ABC (§ 560); hence the proposition.

689. Theorem. *The six projections of the vertices of the orthic triangle upon the sides of the given triangle are concyclic.*

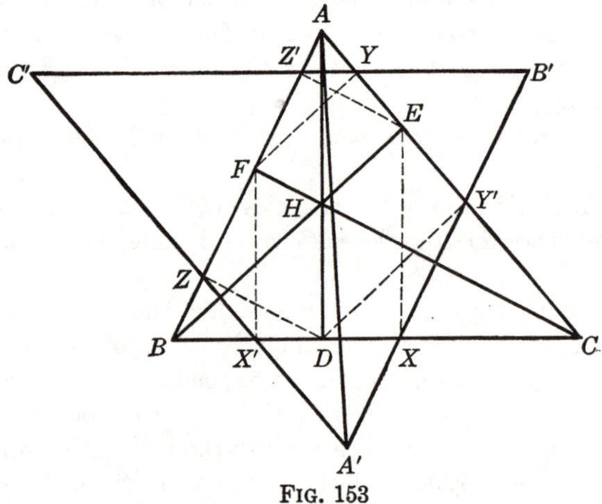

FIG. 153

Let H be the orthocenter and DEF the orthic triangle of ABC (Fig. 153). Let X, X' be the projections of E, F upon BC; Y', Y the projections of D, F upon CA; and Z, Z' the projections of D, E upon AB.

In the triangle AEF the line YZ' joins the feet of the altitudes EZ',

FY; hence YZ' is antiparallel to EF, and since EF is antiparallel to BC, the line YZ' is parallel to BC. Similarly for ZX' and XY'. Thus the lines YZ', ZX', XY', extended, form a triangle $A'B'C'$ homothetic to ABC.

The line AA' bisects $Y'Z$, the two lines being the diagonals of the parallelogram $AZA'Y'$. Now from the two similar triangles $DY'Z$, HEF it follows that $Y'Z$ is parallel to EF and therefore antiparallel to BC; hence AA' is a symmedian of the triangle ABC. Similarly for BB' and CC'.

Thus the homothetic center of the two triangles ABC, $A'B'C'$ coincides with the symmedian point K; hence the six points X, X', Y, Y', Z, Z' lie on a circle (§ 684).

690. Definition. The circle (§ 689) is called the *Taylor circle* of the triangle.

It follows from the proof that the Taylor circle belongs to the circles of the Tucker group.

G. THE ORTHOPOLE

691. Theorem. *The perpendiculars dropped upon the sides of a triangle from the projections of the opposite vertices upon a given line are concurrent.*

Let A', B', C' (Fig. 154) be the projections of the vertices A, B, C of the given triangle ABC upon a given line m, and $A'A''$, $B'B''$, $C'C''$ the perpendiculars to the sides BC, CA, AB. If the lines $B'B''$, $C'C''$ meet $A'A''$ in M, N, and AA' meets BC in K, the two pairs of triangles KAC and $A'B'M$, KAB and $A'C'N$ have their sides respectively perpendicular; hence:

$$A'M:A'B' = CK:AK, \quad A'C':A'N = AK:BK.$$

Now the lines AA', BB', CC' are parallel; hence:

$$A'B':A'C' = BK:CK.$$

Multiplying the three proportions, we obtain $A'M = A'N$; hence N coincides with M, which proves the proposition.

692. Definition. The point M (§ 691) is the *orthopole* of the line m with respect to, or, more briefly, for the triangle ABC.

693. ***Remark.*** The orthopoles of two parallel lines lie on a common perpendicular to the two lines, and the distance between the orthopoles is equal to the distance between the parallel lines.

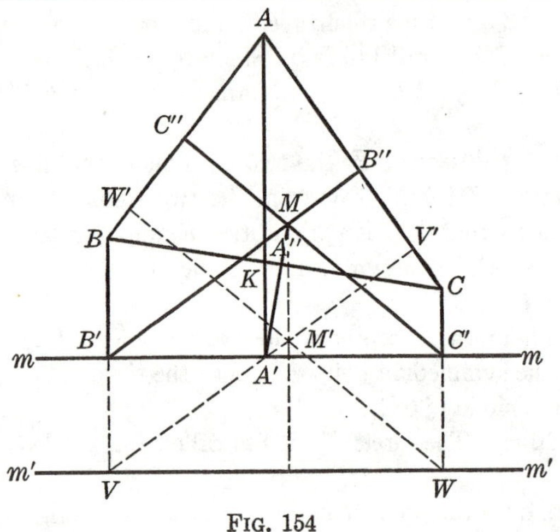

FIG. 154

Indeed, the triangle $M'VW$ (Fig. 154) may be obtained from the triangle $MB'C'$ by moving the vertices B', C' in the direction $B'V$ a distance equal to $B'V = C'W$.

It may also be observed that the distances of the two lines m, m' from their respective orthopoles M, M' are equal.

694. Theorem. *The orthopole, for a given triangle, of a circumdiameter of that triangle lies on the nine-point circle of the triangle.*

Let the perpendiculars AA', BB', CC' from the vertices of the triangle ABC upon the circumdiameter d of ABC meet the circumcircle (O) in the points A_1, B_1, C_1 (Fig. 155). The three perpendiculars from A_1, B_1, C_1 to the sides BC, CA, AB meet (O) in the same point P, the pole of the Simson line perpendicular to d (§ 288).

The midpoint A' of the chord AA_1 is equidistant from the altitude AA_h of ABC and the perpendicular A_1P from A_1 to BC; hence the perpendicular p from A' to BC bisects, in D, the segment PH joining P to the orthocenter H of ABC. The perpendiculars q, r, from B', C' to CA, AB pass through D, for analogous reasons; hence the point $D = pqr$ is the orthopole of the diameter d. Now D lies on the nine-point circle of ABC (§ 291); hence the proposition.

It should be observed that we have incidentally proved the existence of the orthopole of the diameter d, and therefore (§ 693) of any line.

695. COROLLARY. The line s through D parallel to A_1A, or, what is the same thing, perpendicular to d (Fig. 155), is the Simson line of the point P (§ 289). Now any line parallel to d will have its orthopole on s (§ 693); hence: *The orthopole of a line for a given triangle lies on the Simson line perpendicular to the given line.*

696. *Remark.* (a) The orthopole D (Fig. 155) of a circumdiameter d of a triangle ABC may be constructed as follows.

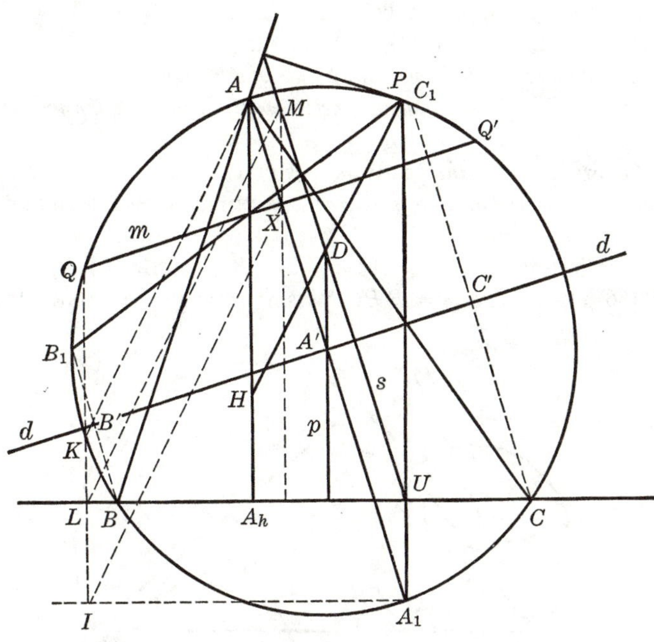

FIG. 155

If the perpendicular from A to d meets d in A' and the circumcircle in A_1, the parallel s to AA_1 through the projection U of A_1 upon BC meets the perpendicular to BC through A' in the required orthopole D.

(b) If m is any line parallel to d, its orthopole M is the trace, on s (§ 693), of the perpendicular to BC through the trace X of m on AA_1.

697. Theorem. *If a line meets the circumcircle of a triangle, the Simson lines of the points of intersection with the circle meet in the orthopole of the line for the triangle.*

Let the line m meet the circumcircle (O) of ABC in Q, Q' (Fig. 155), and let the perpendicular QL from Q to BC meet (O) in K and the parallel IA_1 through A_1 to BC in I.

Considering the two cyclic quadrilaterals A_1XQI, AA_1KQ, we have successively:

$$\angle IXA_1 = IQA_1 = KQA_1 = KAA_1;$$

hence XI is parallel to AK.

On the other hand, the segments XM, LI being equal and parallel to A_1U, the line LM is parallel to XI; hence LM is the Simson line of Q (§ 289). Similarly for Q'. Hence the proposition.

698. Corollary. *The Feuerbach points of a triangle are the orthopoles, for that triangle, of the circumdiameters passing through the tritangent centers* (§ 648).

699. Theorem. *If a line passes through the orthocenter of a triangle, the symmetric of the orthopole of this line with respect to the line lies on the nine-point circle.*

The projections B', C' of the vertices B, C of the triangle ABC (Fig. 156) upon a line $m = B'HC'$ passing through the orthocenter H

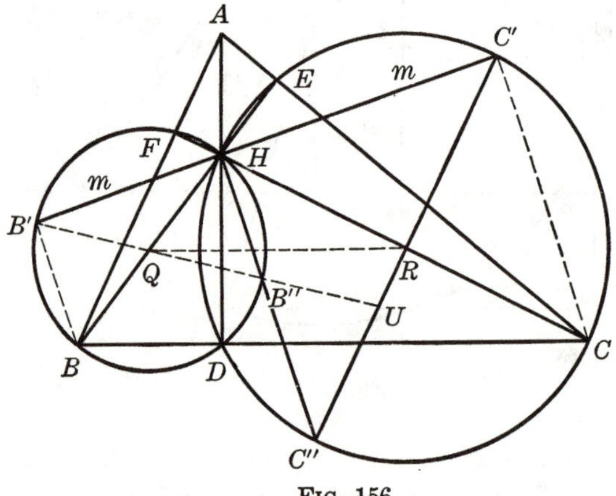

Fig. 156

of ABC lie on the circles (Q), (R) having BH, CH for diameters. If M is the orthopole of m, the lines $B'M$, $C'M$ are perpendicular to AC, AB (§ 689) and therefore respectively parallel to BH, CH, i.e., $B'M$, $C'M$ are the symmetrics, with respect to m, of the lines $B'Q$, $C'R$ joining B', C' to the centers Q, R of (Q), (R).

Let us now show that the point $U = (B'Q, C'R)$ lies on the nine-point circle (N) of ABC. The diametrical opposites B'', C'' of B', C' on (Q), (R) lie on the perpendicular to m at H, and we have:

$$\angle QUR = B''UR = UC''B'' + UB''C'' = UC''B'' + QB''H$$
$$= RC''H + QB''H = RHC'' + QHB'' = QHR = QDR;$$

hence the four points U, D, Q, R are concyclic. But the foot D of the altitude AD and the Euler points Q, R lie on the nine-point circle (N) of ABC; hence the proposition.

EXERCISES

1. Show that the orthocenter of a triangle is the orthopole of the sides with respect to the triangle.
2. Through the orthogonal projections of the vertices of a triangle ABC upon a given line m parallels are drawn to the respective sides of ABC forming a triangle $A'B'C'$. Show that the orthopole of m for ABC is concyclic with the points A', B', C'.
3. The line joining the orthopoles M, M_1 of a line m with respect to a triangle and its medial triangle $A_1B_1C_1$ meets m in a point G' such that $MG' = 2\,G'M_1$. Show that the point G' is the projection upon m of the centroid G of ABC.
4. If A', B', C' are the projections of the vertices of a triangle ABC upon a given line, show that the perpendiculars from the midpoints of the segments $B'C'$, $C'A'$, $A'B'$ upon the respective sides of ABC are concurrent.
5. Show that (a) the distance between the orthopoles of the two bisectors of the same angle of a triangle is equal to the circumdiameter; (b) if a chord PQ of the circumcircle (O) of a triangle ABC is perpendicular to BC, the distance between the orthopoles of the lines AP, AQ is equal to PQ.
6. Show that the orthopole of a line for a triangle is the radical center of the three circles tangent to the given line and having for centers the vertices of the anticomplementary triangle of the given triangle.

SUPPLEMENTARY EXERCISES

1. P and Q are any pair of isogonal conjugates with respect to a triangle ABC. If AQ cuts the circumcircle in O, and OP cuts BC in R, show that QR is parallel to AP.
2. If O and G are the circumcenter and the centroid of the triangle ABC, and O_a, O_b, O_c the circumcenters of the triangles GBC, GCA, GAB, show that the points O and G are respectively the centroid and the Lemoine point of the triangle $O_aO_bO_c$.
3. If $B'C'$ is the antiparallel to BC, in the triangle ABC, passing through the foot of the symmedian issued from A, show that the polar of A for the circle having BC for diameter passes through the orthocenters of the triangles ABC, $AB'C'$.
4. If DEF is the orthic triangle of the triangle ABC, and X', Y', Z' are the symmetrics, with respect to the symmedian point K of ABC, of the projections X, Y, Z of K upon BC, CA, AB, show that X', Y', Z' are the Lemoine points of the triangles AEF, BFD, CDE.

5. A variable triangle has a fixed circumcircle and a fixed centroid. Show that the locus of the Lemoine point is a circle.

6. A variable triangle has a fixed circumcircle and a fixed symmedian point. Show that the locus of the centroid is a circle.

7. If P and P', Q and Q' are pairs of isogonal conjugate points, show that the points $R = (PQ, P'Q')$, $R' = (PQ', P'Q)$ are also isogonal conjugates.

8. Show that the isogonal conjugate of the Nagel point is collinear with the circumcenter and the incenter of the triangle.

9. Show that the reciprocal transversals of two rectangular circumdiameters of a triangle meet on the polar of the circumcenter with respect to the nine-point circle.

10. Show that the symmetric, with respect to a side of a given triangle, of the corresponding vertex of the cosymmedian triangle is the isogonal conjugate of the foot of the perpendicular from the circumcenter upon the symmedian considered.

11. Construct a triangle given, in position, the traces of its medians on the circumcircle.

12. Construct a triangle given, in position, its second Brocard triangle.

13. Show that the symmetrics, with respect to the sides of a triangle, of the traces of the corresponding medians on the circumcircle form a triangle homothetic to the second Brocard triangle of the given triangle.

14. Show that the lines joining the corresponding vertices of the two Brocard triangles of a given triangle meet in the centroid of the given triangle.

15. M is any point in the plane of a triangle ABC. If MA, MB, MC cut the circumcircle again in A', B', C', and D, E, F are the projections of M upon BC, CA, AB, show that the triangles $A'B'C'$, DEF are directly similar and that M in $A'B'C'$ corresponds with the isogonal conjugate of M in DEF. Use this proposition to demonstrate the existence of cosymmedian triangles having the sides of each proportional to the medians of the other.

16. Show that the circumcenter of a triangle is the centroid of the antipedal triangle of the Lemoine point of the given triangle.

17. Show that the trilinear polar of a point on the circumcircle of a triangle passes through the symmedian point of the triangle.

18. Show that (a) the line joining the Lemoine point of a triangle to the orthocenter passes through the Lemoine point of the orthic triangle; (b) the four lines joining the symmedian points of an orthocentric group of triangles to the corresponding orthocenters are concurrent.

19. The midpoint A' of the side BC of the triangle ABC is also the midpoint of the segment MN intercepted on a line through A' by the two isogonal lines AM, AN. If the lines BN, CN meet AM in E, F, show that the tangents at B and C to the circles ABE, ACF meet on AN.

20. Let D, E be the projections of the vertex A of the triangle ABC upon the side BC and upon the mediator $A'E$ of this side. Show that the line DE passes through the vertex A_2 of the second Brocard triangle and that $4\,EA_2 \cdot AA' = AB^2 + AC^2$.

21. The Brocard points of a triangle are fixed and the value of the Brocard angle is equal to one-half of an angle of the triangle. Show that the vertices of the triangle describe straight lines.

22. Construct an isosceles triangle given, in position, any three of the following five points: a vertex, the centroid, the Lemoine point, the two Brocard points.

23. If a variable point P describes a circle concentric with a given equilateral triangle ABC, show that the Brocard angle of the pedal triangle of P for ABC is constant.

24. If the orthocenter of a triangle lies on the Brocard circle, show that the triangle is equilateral.

25. P, Q, R and P', Q', R' are the images of the Brocard points on the sides of the triangle ABC. Prove that (a) the triangles PQR, $P'Q'R'$ have a common Brocard circle, whose center is the symmedian point of ABC; (b) they also have a common first Lemoine circle which is identical with the second Lemoine circle of ABC; (c) the radical axes of the circumcircles of these three triangles are their Brocard diameters.

26. If a line through the symmedian point K of the triangle ABC meets the sides BC, BA in the points D, E so that $DK = KE$, and Q is the second point of intersection of the circumcircle of the triangle DBE with the circle having AB for diameter, prove that the points B, C, Q, and one of the Brocard points of ABC are concyclic.

HISTORICAL AND BIBLIOGRAPHICAL NOTES

In these notes the following abbreviations (in bold italics) refer to the abbreviations listed.

A. BOOKS

Chl. Michel Chasles (1793–1880), *Traité de géométrie supérieure* (Paris, 1852; 2nd ed., 1889).

Cn. E. Catalan (1814–1894), *Théorèmes et problèmes de géométrie élémentaire* (6th ed.; Paris, 1879. 1st ed., 1852).

Crt. L. N. M. Carnot (1753–1823), *Géométrie de position* (Paris, 1803).

Ech. A. Emmerich (1856–1915), *Die Brocardschen Gebilde* (Berlin, 1891).

Gy. William Gallatly, *The Modern Geometry of the Triangle* (2nd ed.; London, 1922).

LHr. Simon A. J. LHuilier (1750–1840), *Elémens d'analyse géométrique et d'analyse algebrique appliquées à la recherche des lieux géométriques* (Paris and Geneva, 1809).

Pt. Jean Victor Poncelet (1788–1867), *Traité des propriétés projectives des figures* (Paris, 1822; 2nd ed., 1865).

Rouché. Rouché et Comberousse, *Traité de géométrie*, I (7th ed., 1900; 1st ed., Paris, 1864). The note on the geometry of the triangle (pp. 445–515) was contributed by J. Neuberg.

B. PERIODICALS

Bull. *Bulletin des sciences mathématiques et physiques élémentaire* (Paris, 1895–1906). Founded by B. Niewenglowski, edited by L. Gerard and Ch. Michel.

Edinb. *Proceedings of the Edinburgh Mathematical Society* (1894–).

ET. Mathematical questions and solutions from the *Educational Times* (Reprints, London, 1863–1918).

GAM. *Annales de mathématiques* (1810–1832). Founded and edited by J. D. Gergonne (1771–1859).

JME. *Journal de mathématiques élémentaires* (1877–1897).

JMS. *Journal de mathématiques spéciales* (1882–1897). This periodical and the preceding one were founded by J. Bourget (1822–1887) and continued by G. de Longchamps (1842–1906).

Mathesis. *Mathesis* (1881–). Founded and edited by Paul Mansion (1844–1919) and J. Neuberg (1840–1926). Continued by Ad. Mineur (1867–1950) and Ronald Deaux (1893–). Ghent, Brussels, Mons.

Monthly. *American Mathematical Monthly* (1894–). Founded by B. F. Finkel (1865–1947). Since 1916 the official organ of the Mathematical Association of America.

NA. *Nouvelles annales de mathématiques* (1842–1927). Founded by O. Terquem (1782–1862) and C. Gerono (1799–1892).

NC. *Nouvelle correspondance mathématique* (1874–1880). Founded and edited by E. Catalan (1814–1894).

(*The numbers in these notes correspond to those of the articles in this book.*)

4. The analytical method of solving construction problems was developed by the Greek philosopher Plato and his school. The credit is usually given to the master.

6. *JME.* (1880), p. 19, Q. 153.

7. E. N. Barisien, *Mathesis*, 1922, p. 43, Q. 2047.

8. Cn., p. 57, prob. 3.

11. The notion of a geometrical locus arose and developed in Plato's school in connection with the study of construction problems. See Note 4.

Apollonius of Perga wrote a book *Plane Loci* in which all the loci mentioned in this article were included. This book has not come down to us. It was restored by Robert Simson (1687–1768) under the title, *The loci plani of apollonius restored*, Edinburgh, 1749, according to the indications found in the *Mathematical Collection* of Pappus (300 A.D.).

11, locus 4. The term "mediator" was introduced by the Belgian mathematician J. Neuberg.

11, locus 11. Pappus in his *Collection* states that Apollonius made use of this circle in his book *Plane Loci;* hence the name *circle of Apollonius* or *Apollonian circle.*

12. Ignacio Beyens, *JME.* (1888), p. 238, Q. 255.

13. Cn., p. 57, prob. 4.

14. Ignacio Beyens, *JME.* (1888), p. 236, Q. 252.

15. LHr., p. 221, ex. 8.
Cn., p. 59, prob. 9.

17. LHr., p. 212, art. 132.
Cn., p. 59, prob. 9.

18. LHr., p. 218.
Cn., p. 60, prob. 10.

24–30 (Chapter II, A). The builders of the pyramids made practical use of similar figures in connection with their constructions. Legend and tradition credit Thales (600 B.C.) with the use of similar figures in indirect measurements of inaccessible heights. The first theoretical considerations of similar figures go back to Hippocrates of Chios (440 B.C.).

28. *JME.* (1882), p. 39, Q. 368.

31–53 (Chapter II, B). Similar figures similarly placed were considered by J. V. Poncelet.

The term *homothetic* is attributed to Michel Chasles, *GAM.*, XVIII (1827–1828), p. 220.

42. W. J. Greenstreet, *ET.*, LV (1891), p. 52, Q. 10,672.

43. I. Alexandrov, *Geometrical Construction Problems* (Russian, first ed., 1882; French translation, 1899; German translation, 1903), p. 78, art. 289. (1934).

J. S. MacKay (1843–1914), Edinb. II (1883–1884), pp. 5 and 27.

Frank Morley (1860–1937), *ET.*, LXIX (1898), p. 93, Q. 9367.

44. I. Alexandrov, *op. cit.*, p. 99, art. 371. *Mathesis* (1929), p. 379, note 37.

47. LHr., p. 39, art. 28.

51–52. LHr., p. 45, art. 29. Cn., p. 81, prob. 48.

53. LHr., p. 44, art. 30. Cochez, *ET.*, III (1903), p. 50, Q. 15017.

54. This construction is given by Pappus of Alexandria in *Mathematical Collection* (third century A.D.).

59. If the reciprocals of three numbers a, b, c form an arithmetic progression, then b is said to be the harmonic mean of a and b. It is readily verified that the segment CD is the harmonic means of segments AD, BD (art. 346); hence the term "harmonic points." The term "harmonic" was used by the Greeks.

62. Pt., Sec. I., chap. I, art. 30, p. 16.

JME. (1881), p. 311, Q. 290.

67. LHr., p. 199, ex. 8. The geometrical method of solving quadratic equations was used by the Greeks.

70. Sanjana. *ET.*, V (1904), p. 61, Q. 15238.

73. J. Mention, *NA.*, IX (1850), p. 326.

76. Cn., p. 65, prob. 20. R. Ivengur, *ET.*, LXI (1894), p. 64, Q. 11986.

77. A. Emmerich, *Mathesis* (1901), p. 162, note 14.

78. *Bull.*, IX (1908–1909), pp. 297–299, Q. 1444.

81. E. N. Barisien, *Bull.*, III (1897–1898), p. 114, Q. 342.

83. *JME.* (1882), p. 15, Q. 352.

84. Considered by Pappus in his *Mathematical Collection.* LHr., p. 241, remark 2.

85, 86. These formulas occur in the works of the Hindu astronomer Brahmagupta (7th century).

88. The property was known to Archimedes. It is given explicitly by Heron of Alexandria in his *Mechanica*, p. 190. (First century A.D.)

90. Cn., p. 13, prob. 5. A. E. A. Williams, *ET.*, LXIII (1895), p. 49, Q. 12550.

91, 92. *NA.* (1863), pp. 93 and 419.

93. Anonymous, *Bull.*, XI (1905–1906), p. 232, Q. 1669.

95. O. Delhez, *Mathesis* (1926), p. 95, note 10.

98. The term "complementary point" was proposed by Hain, Grunert's Archiv (1885), p. 214.

100, 101. Ch. H. Nagel (1803–1882), *Untersuchungen ueber die wichtigsten zum Dreieck gehaerigen Kreise* (1836).

The term "anticomplementary point" was proposed by Longchamps, *JME.* (1886), p. 131.

104. Charruit, *Bull.*, VIII (1902–1903), p. 119, Q. 437.

108. Crt., p. 321, art. 281.

109. A. M. Nesbit, *ET.* XXIII (1913), pp. 36–37, Q. 17309. This is a special case of a more general formula. See Crt., pp. 316–317, art. 274, th. 26.

110. Georges Dostor, *NA.* (1883), p. 369.

112, 113. A. Lascases, *NA.* (1859), pp. 171, 265, Q. 477. *ET.*, LXIV (1896), p. 58, Q. 12801.

114. L. M. Kelly, *Monthly,* LI (1944), p. 390, QE., 613.

115. This problem was proposed in 1840 by D. C. Lehmus (1780–1863) to J. Steiner (1796–1867). See N. A.–C., *Mathematical Gazette,* XVIII (1934), p. 120. V. Thébault (1882–), *Scripta Mathematica,* XV (1949), pp. 87–88.

116. The proposition was known in the Pythagorean school (600 B.C.).

118–120, 121–124. Cn., p. 46, th. 21. N. A.–C., *Monthly,* XXV (1918), pp. 241–242.

127. G. H. Hopkins, *ET.,* XVI (1872), p. 59, Q. 3500. C. Bourlet, *Bull.,* II (1896–1897), p. 187, Q. 196.

130. *JME.* (1880), p. 542, Q. 229.

131–174. The inscribed circle is considered in Euclid's *Elements,* but the interest in the escribed circles awakened only in the beginning of the 19th century. Thus LHr., p. 198, rem. I, mentions all three escribed circles.

The term "tritangent circles" came into use during the second quarter of the 20th century.

132. The formula is given by Heron of Alexandria (first century) in his *Mechanica.*

134. Mahieu in 1807. LHr., p. 224.

135–137. Anonymously, *GAM.,* XIX (1828–1829), p. 211. W(illiam) W(allace), *Cambridge Math. J.,* I (1839), pp. 21–22. O. Terguem, *NA.,* II (1843), p. 554.

145. Karl Wilhelm Feuerbach (1800–1834), *Eigenschaften einiger merkwuerdigen Punkte des geradelinigen Dreiecks und mehrerer durch Sie bestimmter Linien und Figuren* (Nuernberg, 1822, p. 4. Sec. ed., Berlin, 1908.)

146. Crt., p. 168.

149. Anonymous, *Bull.,* VI (1900–1901), p. 235, Q. 904.

150. Anonymous, *Mathesis* (1898), p. 79, QE. 824.

151. LHr., p. 219. N. A.-C., *Mathesis* (1929), p. 251, note 19. Bibliography.

153. Simon LHuilier, *GAM.,* I (1810–1811), p. 157.

157–158. LHr., pp. 214–215.

159–163. Cn., pp. 44–46, th. 20.

166. Cf. Cn., p. 47, prop. 22 (proof).

168. Anonymous, *Bull.,* X (1904–1905), p. 220, Q. 1594.

169. Cochez, *ET.,* I (1902), p. 36, Q. 14772.

175. The proposition is not included in the *Elements* of Euclid. It is found in the writings of Archimedes (287–212 B.C.) in an indirect form, and explicitly in Proclus (410–485), a commentator of Euclid.

176. The term "orthocenter" was first used by W. H. Besant in *Conic Sections Treated Geometrically,* 1869.

177. Crt., p. 187, art. 153.

178–180. Carnot, *De la correlation des figures gêométriques* (Paris, 1801), pp. 101 ff.

181. J. Steiner, *Gesammelte Werke,* I, p. 345.

182. J. Ardueser, *Geometriae theoricae et practicae* (1627), p. 149.
Cn., p. 195, prob. 15.

184. Crt., p. 187, art. 151.

185. The term "antiparallel" in its present sense is used by A. Arnauld (1612–1694) in his *Nouveaux Éléments de Gêométrie* (1667).
Crt., p. 187, art. 151.

186. L. N. M. Carnot, *De la correlation des figures gêometriques,* p. 101, Paris, 1801.

192. Ph. Naudé (1654–1729) in 1737.
Cn., p. 27, Pr. 1.

195. Crt., p. 163, art. 131.

196. H. D. Drury, *ET.*, XXVIII (1915), p. 59, Q. 17521.

199. James Booth, *JME.* (1879), p. 236. A. Causse, *Bull.*, I (1895–1896), pp. 23–24, gives corresponding formulas for an obtuse angled triangle.

200. Janculescu, *Mathesis* (1913), p. 111, Q. Ex. 1625.

201. Leonhard Euler (1707–1783), *Proceedings of the St. Petersburg Academy of Sciences* (1765). The proof is analytic. First geometric proof Crt., pp. 163–164, art. 131.

203. G. Dostor, *NA.* (1883), p. 369.

205. A. Gob, "Sur la droite et le cercle d'Euler," *Mathesis* (1889), Supplement, p. 1, art. 1.

207, 208. In 1765, in the *Proceedings of St. Petersburg* (*Leningrad*) *Academy of Sciences*, Leonhard Euler showed that, in modern terminology, the circumcircle of the orthic triangle of a given triangle is also the circumcircle of the medial triangle.

In a joint paper by Ch. J. Brianchon and J. V. Poncelet, published in *GAM.*, XI (1820–1821), pp. 205 ff., the authors not only rediscovered the Euler circle (th. IX, p. 215), but showed that this circle also passes the Euler points of the triangle. See D. E. Smith (Editor), *A Source Book in Mathematics* (New York, 1929), pp. 337–338.

Many writers rediscovered Euler's circle, both before and after Brianchon and Poncelet. Among them was Feuerbach. (See note 145.) He missed the Euler points.

Poncelet calls this circle the *nine-point circle*, and that is the name commonly used in the English-speaking countries. Most of the French writers, however, prefer to call it the *Euler circle*. In Germany the circle is consistently referred to as the *Feuerbach circle.*

211. A. Gob, "Sur la droite et le cercle d'Euler," *Mathesis* (1889), Supplement, p. 2, art. 2.

212. O. Delhez, *Mathesis* (1915), p. 110, note 10.

214. A. Droz-Farny, *ET.*, XVI (1894), p. 42, Q. 12157. *JME.* (1894), p. 214, Q. 537.

215, 216, 217. Feuerbach, art. 57 (see note 145). The proof is by computation, involving trigonometry. See note 207, D. E. Smith, pp. 339–345.
E. H. Neville, *Mathematical Gazette*, XXI (1937), p. 127.

218. Mascart, *ET.*, LXIII (1895), p. 46, Q. 10228.

219. *Mathesis* (1902), p. 48, Q. 1309.

220. Crt., p. 162, art. 130.

The term "orthocentric group of points" was introduced by Longchamps, *Mathesis* (1888), p. 210.

223. Crt., p. 163.

224, 225. Sarkar, *ET.*, LVII (1892), p. 82, Q. 11269. *Mathesis* (1928), p. 436, Q. Ex. 52.

225–228. N. A.-C., *Monthly*, XXVII (1920), pp. 199–202.

233. E. N. Barisien, *NA.*, pp. 485–486, Q. 1538.

235. William Gallatly, *ET.*, XVI (1909), p. 46.

236. R. Goormaghtigh, *Mathesis* (1933), pp. 120–121, note 10.

240. Pierre Varignon, *Eléments de mathématiques* (1731; 2nd ed., 1734, p. 62).
Cn., p. 9, th. 11.

242. J. D. Gergonne, *GAM.*, I (1810–1811), pp. 232, 311.
Cn., p. 9, th. 12.

246. Crt., p. 328. Carnot attributes the proposition to Euler.

248. Crt., p. 327, art. 290, th. 29.

249. Cn., p. 14, prob. 8.
W. J. C. Miller, *ET.*, LXIV (1896), p. 70, Q. 12649.

250. Cn., p. 69, prob. 27.

255. The proposition is included in the *Almagest*, a famous work on astronomy by Claudius Ptolemy.
Crt., p. 270, art. 212, 1°.

256. A special case of this problem preoccupied Regiomontanus (Johannes Meuller, 1436–1476) in 1464.

The first ruler and compass construction was given by Fr. Viète in 1540. The problem has been the object of considerable attention.

LHr., p. 255, art. 140.

257. Part b. was known to Regiomontanus (see note 256).
Crt., p. 270, art. 212, 2°.
Part d. G. Fontené, *Bull.*, IX (1903–1904), p. 147.

258. H. D. Drury. *ET.*, LXV (1896), p. 107, Q. 13127. Auguste Deteuf, *NA.* (1908), p. 443, prop. 2°.

260. J. Neuberg, *Mathesis* (1924), p. 314, art. 38.

261. Jules Mathot, *Mathesis* (1901), pp. 25–26. A. Deteuf, *NA.* (1908), p. 443, prop. 3°.

263. E. Lemoine, *NA.* (1869), pp. 47, 174, 317. Jules Mathot, *Mathesis* (1901), pp. 25–26. A. Deteuf, *NA.* (1908), p. 447.

264. Thomas Dobson, *ET.*, I (1864), p. 34, Q. 1431.

265. J. Neuberg, *NC.*, I (1874), pp. 96, 198. Frank V. Morley, *Monthly* (1917), p. 124, Q. 511.

266, 267. N. A.-C., *Monthly*, XXV (1918), pp. 243–246. L. Droussent, "Sur le quadrilatere inscriptible," *Mathesis*, LXVI (1947), pp. 93–95.

268–269. The circumscriptible quadrilateral was first considered in the thirteenth century by Jordanus Nemorarius.

270. J. B. Durrande, *GAM.*, VI (1815–1816), p. 49.

278–280. Cn., p. 134, th. 47; p. 129, th. 61; pp. 135, 136, th. 48.

281. Crt., p. 165, art. 132. Cn., p. 129, th. 61 corollary.

282, 283. The proposition was given by William Wallace (1768–1873) in 1799, in the *Mathematical Repository*. F. J. Servois, in 1814, referred to the line as the Simson line (after Robert Simson, 1687–1768). This was obviously a slip, as is apparent from the dates involved. Moreover, the line has not been found in the writings of Simson, at least not up to the present. But the name Simson line remains in use, by tradition and force of habit.

284. F. J. Servois, *GAM.*, IV (1813–1814), p. 250. The proof is analytic.

286, 287. J. G., *NA.* (1871), p. 206.

288, 289. Heinen, *Crelle's Journal*, III (1828), pp. 285–287.

290. J. Steiner, *GAM.*, XVIII (1827–1828). No proof. W. F. Walker, *Quarterly Journal of Mathematics,* VIII (1867), p. 47.

292. Philippe de la Hire, *Nouveaux éléments des sections coniques*, p. 196. Solved by the use of conics.

LHr., pp. 220–221. This solution involves trigonometric computations.

N. A.–C., *Monthly* (1927), p. 161, rem. IV, bibliography.

293. J. Steiner, *Crelle's Journal*, LIII (1857), p. 237. W. H. Besant, *Quarterly Journal of Mathematics,* X (1870), p. 110.

294. C. E. Youngman, *ET.*, LXXV (1901), p. 107.

296, 297. J. Alison, *Edinb.*, III (1885), p. 86.

298. Ad. Mineur, *Mathesis* (1929), p. 330, note 34.

299. J. Steiner, *GAM.*, XVIII (1827–1828), p. 302. No proof.

302, 303. J. G., *NA.* (1871), p. 206.

304. E. Lemoine, *NA.* (1869), pp. 47, 174, 317.

Weill, *JMS.* (1884), p. 13, th. III.

W. F. Beard, *ET.*, VI (1904), p. 57, note.

306, 307. Negative segments came into use in geometry during the 17th century due to Albert Girard (1595–1632), René Descartes (1596–1650), and others. The systematic use of negative quantities in geometry was established by L. N. M. Carnot, and by A. F. Moebius, *Barycentrische Calcul* (1827).

The term "transversal" was introduced by Carnot.

308. Matthew Stewart (1717–1785), *Some General Theorems of Considerable Use in the Higher Parts of Mathematics.* Edinburgh, 1746. No proof is given. The proposition was rediscovered and proved by Thomas Simpson (1710–1761) in 1751, by L. Euler in 1780, and by L. N. M. Carnot in 1803 (Crt., p. 265, th. 2).

310. Menelaus, a Greek astronomer of the first century A.D., wrote a book of spherics in which this proposition was included. But the Greek original was lost, and the book came down to us only in the Hebrew and Arabic translations. However, the proposition became known through Ptolemy's *Almagest* (see note 255), where it was quoted.

Crt., p. 276, art. 219.

321, 322. G. de Longchamps, *NA.* (1866), pp. 118–119.

323. E. Lemoine, *JME.* (1890), p. 47, Q. 300.

324. Ernesto Cesàro, *NC.*, VI (1880), p. 472, Q. 573. Sanjana, *ET.*, XXVIII (1915), p. 33, Q. 17036; p. 35, Q. 17539.

326. Giovanni Ceva, *De lineis rectis se invicem secantibus* (Milan, 1678).

Crt., p. 281, th. 7.

330, 331. J. D. Gergonne, *GAM.*, IX (1818–1819), p. 116.

332, 333. Ch. Nagel (see note 100), p. 32, art. 99.

334. E. Vigarié, *JME.* (1886), p. 158.

335. G. de Longchamps, *NA.* (1866), p. 124, prop. 4.

336. The term is due to John Casey, *Mathesis* (1889), p. 5, Sec. I., prob. 3.

337, 338. G. de Longchamps, *JME.*, p. 92, Q. 94.

340, 341. J. D. Gergonne, *GAM.*, IX (1818–1819), p. 277.

342. H. Van Aubel, *Mathesis* (1882), Q. 128. See also N. Plakhowo, *Bull.*, V (1899–1900), p. 289.

343. Girard Desargues (1593–1662). The proposition is found in the writings (1648) of a follower of Desargues and is attributed by the pupil to the master. D. E. Smith (see note 207), p. 307.

The proposition is given, without indication of source, by F. J. Servois in his "Solutions peu connues . . . ," p. 23 (Paris, 1804), and by Poncelet, Pt., p. 89, who attributes it to Desargues.

344. The term "homological" was proposed by Poncelet.

345. See chap. iii, arts. 54–62.

346, 347. Pt., Sec. I, chap. i, art. 30, p. 16.

348. See notes 59, 54.

351, 352. Pt., Sec. I, chap. i, art. 27, p. 14.

353. LHr., p. 94, rem. 1.

354. The term *harmonic pencil* was used by Brianchon in 1817.

355. Pappus' *Collection.*

358. Pt., Sec. I, chap. ii., art. 76, pp. 41–42.

360. Cn., p. 104, th. 21.

362. P. H. Schoute, *Academy of Amsterdam,* III, p. 59.
J. Griffiths, *ET.*, LX (1894), p. 113, Q. 12113.

363. Orthogonal circles were considered in the early nineteenth century by Gaultier, Poncelet, Durrande, Steiner, and others.

368. Pt., p. 43, art. 79.

372. N. A.–C., *Monthly* (1934), p. 500.

373–392. The theory of poles and polars had its beginnings in the study of harmonic points and pencils. Some properties of polar lines may be found in the writings of both Apollonius and Pappus.

This theory was developed in considerable detail by Girard Desargues in his treatise on conic sections entitled *Brouillon projet d'une atteinte aux évênemens des rencontres d'un cone avec un plan* (Paris, 1639), and by his follower Philippe de la Hire.

The major elaboration of the theory took place in the first half of the nineteenth century in connection with the study of conic sections in projective geometry.

The term *pole* (art. 372) was first used by F. J. Servois, *GAM.*, I (1810–1811), p. 337. Two years later J. D. Gergonne suggested the term *polar line* in his own *GAM.*, III (1812–1813), p. 297.

386, 387. J. B. Durrande, *GAM.*, XVI (1825–1826), p. 112.

393, 394, 395. LHr., p. 41, rem. 8. The ancient Greeks were familiar with the centers of similitude of two circles. See R. C. Archibald, *Monthly* (1915), pp. 6–12; (1916), pp. 159–161.

397. Pt., p. 130, art. 242.

398. Rebuffel, *Bull.*, V (1899–1900), p. 113.

400, 401. Pt., pp. 139, 140, art. 262.
Chl., pp. 504–505, art. 725.

402, 404. Chl., p. 520, art. 743. See also *GAM.*, XI (1820–1821), p. 364, and XX (1829–1830), p. 305.

405, 406, 407. Gaspar Monge, *Géomêtrie descriptive,* pp. 54–55 (1798).

409. Chl., p. 541, art. 770.

410. The power of a point with respect to a circle was first considered by Louis

Gaultier in a paper published in the *Journal de L'École Polytechnique, Cahier* 16 (1813), pp. 124–214. Gaultier uses here for the first time the terms *radical axis* (p. 147) and *radical center* (p. 143). However, the term *power* (Potenz) was first used by J. Steiner.

412. Steiner. See note 410.

421, 422. Gaultier. See note 410.

425. Crt., p. 347, art. 305. Pt., Sec. I, chap. ii, art. 71, p. 40, footnote. Poncelet ascribes the proposition to his teacher Gaspar Monge.

429. Pt., Sec. I, chap. ii, art. 82, p. 44.

434. Pt., p. 123, art. 249, 2°.

436. Pt., pp. 133–134, art. 250, 2°.

437. Cf. Pt., p. 134, art. 251.

438. N. A.–C., *Monthly* (1932), p. 193.

439. Chl., p. 501, art. 716.

440. Coaxal circles were considered early in the nineteenth century by Gaultier, Poncelet, Steiner, and others.

447. Pt., Sec. I, chap. ii, art. 76, pp. 41, 42.

448. Pt., Sec. I, chap. ii, art. 79, p. 43.

453–457. Gaultier. See note 410. Pt., Sec. I, chap. ii, arts. 73, 74, p. 41.

460. Pt., Sec. I, chap. ii, art. 87, p. 44.

463. N. A.–C., *Mathesis* XLII (1828), p. 158.

464–466. N. A.–C., *Annals of Math.*, XXIX (1928), p. 369, art. 2.

471, 472. John Casey, *A Sequel to Euclid* (6th ed., 1900), VI, Sec. V, pp. 113–114. This is a very useful proposition. A. Droz-Farny shows in *JME.* (1895), pp. 242–245, that most of the properties of the radical axis given in Chl. are corollaries of Casey's theorem.

477, 480. Chl., p. 520, art. 747.

481. N. A.–C., *Monthly*, XXXIX (1932), p. 56, Q. 3477.

491, 492. Gaultier. See note 410.

494. J. B. Durrande, *GAM.*, XVI (1825–1826), p. 112.

495. Cf. J. Neuberg, *Mathesis* (1925), p. 37, Q. 2225.

496. Vecten and Durrande, *GAM.*, XI (1820–1821), p. 364.

499–501. See note 438.

502, 503. Cf. N. A.–C., *Modern Pure Solid Geometry* (New York, 1935), pp. 205, 206, arts. 641, 642.

505. Cl. Servais, *Mathesis* (1891), p. 238, Q. 717

508. E. Lemoine, *JME.* (1874), p. 21.

516. The problem was included in the book *De tactionibus* by Apollonius. This book has not come down to us. It was restored by François Viète (Francis Vieta). The solution of the problem of Apollonius as given in this restoration is reproduced here.

This problem has interested many great mathematicians, including Descartes and Newton.

518. Inverse points were known to François Viète. Robert Simson in his restoration of a lost book of Apollonius on geometrical loci (see note 11) included one of the basic theorems of the theory of inversion (art. 523). LHuilier in LHr. gives the two special cases of this theorem (arts. 521, 523).

The theory of inversion as a method of studying and transforming geometrical figures is largely a product of the second quarter of the nineteenth century. Among the outstanding contributors to the elaboration of this theory may be named Poncelet, Steiner, Quetelet, Pluecker (1801–1868), Moebius, Liouville, and William Thompson (Lord Kelvin).

533. In 1864 in a letter to the editors of the *Nouvelles annales de mathématiques* (p. 414) A. Peaucellier (1832–1913) suggested to the readers of that periodical the idea of finding a linkage (compas composé) which would draw rigorously a straight line by a continuous motion. From the remarks made in that letter it is readily inferred that Peaucellier himself was in possession of such an instrument.

In 1867 A. Mannheim (1831–1906) formally announced, without description, Peaucellier's invention at a meeting of the Paris Philomathic Society. The first paper on the linkage was not published by Peaucellier until 1873 (*NA*., pp. 71–73).

In the meantime, Lipkin, a young man of St. Petersburg (Leningrad), rediscovered the linkage and a description of it appeared in the *Bulletin* of the St. Petersburg Academy of Sciences, XVI (1781), pp. 57–60. See R. C. Archibald, *Outline of the History of Mathematics*, p. 99, note 280, published in 1949 by the Mathematical Association of America.

535. J. Tummers, *Mathesis* (1929), p. 130, Q. 2515.

536. See note 255.

537. R. C. J. Nixon, *ET*., LV (1891), p. 107, Q. 10693.

547, 548. J. J. A. Mathieu, *NA*. (1865), p. 399. E. Lemoine, Association Française pour l'Avancement des Sciences (1884).

The term *trilinear polar* is due to Mathieu. The term *harmonic polar* was proposed by Longchamps, *JMS*. (1886), p. 103.

558–586 (Lemoine Geometry). In 1873, Emile Lemoine read a paper before the Association Française pour l'Avancement des Sciences entitled "Sur quelques propriétés d'un point remarquable du triangle." The paper appeared in the *Proceedings*, pp. 90–91, and was also published in *NA*. (1873), pp. 364–366. It included the articles 561, 572, 573, 588, 589, 592, 593, 594. No proofs were given. The paper may be said to have laid the foundations not only of Lemoinian Geometry, but also of the modern geometry of the triangle as a whole.

The basic point of Lemoinian geometry (art. 570) Lemoine called the center of antiparallel medians; J. Neuberg named it the Lemoine point. (*Mémoire sur le tétraèdre*, The Belgian Academy, 1884, p. 3; *Supplement* to *Mathesis*, 1885). This point was met with by a number of writers before Lemoine, in connection with its various properties. LHuilier encountered it in 1809 (LHr., p. 296).

In 1847 Grebe was led to this point through the property of ex. 10, p. 257. The proposition of art. 583 was found by Catalan in 1852. In 1862, O. Schloemilch chanced upon the theorem art. 586.

Lemoine through his numerous papers brought to light the great abundance of geometrical properties connected with this point and this invested it with an importance rivaling that of the centroid and the orthocenter in the study of the geometry of the triangle.

558. The term *symmedian* was proposed by Maurice d'Ocagne, *NA*. (1883), p. 451, as a substitute for the term *antiparallel median* used by Lemoine, *NA*. (1873), p. 364.

585. Ad. Mineur, *Mathesis* (1934), p. 257. A. Droz-Farny, *ET.*, LXI (1894), p. 90, Q. 12223.

586. Schloemilch. See note 558.

588. The method of proof used in this article originated with Brianchon and Poncelet. See D. E. Smith (note 207), pp. 337, 338.

592. R. Tucker, *Proceedings of the London Math. Soc.*, XVIII (1886–1887), p. 3; *Mathesis* (1887), p. 12.

E. Lemoine (1873), *NA.*, p. 365, prop. 4°.

596. E. Lemoine, *JME.* (1894), p. 112, Q. 442.

597. See note 11, locus 11. J. Neuberg, *Mathesis* (1885), p. 204.

600. Vecten, *GAM.*, X (1819–1820), p. 202.

601. J. Neuberg, *Mathesis* (1885), p. 204. Rouché, p. 475.

604, 605. N. A.–C., *Tohoku Mathematical Journal*, XXXIX (1934), Part 2, p. 264, art. 1.

609–614. N. A.–C., "On the Circles of Apollonius," *Monthly*, XXII (1915), pp. 304–305. *Mathesis* (1926), p. 144.

621, 622. Sollertinski, *Mathesis* (1894), p. 116, Q. 841.

624. E. Lemoine, *Mathesis* (1902), p. 147. N. A.–C., *Mathesis* (1927), note 49, 2°.

625. The isogonal relation was first considered by J. J. A. Mathieu, *NA.* (1865), pp. 393 ff.

Most of the arts. 625–644 are given by Émile Vigarié in *JME.* (1885), pp. 33 ff.

635. Mathieu (see note 625), p. 400.

638. *Bull.*, I (1895–1896), p. 73, Q. 77.

644. Cf. N. A.–C., *Modern Pure Solid Geometry*, p. 244, art. 750.

646, 647, 648. K. Sivaraj, *Mathematics Student*, XIII (1945), p. 69. V. Thébault, *Mathesis*, LVI (1947), p. 31. See note 215.

649. In 1875 (*NA.*, pp. 192, 286, Q. 1166) H. Brocard stated a characteristic property of the two points which now bear his name. As in the case of the Lemoine point (see note 558) the Brocard points had been encountered by previous writers. But in 1881 Brocard read a paper before the Association Française pour l'Avancement des Sciences under the title *Nouveau cercle du plan du triangle* (art. 666) which brought out numerous properties of the Brocard points, and thus established Brocard's reputation as the co-founder of the modern geometry of the triangle.

The name of J. Neuberg (Liège, Belgium) is generally associated with those of Lemoine and Brocard as the third co-founder of this branch of geometry.

652. The name *Brocard points* is attributed to A. Morel, *JME.* (1883), p. 70.

656, 659, 661, 663. Ech., p. 25, art. 12.3°; p. 77, art. 3; p. 29, art. 15.1° p. 36, art. 16.1°.

666. The term *Brocard circle* was suggested by A. Morel, *JME.* (1883), p. 70.

669. H. Brocard, *ET.*, LIX (1893), p. 50, Q. 11638; *JME.* (1893), p. 69, Q. 437.

670a. Gy., p. 106, art. 150.

670b. Neuberg, Rouché, p. 480, art. 42.

671. C. Jonescu-Bujor, *Mathesis* (1938), p. 360, note 36, 3°.

672, 673, 674. Neuberg, Rouché, p. 480, art. 42.

675. Ech., p. 90, art. 43, 1°.

676. *Mathesis* (1889), p. 243, no. 1. A. Emmerich, *ET.*, XIII (1908), p. 88, Q. 9992.

677. Neuberg, Rouché, p. 480, art. 41.

678, 679. The proposition is due to J. Neuberg, "Sur le point de Steiner," *JMS.* (1886), p. 29, 1°, where he identified the point *R* with one which J. Steiner had considered in a different connection (Crelle, XXXII, 1846, p. 300).

The name *Steiner point* is due to Neuberg, *Mathesis* (1885), p. 211.

680. Gaston Tarry, *Mathesis* (1884), Supplement.

681. The term was introduced by J. Neuberg, *Mathesis* (1886), p. 5.

684, 685. R. Tucker (1832–1905), "On a Group of Circles," *Quarterly Journal of Mathematics*, XX (1885), p. 57.

Some circles of this group were found by other writers, before Tucker. Lemoine (arts. 588, 594), Taylor (art. 689).

The name *Tucker circles* is due to Neuberg, "Sur les cercles de Tucker," *ET.*, XLIII (1885), pp. 81–85.

689. H. M. Taylor, "On a Six-Point Circle Connected with a Triangle," *Messenger of Math.*, LXXVII (1881–1882), pp. 177–179. However, Taylor was anticipated by others. See *NC.*, VI (1880), p. 183; *National Math. Magazine*, XVIII (1943–1944), p. 40–41.

691. J. Neuberg, *NC.*, II (1875), pp. 189, 316, Q. 111; IV (1878), p. 379; E. Lemoine, *JME.* (1884), p. 51.

692. The term *orthopole* was proposed by Neuberg, *Mathesis* (1911), p. 244.

694. M. Soons, *Mathesis* (1896), p. 57.

697. Gy., p. 49, art. 75.

LIST OF NAMES

This list covers both the text and the notes. Numbers refer to sections in this book. See also the general index.

A

Alexandrov, I. 43, 44.
Alison, J. 296, 297.
Apollonius of Perga (?–225 B.C.) 11, 373, 516, 518.
Archibald, R. C. (1875–) 393, 394, 395, 533.
Archimedes (287–212 B.C.) 88, 175.
Arduesser, J. (1584–1665) 182.
Arnauld, A. (1612–1694) 185.

B

Barisien, E. N. 7, 81, 233.
Beard, W. F. 304.
Besant, W. H. 176, 293, 295.
Beyens, Ignacio 12, 14.
Booth, James 199.
Bourlet, Carlo (1866–1913) 127.
Brahmagupta (7th century) 85, 86.
Brianchon, Ch. J. (1785–1864) 207, 208, 354, 588.
Brocard, H. (1845–1922) 649, 669.

C

Carnot, L. N. M. (1753–1823) 108, 109, 146, 177, 178–180, 184, 185, 186, 195, 201, 220, 223, 246, 248, 255, 257, 281, 306, 307, 308, 310, 326, 425.
Casey, John (1820–1891) 336, 471, 472.
Catalan, E. (1814–1894) 8, 13, 15, 17, 18, 51, 52, 76, 90, 118, 119, 120, 122–126, 159–164, 166, 182, 192, 240, 249, 250, 278, 279, 280, 281, 360.
Causse, A. 199.
Cesàro, Ernesto (1859–1906) 324.
Ceva, Giovanni (1647–1734) 326.
Charruit 104.
Chasles, Michel (1793–1880) 31, 400, 401, 402, 404, 411, 439, 477, 480, 558.
Cochez, O. 53, 169.

D

Delhez, O. 95, 212.
Desargues, Gérard (1593–1662) 343.
Descartes, René (1596–1650) 306, 307, 516.
Deteuf, Auguste 258, 261, 262.
Dobson, Thomas 264.
Dostor, Georges (1868–?) 110, 203.
Droussent, L. (1907–) 266, 267.
Droz-Farny, A. 214, 470, 471, 585.
Drury, H. D. 196, 258
Durrande, J. B. (1793–1825) 270, 363, 386, 387, 494, 496.

E

Emmerich, A. (1856–1915?) 77, 656, 659, 661, 663, 675, 676.
Euclid (c. 300 B.C.) 129, 173.
Euler, Leonhard (1707–1783) 201, 207, 208, 211, 246, 308.

F

Feuerbach, K. W. (1800–1834) 145, 207, 208, 215, 216, 217, 698
Fontené, G. 257.

G

Gallatly, William 235, 670a, 697.
Gaultier, Louis 363, 410, 421, 422, 440, 453–457, 491, 492 (See note 410).
Gergonne, J. D. (1771–1859) 242, 330, 331, 340, 341, 374.
Girard, Albert (1595–1632) 306, 307.
Gob, A. 205, 211.
Goormaghtigh, R. (1893–) 236.
Griffiths, J. 362.
Grebe, 556.
Greenstreet, W. J. 42.

INDEX

This index covers both the text and the exercises.

References are to pages. A number in italics indicates the page where the term is defined.

A page frequently contains the same term more than once.

Consult also the Table of Contents and the List of Names.